ALTERNATIVES FOR MANAGING THE NATION'S COMPLEX CONTAMINATED GROUNDWATER SITES

Committee on Future Options for Management in the Nation's Subsurface Remediation Effort

Water Science and Technology Board

Division on Earth and Life Studies

NATIONAL RESEARCH COUNCIL
OF THE NATIONAL ACADEMIES

THE NATIONAL ACADEMIES PRESS
Washington, D.C.
www.nap.edu

THE NATIONAL ACADEMIES PRESS 500 Fifth Street, NW Washington, DC 20001

NOTICE: The project that is the subject of this report was approved by the Governing Board of the National Research Council, whose members are drawn from the councils of the National Academy of Sciences, the National Academy of Engineering, and the Institute of Medicine. The members of the panel responsible for the report were chosen for their special competences and with regard for appropriate balance.

This study was supported by Contract Number W911SR-09-1-0004 between the National Academy of Sciences and the U.S. Department of the Army. Any opinions, findings, conclusions, or recommendations expressed in this publication are those of the author(s) and do not necessarily reflect the views of the organizations or agencies that provided support for the project.

International Standard Book Number-13: 978-0-309-27874-4
International Standard Book Number-10: 978-0-309-27874-0
Library of Congress Catalog Card Number 2012955130

Cover: The plume maps represent the distribution of TCE in one groundwater zone at the MEW Superfund site in California, before and after 17 years of applying pump and treat technology.

Additional copies of this report are available for sale from the National Academies Press, 500 Fifth Street, NW, Keck 360, Washington, DC 20001; (800) 624-6242 or (202) 334-3313; http://www.nap.edu.

Printed and bound in Great Britain by Marston Book Services Ltd, Oxfordshire

THE NATIONAL ACADEMIES

Advisers to the Nation on Science, Engineering, and Medicine

The **National Academy of Sciences** is a private, nonprofit, self-perpetuating society of distinguished scholars engaged in scientific and engineering research, dedicated to the furtherance of science and technology and to their use for the general welfare. Upon the authority of the charter granted to it by the Congress in 1863, the Academy has a mandate that requires it to advise the federal government on scientific and technical matters. Dr. Ralph J. Cicerone is president of the National Academy of Sciences.

The **National Academy of Engineering** was established in 1964, under the charter of the National Academy of Sciences, as a parallel organization of outstanding engineers. It is autonomous in its administration and in the selection of its members, sharing with the National Academy of Sciences the responsibility for advising the federal government. The National Academy of Engineering also sponsors engineering programs aimed at meeting national needs, encourages education and research, and recognizes the superior achievements of engineers. Dr. Charles M. Vest is president of the National Academy of Engineering.

The **Institute of Medicine** was established in 1970 by the National Academy of Sciences to secure the services of eminent members of appropriate professions in the examination of policy matters pertaining to the health of the public. The Institute acts under the responsibility given to the National Academy of Sciences by its congressional charter to be an adviser to the federal government and, upon its own initiative, to identify issues of medical care, research, and education. Dr. Harvey V. Fineberg is president of the Institute of Medicine.

The **National Research Council** was organized by the National Academy of Sciences in 1916 to associate the broad community of science and technology with the Academy's purposes of furthering knowledge and advising the federal government. Functioning in accordance with general policies determined by the Academy, the Council has become the principal operating agency of both the National Academy of Sciences and the National Academy of Engineering in providing services to the government, the public, and the scientific and engineering communities. The Council is administered jointly by both Academies and the Institute of Medicine. Dr. Ralph J. Cicerone and Dr. Charles M. Vest are chair and vice chair, respectively, of the National Research Council.

www.national-academies.org

COMMITTEE ON FUTURE OPTIONS FOR MANAGEMENT IN THE NATION'S SUBSURFACE REMEDIATION EFFORT*

MICHAEL C. KAVANAUGH, *Chair*, Geosyntec, Oakland, California
WILLIAM A. ARNOLD, University of Minnesota, Minneapolis
BARBARA D. BECK, Gradient, Cambridge, Massachusetts
YU-PING CHIN, The Ohio State University, Columbus
ZAID CHOWDHURY, Malcolm Pirnie, Phoenix, Arizona
DAVID E. ELLIS, DuPont Engineering, Newark, Delaware
TISSA H. ILLANGASEKARE, Colorado School of Mines, Golden
PAUL C. JOHNSON, Arizona State University, Tempe
MOHSEN MEHRAN, Rubicon Engineering, Irvine, California
JAMES W. MERCER, Tetra Tech GEO, Sterling, Virginia
KURT D. PENNELL, Tufts University, Medford, Massachusetts
ALAN J. RABIDEAU, State University of New York, Buffalo
ALLEN M. SHAPIRO, U.S. Geological Survey, Reston, Virginia
LEONARD M. SIEGEL, Center for Public Environmental Oversight, Mountain View, California
WILLIAM J. WALSH, Pepper Hamilton LLP, Washington, DC

NRC Staff

LAURA J. EHLERS, Study Director
STEPHANIE E. JOHNSON, Senior Staff Officer
KERI SCHAFFER, Research Associate
JEANNE AQUILINO, Senior Administrative Associate
ELLEN DEGUZMAN, Research Associate, *through June 2011*
ANITA HALL, Senior Program Associate

---

*Kevin J. Boyle, Virginia Polytechnic Institute and State University, Blacksburg, was a member of the Committee from February 2010 to June 2012.

WATER SCIENCE AND TECHNOLOGY BOARD

DONALD I. SIEGEL, *Chair*, Syracuse University, Syracuse, New York
LISA ALVAREZ-COHEN, University of California, Berkeley
EDWARD J. BOUWER, Johns Hopkins University
YU-PING CHIN, The Ohio State University, Columbus
M. SIOBHAN FENNESSY, Kenyon College, Gambier, Ohio
BEN GRUMBLES, Clean Water America Alliance, Washington, DC
GEORGE R. HALLBERG, The Cadmus Group, Inc., Watertown, Massachusetts
KENNETH R. HERD, Southwest Florida Water Management District, Brooksville
GEORGE M. HORNBERGER, Vanderbilt University, Nashville, Tennessee
CATHERINE L. KLING, Iowa State University, Ames
DEBRA S. KNOPMAN, The Rand Corporation, Washington, DC
LARRY LARSON, Association of State Floodplain Managers, Madison, Wisconsin
RITA P. MAGUIRE, Maguire & Pearce PLLC, Phoenix, Arizona
DAVID H. MOREAU, University of North Carolina, Chapel Hill
ROBERT SIMONDS, The Robert Simonds Company, Culver City, California
FRANK H. STILLINGER, Princeton University, Princeton, New Jersey
MARYLYNN V. YATES, University of California, Riverside
JAMES W. ZIGLAR, SR., Van Ness Feldman, Washington, DC

NRC Staff

JEFFREY JACOBS, Director
LAURA J. EHLERS, Senior Staff Officer
LAURA J. HELSABECK, Senior Staff Officer
STEPHANIE JOHNSON, Senior Staff Officer
JEANNE AQUILINO, Financial and Administrative Associate
ANITA HALL, Senior Program Associate
MICHAEL STOEVER, Research Associate
SARAH BRENNAN, Senior Program Assistant

Preface

Despite nearly 40 years of intensive efforts in the United States as well as in other industrialized countries worldwide, restoration of groundwater contaminated by releases of anthropogenic chemicals to a condition allowing for unlimited use and unrestricted exposure remains a significant technical and institutional challenge. Recent (2004) estimates by the U.S. Environmental Protection Agency (EPA) indicate that expenditures for soil and groundwater cleanup at over 300,000 sites through 2033 may exceed $200 billion (not adjusted for inflation), and many of these sites have experienced groundwater impacts.

One dominant attribute of the nation's efforts on subsurface remediation efforts has been lengthy delays between discovery of the problem and its resolution. Reasons for these extended timeframes are now well known: ineffective subsurface investigations, difficulties in characterizing the nature and extent of the problem in highly heterogeneous subsurface environments, remedial technologies that have not been capable of achieving restoration in many of these geologic settings, continued improvements in analytical detection limits leading to discovery of additional chemicals of concern, evolution of more stringent drinking water standards, and the realization that other exposure pathways, such as vapor intrusion, pose unacceptable health risks. A variety of administrative and policy factors also result in extensive delays, including, but not limited to, high regulatory personnel turnover, the difficulty in determining cost-effective remedies to meet cleanup goals, and allocation of responsibility at multiparty sites.

Over the past decade, however, remedial technologies have shown increased effectiveness in removing contaminants from groundwater, and the use of more precise characterization tools and other diagnostic technologies have improved our ability to achieve site-specific remedial action objectives within a reasonable time frame at an increasing number of sites. For example, of the over 1,700 National Priority List sites, the EPA has deleted over 360 (as of March 2012), including some that have reported achieving restoration goals for groundwater, usually defined as drinking water standards. Other regulatory programs at both the federal and state level report closures of many sites with contaminated groundwater, although "closure" is often defined by site-specific conditions, such as the need for long-term institutional controls. Such trends and financial pressures have prompted the Department of Defense (DoD) to set very aggressive goals for significantly reducing the expenditures for the Installation Restoration Program (IRP) within the next few years.

There is general agreement among practicing remediation professionals, however, that there is a substantial population of sites, where, due to inherent geologic complexities, restoration within the next 50 to 100 years is likely not achievable. Reaching agreement on which sites should be included in this category, and what should be done with such sites, however, has proven to be difficult. EPA recently summarized the agency's recommended decision guidance (July 2011) for these more complex sites, presenting a Road Map for groundwater restoration that targets both Superfund and Resource Conservation and Recovery Act Corrective Action sites. A key decision in that Road Map is determining whether or not restoration of groundwater is "likely." If not, alternative strategies must be evaluated to achieve the remedial action objectives, including possible modification of these objectives or the points of compliance. The National Research Council (NRC) has also addressed the issue of complex and difficult sites. Since 1987, there have been at least six NRC studies to evaluate barriers to achieving the goal of groundwater restoration. These reports addressed both technical and institutional barriers to restoration, but in general, the reports have concluded that some fraction of sites will require containment and long-term management and the number of such sites could be in the thousands. Other organizations have also undertaken in-depth assessments of barriers to restoration at more complex sites including the Interstate Technology and Regulatory Council.

In this context, the U.S. Army Environmental Command (AEC) agreed to support an NRC study to address the technical and management issues arising from barriers to restoration of contaminated groundwater at these complex sites. In particular, the AEC was concerned that delays in decision making on the final remedies at many of their more complex sites could diminish their ability to achieve DoD goals for the IRP. For the Army, one significant goal is achieving the remedy-in-place or response-complete milestones for 100 percent of their IRP sites at active installations by 2014. This study was established under the Water Science and Technology Board (WSTB) of the NRC with the title "Future Options for Management in the Nation's Subsurface Remediation Effort." The Committee included fifteen individuals representing expertise in all areas relevant to the statement of task, including various scientific and technical disciplines, resource economics, environmental policy, risk assessment, and public stakeholder issues. Seven meetings were held over the past two years, with presentations from a wide range of interested parties. I would like to thank the following individuals for giving presentations to the committee during one or more of its meetings: Laurie Haines-Eklund, Army Environmental Command; Jim Cummings, EPA Superfund Office; Adam Klinger, EPA Underground Storage Tank Office; Jeff Marquesee and Andrea Leeson, Strategic Environmental Research and Development Program; Brian Looney, Department of Energy Environmental Management; John Gillespie, Air Force Center for Environmental Excellence; Anna Willett, Interstate Technology and Regulatory Council; Alan Robeson, American Water Works Association; Jill Van Dyke, National Groundwater Association; Ira May, May Geoenvironmental Services; Roy Herndon, Orange County Water District; Milad Taghavi, Los Angeles Department of Water and Power; Carol Williams, San Gabriel Water Supply; Gil Borboa, City of Santa Monica; David Lazerwitz, Farella Braun + Martel, LLP; James Giannopoulos, California State Water Quality Control Board; Herb Levine, EPA Region 9; Alec Naugle, California Region 2 Water Board; David Sweeney, New Jersey Department of Environmental Protection; Rula Deeb, Malcolm Pirnie; Amy Edwards, Holland & Knight LLP; Brian Lynch, Marsh Environmental Practice; Richard Davies, Chartis; Henry Schuver and Helen Dawson, EPA; Tushar Talele, Arcadis; Anura Jayasumana, Colorado School of Mines; Deborah Morefield, Office of the Deputy Undersecretary of Defense; Alana Lee, EPA Region 9; Betsy Southerland and Matt Charsky, EPA; Mike Truex, Pacific National Lab; and Jim Gillie, Versar/Joint Base Lewis McChord.

I wish to acknowledge the herculean efforts of Laura Ehlers and her colleagues at the WSTB for organizing our meetings, managing multiple tasks, and finally completing the editing of contributions from committee members, a task that requires both editing and substantial technical expertise and diplomacy in helping a diverse committee reach consensus. I am indebted to Laura for her efforts on completing this report. I also want to send special thanks to all the Committee members who so diligently participated in long sessions at our meetings, produced comprehensive summaries of the state of the science in subsurface remediation, and who wrestled with the complexities of addressing the challenges of better decision making. The contributions of those who worked on the final chapter are especially appreciated, and particularly those individuals who joined the committee later in deliberations to fill in for vacancies caused by unanticipated changes in the committee roster.

This report has been reviewed in draft form by individuals chosen for their diverse perspectives and technical expertise, in accordance with procedures approved by the National Research Council's Report Review Committee. The purpose of this independent review is to provide candid and critical comments that will assist the institution in making its published report as sound as possible and to ensure that the report meets institutional standards for objectivity, evidence, and responsiveness to the study charge. The review comments and draft manuscript remain confidential to protect the integrity of the deliberative process. We wish to thank the following individuals for their review of this report: Lisa Alvarez-Cohen, University of California, Berkeley; Linda Lee, Purdue University; Jacqueline MacDonald Gibson, University of North Carolina, Chapel Hill; David Nakles, Carnegie Mellon University; Stavros Papadopulos, S.S. Papadopulos & Associates, Inc.; Tom Sale, Colorado State University; Rosalind Schoof, Environ International Corporation; Hans Stroo, HydroGeoLogic, Inc.; and Marcia E. Williams, Gnarus Advisors, LLC.

Although the reviewers listed above have provided many constructive comments and suggestions, they were not asked to endorse the conclusions or recommendations nor did they see the final draft of the report before its release. The review of this report was overseen by Susan L. Brantley, Pennsylvania State University, and Mitchell Small, Carnegie Mellon University. Appointed by the National Research Council, they were responsible for making certain that an independent examination of this report was carried out in accordance with institutional procedures and that all review comments were carefully considered. Responsibility for the final content of this report rests entirely with the authoring committee and the institution.

Michael C. Kavanaugh, *Chair*
Committee on Future Options for Management
in the Nation's Subsurface Remediation Efforts

Contents

*Pages 321-408 of Appendix C are available online at www.nap.edu/catalog.php? record_ID=14668.

Summary

At hundreds of thousands of hazardous waste sites across the country, groundwater contamination remains in place at levels above cleanup goals. The most problematic sites are those with potentially persistent contaminants including chlorinated solvents recalcitrant to biodegradation, and with hydrogeologic conditions characterized by large spatial heterogeneity or the presence of fractures. While there have been success stories over the past 30 years, the majority of hazardous waste sites that have been closed were relatively simple compared to the remaining caseload. In 2004, the U.S. Environmental Protection Agency (EPA) estimated that more than \$209 billion would be needed to mitigate these hazards over the next 30 years—likely an underestimate because this number did not include sites where remediation was already underway or where remediation had transitioned to long-term management.

The Department of Defense (DoD) exemplifies a responsible party that has made large financial investments (over \$30 billion) in hazardous waste remediation to address past legacies of their industrial operations. Although many hazardous waste sites at military facilities have been closed with no further action required, meeting goals like drinking water standards in contaminated groundwater has rarely occurred at many complex DoD sites. It is probable that these sites will require significantly longer remediation times than originally predicted and, thus, continued financial demands for monitoring, maintenance, and reporting.

In this context, the Water Science and Technology Board, under the auspices of the National Research Council (NRC), convened a committee to assess the future of the nation's groundwater remediation efforts focusing on the technical, economic, and institutional challenges facing the Army and other responsible parties as they pursue site closure. Previous NRC reports concluded that complete restoration of contaminated groundwater is unlikely to be achieved for many decades for a substantial number of sites, in spite of the fact that technologies for removing contaminants from groundwater have continued to evolve and improve. Since the most recent NRC report in 2005, better understanding of technical issues and barriers to achieving site closure have become evident. The following questions comprised the statement of task for this Committee, which considered both public and private hazardous waste sites.

Size of the Problem. At how many sites does residual contamination remain such that site closure is not yet possible? At what percentage of these sites does residual contamination in groundwater threaten public water systems?

Current Capabilities to Remove Contamination. What is technically feasible in terms of removing a certain percentage of the total contaminant mass? What percent removal would be needed to reach unrestricted use or to be able to extract and treat groundwater for potable reuse? What should be the definition of "to the extent practicable" when discussing contaminant mass removal?

Correlating Source Removal with Risks. How can progress of source remediation be measured to best correlate with site-specific risks? Recognizing the long-term nature of many problems, what near-term endpoints for remediation might be established? Are there regulatory barriers that make it impossible to close sites even when the site-specific risk is negligible and can they be overcome?

The Future of Treatment Technologies. The intractable nature of subsurface contamination suggests the need to discourage future contaminant releases, encourage the use of innovative and multiple technologies, modify remedies when new information becomes available, and clean up sites sustainably. What progress has been made in these areas and what additional research is needed?

Better Decision Making. Can adaptive site management lead to better decisions about how to spend limited resources while taking into consideration the concerns of stakeholders? Should life cycle assessment become a standard component of the decision process? How can a greater understanding of the limited current (but not necessarily future) potential to restore groundwater be communicated to the public?

MAGNITUDE OF THE PROBLEM

Chapter 2 presents information on the major federal and state regulatory programs under which hazardous waste is cleaned up to determine the size and scope of these programs. The Committee sought to determine (1) the number of sites that have not yet reached closure, (2) principal chemicals of concern, (3) remediation costs expended to date, (4) cost estimates for reaching closure, and (5) the number of sites affecting local water supplies. Information was gathered for sites in the EPA's Comprehensive Environmental Response, Compensation, and Liability Act (CERCLA), Resource Conservation and Recovery Act (RCRA), and Underground Storage Tank (UST) programs; sites managed by the DoD, the Department of Energy (DOE) and other federal agencies; and sites under state purview (e.g., state Superfund, voluntary cleanup programs, and Brownfields programs). The metrics and milestones across all these programs differ, making comparisons and the elimination of overlap difficult. Nonetheless, the Committee used these data to estimate the number of complex sites, the likelihood that sites affect a drinking water supply, and the remaining costs associated with remediation.

At least 126,000 sites across the country have been documented that have residual contamination at levels preventing them from reaching closure. This number is likely to be an underestimate of the extent of contamination in the United States for many reasons. For example, the CERCLA and RCRA programs report the number of facilities, which are likely to have multiple sites. The total does not include DoD sites that have reached *remedy in place* or *response complete*, although some such sites may indeed contain residual contamination. Although there is overlap between some of the categories, in the Committee's opinion it is not significant enough to dismiss the conclusion that the total number of 126,000 is an underestimate.

No information is available on the total number of sites with contamination in place above levels allowing for unlimited use and unrestricted exposure, although the total is certainly greater than 126,000. For the CERCLA program, many facilities have been delisted with contamination remaining in place at levels above unlimited use and unrestricted exposure. Depending on state closure requirements, USTs are often closed with contamination remaining due to the biodegradability of petroleum hydrocarbons. Most of the DOE sites, including those labeled as "completed," contain recalcitrant contamination that in some cases could take hundreds of years to reach levels below those allowing for unlimited use and unrestricted exposure.

A small percentage (about 12,000 or less than 10 percent) of the 126,000 sites are estimated by the Committee to be complex from a hydrogeological and contaminant perspective. This total represents the sum of the remaining DoD, CERCLA, RCRA, and DOE sites and facilities, based on the assumption that many of the simpler sites in these programs have already been dealt with.

Approximately 10 percent of CERCLA facilities affect or significantly threaten public water supply systems, but similar information from other programs is largely unavailable. Surveys of groundwater quality report that 0.34 to 1 percent of raw water samples from wells used for drinking water (including public supply and private wells) contain mean volatile organic compound (VOC) concentrations greater than the applicable drinking water standard, although there are no data linking these exceedances to specific hazardous waste sites. The percentage of drinking water wells with samples containing low-level VOC concentrations is likely to be higher for areas in close proximity to contaminated sites, for urban rather than rural areas, and in shallow unconfined sandy aquifers.

Information on cleanup costs incurred to date and estimates of future costs are highly uncertain. Despite this uncertainty, the estimated "cost to complete" of $110-127 billion is likely to be an underestimate of future liabilities. Remaining sites include some of the most difficult to remediate sites, for which the effectiveness of planned remediation remains uncertain given their complex site conditions. Furthermore, many of the estimated costs do not fully consider the cost of long-term management of sites that will have contamination remaining in place at levels above those allowing for unlimited use and unrestricted exposure for the foreseeable future.

The nomenclature for the phases of site cleanup and cleanup progress are inconsistent between federal agencies, between the states and federal government, and in the private sector. Partly because of these inconsistencies, members of the public and other stakeholders can and have confused the concept of "site closure" with achieving unlimited use and unrestricted exposure goals for the site, such that no further monitoring or oversight is needed. In fact, many sites thought of as "closed" and considered as "successes" will require oversight and funding for decades and in some cases hundreds of years in order to be protective. CERCLA and other programs have reduced public health risk from groundwater contamination by preventing unacceptable exposures in water or air, but not necessarily by reducing contamination levels to drinking water standards throughout the affected aquifers.

REMEDIAL OBJECTIVES, REMEDY SELECTION, AND SITE CLOSURE

Chapter 3 focuses on the remedial objectives dictated by the common regulatory frameworks under which groundwater cleanup generally occurs because such objectives are often a substantial source of controversy. This is particularly true for complex sites, where the remedial objectives are drinking water standards (denoted as maximum contaminant levels or MCLs) and hence are typically difficult, if not impossible, to attain for many decades. Faced with shrinking budgets and a backlog of sites that include an increasing percentage of complex sites, some states (e.g., California) have proposed closing large numbers of petroleum underground storage tank sites deemed to present a low threat to the public, despite the affected groundwater not meeting remedial goals at the time of closure. Other states (New Jersey and Massachusetts) have sought to privatize parts of the remediation process in order to unburden state and local regulatory agencies.

EPA's current remediation guidance provides substantial flexibility to the remedy selection process in a number of ways, although there are legal and practical limits to this flexibility. There are several alternatives to traditional cleanup goals, like technical impracticability waivers, that can allow sites with intractable contamination to move more expeditiously through the phases of cleanup while still minimizing risks to human health and the environment. The chapter also discusses sustainability concepts, which have become goals for some stakeholders and could impact the remedy selection process. The following conclusions and recommendations discuss the value of exploring goals and remedies based on site-specific risk, sustainability, and other factors.

By design (and necessity), the CERCLA process is flexible in (a) determining the beneficial uses of groundwater; (b) deciding whether a regulatory requirement is an applicable or relevant and appropriate requirement (ARAR) at a site; (c) using site-specific risk assessment to help select the remedy; (d) using at least some sustainability factors to help select the remedy; (e) determining what is a reasonable timeframe to reach remedial goals; (f) choosing the point of compliance for monitoring; and (g) utilizing alternate concentration limits, among others. **These flexible approaches to setting remedial objectives and selecting remedies should be explored more fully by state and federal regulators, and EPA should take administrative steps to ensure that existing guidance is used in the appropriate circumstances.**

To fully account for risks that may change over time, **risk assessment at contaminated groundwater sites should compare the risks from taking**

"no action" to the risks associated with the implementation of each remedial alternative over the life of the remedy. Risk assessment at complicated groundwater sites is often construed relatively narrowly, with an emphasis on risks from drinking water consumption and on the MCL. Risk assessments should include additional consideration of (a) short-term risks that are a consequence of remediation; (b) the change in residual risk over time; (c) the potential change in risk caused by future changes in land use; and (d) both individual and population risks.

Progress has been made in developing criteria and guidance concerning how to consider sustainability in remedy selection. However, in the absence of statutory changes, remedy selection at private sites regulated under CERCLA cannot consider the social factors, and may not include the other economic factors, that fall under the definition of sustainability. At federal facility sites, the federal government can choose, as a matter of policy, to embrace sustainability concepts more comprehensively. Similarly, private companies may adopt their own sustainable remediation policies in deciding which remedial alternatives to support at their sites. **New guidance is needed from EPA and DoD detailing how to consider sustainability in the remediation process to the extent supported by existing laws, including measures that regulators can take to provide incentives to companies to adopt more sustainable measures voluntarily.**

CURRENT CAPABILITIES TO REMOVE/CONTAIN CONTAMINATION

Chapter 4 updates the 2005 NRC report on source removal by providing brief reviews of the major remedial technologies that can be applied to complex hazardous waste sites, particularly those with source zones containing dense nonaqueous phase liquids (NAPLs) like chlorinated solvents and/or large downgradient dissolved plumes. This includes surfactant flushing, cosolvent flushing, *in situ* chemical oxidation, pump and treat for hydraulic containment, physical containment, *in situ* bioremediation, permeable reactive barriers, and monitored natural attenuation. Well-established technologies including excavation, soil vapor extraction/air sparging, and solidification/stabilization are not discussed because they have been presented in prior publications and minimal advancements in these technologies have occurred over the past five to ten years. To address what is technically feasible in terms of removing a certain percentage of the total contaminant mass from the subsurface, the sections discuss current knowledge regarding performance and limitations of the technologies, identify remaining gaps in knowledge, and provide case studies supporting these assessments. The following conclusions and recommendations arise from this chapter.

Significant limitations with currently available remedial technologies persist that make achievement of MCLs throughout the aquifer unlikely at most complex groundwater sites in a time frame of 50-100 years. Furthermore, future improvements in these technologies are likely to be incremental, such that long-term monitoring and stewardship at sites with groundwater contamination should be expected.

The Committee could identify only limited data upon which to base a scientifically supportable comparison of remedial technology performance for the technologies reviewed in Chapter 4. There have been a few well-studied demonstration projects and lab-scale research studies, but adequate performance documentation generated throughout the remedial history at sites either is not available or does not exist for the majority of completed remediation efforts. Furthermore, poor design, poor application, and/or poor post-application monitoring at typical (i.e., non-research or demonstration) sites makes determination of the best practicably achievable performance difficult.

There is a clear need for publically accessible databases that could be used to compare the performance of remedial technologies at complex sites (performance data could be concentration reduction, mass discharge reduction, cost, time to attain drinking water standards, etc.). To ensure that data from different sites can be pooled to increase the statistical power of the database, a standardized technical protocol would be needed, although it goes beyond the scope of this report to provide the details of such a protocol.

Additional independent reviews of source zone technologies are needed to summarize their performance under a wide range of site characteristics. Since NRC (2005), only thermal and *in situ* chemical oxidation technologies have undergone a thorough, independent review. Other source zone technologies should also be reviewed by an independent scientific group. Such reviews should include a description of the state of the practice, performance metrics, and sustainability information of each type of remedial technology so that there is a trusted source of information for use in the remedial investigation/feasibility study process and optimization evaluations.

IMPLICATIONS OF CONTAMINATION REMAINING IN PLACE

Chapter 5 discusses the potential technical, legal, economic, and other practical implications of the finding that groundwater at complex sites is unlikely to attain unlimited use and unrestricted exposure levels for many decades. First, the failure of hydraulic or physical containment systems, as well as the failure of institutional controls, could create new exposures. Second, toxicity information is regularly updated, which can alter drinking water standards, and contaminants that were previously unregulated may become so. In addition, pathways of exposure that were not previously considered can be found to be important, such as the vapor intrusion pathway. Third, treating contaminated groundwater for drinking water purposes is costly and, for some contaminants, technically challenging. Finally, leaving contamination in the subsurface may expose the landowner, property manager, or original disposer to complications that would not exist in the absence of the contamination, such as natural resource damages, trespass, and changes in land values. Thus, the risks and the technical, economic, and legal complications associated with residual contamination need to be compared to the time, cost, and feasibility involved in removing contamination outright. The following conclusions and recommendations are made.

Implementing institutional controls at complex sites is likely to be difficult. Although EPA has developed a number of measures to improve the reliability, enforceability, and funding of institutional controls, their long-term efficacy has yet to be determined. Regulators and federal responsible parties should incorporate a more significant role for local citizens in the long-term oversight of institutional controls. **A national, searchable, geo-referenced institutional control database covering as many regulatory programs as practical as well as all federal sites would help ensure that the public is notified of institutional controls.**

New toxicological understanding and revisions to dose-response relationships will continue to be developed for existing chemicals, such as trichloroethene and tetrachloroethene, and for new chemicals of concern, such as perchlorate and perfluorinated chemicals. The implications of such evolving understanding include identification of new or revised ARARs (either more or less restrictive than existing ones), potentially leading to a determination that the existing remedy at some hazardous waste sites is no longer protective of human health and the environment. **Modification of EPA's existing CERCLA five-year review guidance would allow for more expeditious assessment of the protectiveness of the remedy based on any changes in EPA toxicity factors, drinking water standards, or other risk-based standards.**

Careful consideration of the vapor intrusion pathway is needed at all sites where VOCs are present in the soil or groundwater aquifer. Although it has been recognized for more than a decade that vapor intrusion is a potential exposure pathway of concern, a full understanding of the risks over time and appropriate methods for characterizing them are still evolving. Mitigation strategies such as subslab depressurization can prevent vapor intrusion exposure. As a precautionary measure, vapor mitigation could be built into all new construction on or near known VOC groundwater plumes. Vapor mitigation systems require monitoring over the long term to ensure that they are operating properly.

TECHNOLOGY DEVELOPMENT TO SUPPORT LONG-TERM MANAGEMENT

Despite years of characterization and implementation of remedial technologies, many complex federal and private industrial facilities with contaminated groundwater will require long-term management that could extend for decades or longer. Chapter 6 discusses technological developments that can aid in the transition from active remediation to more passive strategies and provide more cost-effective and protective long-term management of complex sites. In particular, transitioning to and improving long-term management can be achieved through (1) better understanding of the spatial distribution of contaminants, exposure pathways, and processes controlling contaminant mass flux and attenuation along exposure pathways; (2) improved spatio-temporal monitoring of groundwater contamination through better application of conventional monitoring techniques, the use of proxy measurements, and development of sensors; and (3) application of emerging diagnostic and modeling tools. The chapter also explores emerging remediation technologies that have yet to receive extensive field testing and evaluation, and it reviews the state of federal funding for relevant research and development. The following conclusions and recommendations are offered.

Long-term management of complex sites requires an appropriately detailed understanding of geologic complexity and the potential distribution of contaminants among the aqueous, vapor, sorbed, and NAPL phases, as well as the unique biogeochemical dynamics associated with both the source area and downgradient plume. Recent improvements to the understanding of subsurface biogeochemical processes have not been accompanied by cost-effective site characterization methods capable of fully distinguishing between different contaminant compartments. Management of residual contamination to reduce the exposure risks via the vapor intrusion pathway is challenged by the highly variable nature of exposure, as well as uncertain interactions between subsurface sources and indoor background contamination.

Existing protocols for assessing monitored natural attenuation and

other remediation technologies should be expanded to integrate compound-specific isotope analysis and molecular biological methods with more conventional biogeochemical characterization and groundwater dating methods. The development of molecular and isotopic diagnostic tools has significantly enhanced the ability to evaluate the performance of degradation technologies and monitored natural attenuation at complex sites.

Although the Committee did not attempt a comprehensive assessment of research needs, research in the following areas would help address technical challenges associated with long-term management at complex contaminated sites (see Chapter 6 for a more complete list):

- **Remediation Technology Development.** Additional work is needed to advance the development of emerging and novel remediation technologies, improve their performance, and understand any potential broader environmental impacts. A few developing remediation techniques could provide more cost-effective remediation for particular combinations of contaminants and site conditions at complex sites, but they are in the early stages of development.

- **Tools to Assess Vapor Intrusion.** Further research and development should identify, test, and demonstrate tools and paradigms that are practicable for assessing the significance of vapor intrusion, especially for multi-building sites and preferably through short-term diagnostic tests. Development of real-time unobtrusive and low-cost air quality sensors would allow verification of those short-term results over longer times at buildings not needing immediate mitigation.

- **Modeling.** Additional targeted modeling research and software development that will benefit the transition of sites from active remediation to long-term management should be initiated. Particular needs include concepts and algorithms for including the processes of back-diffusion and desorption in screening and plume models, and the development of a larger suite of intermediate-complexity modeling tools to support engineering design for source remediation.

Overall research and development have been unable to keep pace with the needs of practitioners trying to conduct remediation on complex sites. Currently, a national strategy for technology development to support long-term management of complex sites is lacking. It is not clear that the pertinent federal agencies will be capable of providing the funding and other support for the fundamental research and development that is necessary to meet the challenges facing complex sites. A comprehensive assessment of future research needs, undertaken at the federal level and involving coordination between federal agencies, would allow research funding to be allocated in an efficient and targeted manner.

BETTER DECISION MAKING DURING THE LONG-TERM MANAGEMENT OF COMPLEX GROUNDWATER CONTAMINATION SITES

The fact that at most complex groundwater sites drinking water standards will not be attained for decades should be more fully reflected in the decision-making process of existing cleanup programs. Thus, Chapter 7 provides a series of recommendations that will accelerate the transition of sites to one of three possible end states: (1) *closure* in which unlimited use and unrestricted exposure levels have been attained; (2) *long-term passive management* (e.g., using natural attenuation with or without monitoring, physical containment, permeable reactive barriers, and/or institutional controls), and (3) *long-term active management* (e.g., indefinite hydraulic containment using pump and treat). The acceleration of this transition to one of three end states is premised on using remedies that are fully protective of human health and the environment in combination with more rapid acceptance of alternative end states other than clean closure.

An alternative approach for better decision making at complex sites is shown in Figure 7-2. It includes the processes currently followed at all CERCLA facilities and at many complex sites regulated under other federal or state programs (RCRA or state Superfund), but it provides more detailed guidance for sites where recalcitrant contamination remains in place at levels above those allowing for unlimited use and unrestricted exposure. This alternative approach diverges from the status quo by requiring the explicit charting of risk reduction (as indicated by, e.g., contaminant concentration reduction) over time. Specifically, if data indicate that contaminant concentrations are approaching an asymptote, resulting in exponential increases in the unit cost of the remedy, then there is limited benefit in its continued operation. At this point of diminishing returns, it is appropriate to assess whether to take additional remedial action (if legally possible) or whether to transition to more passive long-term management.

If asymptotic conditions have occurred, a *transition assessment* is performed. The transition assessment evaluates each of the relevant alternatives (remedy modification or replacement, passive or active long-term management) based on the statutory and regulatory remedy selection criteria. This includes consideration of the risk from residual contamination in subsurface zones, life-cycle costs and the incremental costs compared to the level of risk reduction achieved, and the likely reaction of stakeholders.

The following conclusions and recommendations about this alternative approach are made.

At many complex sites, contaminant concentrations in the plume remain stalled at levels above cleanup goals despite continued operation of remedial systems. There is no clear path forward to a final end state embodied in the current cleanup programs, such that money continues to be spent, with no concomitant reduction in risks. **If the effectiveness of site remediation reaches a point of diminishing returns prior to reaching cleanup goals and optimization has been exhausted, the transition to monitored natural attenuation or some other active or passive management should be considered using a formal evaluation.** This transition assessment would determine whether a new remedy is warranted at the site or whether long-term management is appropriate.

Five-year reviews are an extremely valuable source of field data for evaluating the performance of remedial strategies that have been implemented at CERCLA facilities and could be improved. To increase transparency and allow EPA, the public, and other researchers to assess lessons learned, more should be done, on a national basis, to analyze the results of five-year reviews in order to evaluate the current performance of implemented technologies. **EPA's technical guidance for five-year reviews should be updated to provide a uniform protocol for analyzing the data collected during the reviews, reporting their results, and improving their quality.**

Public involvement tends to diminish once remedies at a site or facility are in place. No agency has a clear policy for sustaining public involvement during long-term management. Regulators and federal responsible parties should work with members of existing advisory groups and technical assistance recipients to devise models for ongoing public oversight once remedies are in place. Such mechanisms may include annual meetings, Internet communications, or the shifting of the locus of public involvement to permanent local institutions such as public health departments.

Although the cost of new remedial actions may decrease at complex sites if more of them undergo a transition to passive long-term management, there will still be substantial long-term funding obligations. Failure to fund adequately the long-term management of complex sites may result in unacceptable risks to the public due to unintended exposure to site contaminants.

1

Introduction

BACKGROUND OF STUDY

Since the 1970s, hundreds of billions of dollars have been invested by federal, state, and local government agencies as well as responsible parties to mitigate the human health and ecological risks posed by chemicals released to the subsurface environment. Many of the contaminants common to these hazardous waste sites, such as metals and volatile organic compounds, are known or suspected to cause cancer or adverse neurological, reproductive, or developmental conditions. Over the past 30 years, some progress in meeting mitigation and remediation goals at hazardous waste sites has been achieved. For example, of the 1,723 sites ever listed on the National Priorities List (NPL), which are considered by the U.S. Environmental Protection Agency (EPA) to present the most significant risks, 360 have been permanently removed from the list because EPA deemed that no further response was needed to protect human health or the environment (EPA, 2012). Seventy percent of the 3,747 hazardous waste sites regulated under the Resource Conservation and Recovery Act (RCRA) corrective action program have achieved "control of human exposure to contamination," and 686 have been designated as "corrective action completed" (EPA, 2011a). The Underground Storage Tank (UST) program also reports successes, including closure of over 1.7 million USTs since the program was initiated in 1984 (EPA, 2010). The cumulative cost associated with these national efforts underscores the importance of pollution prevention and serves as a powerful incentive to reduce the discharge or release of hazardous substances to the environment, particularly when a groundwater resource is threatened.

Although some of the success stories described above were challenging in terms of contaminants present and underlying hydrogeology, the majority of sites that have been closed were relatively simple (e.g., shallow, localized petroleum contamination from USTs) compared to the remaining caseload. Indeed, hundreds of thousands of sites across both state and federal programs are thought to still have contamination remaining in place at levels above those allowing for unlimited land and groundwater use and unrestricted exposure (see Chapter 2).[1] According to its most recent assessment, EPA estimates that more than $209 billion dollars (in constant 2004 dollars) will be needed over the next 30 years to mitigate hazards at between 235,000 to 355,000 sites (EPA, 2004). This cost estimate, however, does not include continued expenditures at sites where remediation is already in progress, or where remediation has transitioned to long-term management.[2] It is widely agreed that long-term management will be needed at many sites for the foreseeable future, particularly for the more complex sites that have recalcitrant contaminants, large amounts of contamination, and/or subsurface conditions known to be difficult to remediate (e.g., low-permeability strata, fractured media, deep contamination). Box 1-1 describes the characteristics of complex sites, where long-term management is a likely outcome given the difficulty of remediating the groundwater to conditions allowing for unlimited use and unrestricted exposure.

The Department of Defense (DoD) exemplifies a responsible party that has made large financial investments to address past legacies of their industrial operations. According to the most recent annual report to Congress (OUSD, 2011), the DoD currently has almost 26,000 active sites under its Installation Restoration Program where soil and groundwater remediation is either planned or under way. Of these, approximately 13,000 sites are the responsibility of the Army, the sponsor of this report. The estimated cost to complete cleanup at all DoD sites is approximately $12.8 billion. (Note that these estimates do not include sites containing unexploded ordnance.)

DoD has set a procedural goal for each of the Services stating that all sites will reach the *response-complete* or *remedy-in-place* milestone by 2014. *Remedy in place* means that a remedial strategy has been implemented and is in the performance assessment stage of the site's life cycle, while *response complete* means that remedial actions have been completed,

---

[1] "Contamination remaining in place," as used in this report, is consistent with the interagency definition of hazardous substances, pollutants, or contaminants remaining at the site above levels that allow for unlimited use and unrestricted exposure (UU/UE) (EPA, 2001; DoD, 2012).

[2] Long-term management is defined as requiring decades to centuries, well beyond the typical 30 years used to discount remedial costs.

although contamination at levels above those allowing for unlimited use and unrestricted exposure may still remain on-site. In addition, the DoD has directed 90 percent of sites at active installations to achieve *response complete* by the end of FY 2018, and 95 percent by the end of FY 2021 (Conger, 2011). These goals will be extremely challenging to meet because at many of the military's remaining sites that have groundwater contamination, one can anticipate the need for long-term management that may take many decades to resolve.

In this context, the Water Science and Technology Board, under auspices of the National Research Council (NRC), initiated a study to assess the future of the nation's subsurface remediation efforts, with a particular focus on technical, economic, and institutional challenges facing the Army and other responsible parties as they pursue aggressive programmatic goals for site closure. It should be noted that there is no single definition of "site closure," nor was the Committee able to agree on a precise consensus definition of the term that would be applicable to all state and federal programs. The term is often used to mean that "no further action" is required at a site (except for various institutional controls)—a connotation that the Committee is comfortable with. However, "no further action" does not mean that site contaminants have been reduced to levels below those allowing for unlimited use and unrestricted exposure. Whenever possible throughout this report, the term "site closure" is replaced with the more specific designations for success used by the various federal and state remediation programs. Chapter 7 abandons the terms "site closure" and "no further action" entirely and instead presents three end states, one of which all sites will achieve: active long-term management, passive long-term management, and achievement of unlimited use and unrestricted exposure levels. The central theme of this report is how the nation will deal with the complex hazardous waste sites where contamination remains in place at levels above those allowing for unlimited use and unrestricted exposure.

REGULATORY RESPONSE TO GROUNDWATER CONTAMINATION

The federal regulatory regime for responding to groundwater contamination consists of several key statutes and regulations enforced primarily by the EPA's Office of Solid Waste and Emergency Response (see Box 1-2 for an overview of the major U.S. cleanup programs). Designed to address problems related to municipal and industrial waste, RCRA was passed in 1976 and promoted recovery methods and techniques to reduce waste generation while also outlining environmentally sound management of hazardous and nonhazardous wastes. In 1980, Congress passed the Superfund Law (Comprehensive Environmental Response, Compensation,

BOX 1-1
Complex Contaminated Sites

Although progress has been made in remediating many hazardous waste sites, there remains a sizeable population of *complex* sites, where restoration is likely not achievable in the next 50-100 years. Although there is no formal definition of complexity, most remediation professionals agree that attributes include areally extensive groundwater contamination, heterogeneous geology, large releases and/or source zones, multiple and/or recalcitrant contaminants, heterogeneous contaminant distribution in the subsurface, and long time frames since releases occurred. Additional factors that contribute to complexity include restrictions on the physical placement or operation of remedial technologies and challenging expectations (e.g., regulatory requirements, cleanup goals, community expectations). The complexity of a site increases with the number of these characteristics present.

Complexity is most intimately tied to limitations on the fundamental contaminant removal and/or destruction processes inherent to all remediation approaches, and the severity of these limitations at any given site is directly related to geology and contaminant distribution. Thus, the more varied the geologic media or lithology, the more complex the flow patterns of contaminants and injected solutions are. The simplest geology is uniform media, like well-sorted sand (called *homogeneous*), while more complex heterogeneous geology includes such varied media as poorly sorted sand with lenses of silt and clay. Fractured media are often considered the most heterogeneous (see Chapter 6 and NRC, 2005a, for more details on hydrogeologic types). Heterogeneous media not only yield intricate contaminant plumes, but also limit the effectiveness of remedial technologies that

TABLE 1-1 Relative Ease of Remediating Contaminated Aquifers as a Function of Contaminant Chemistry and Hydrology

	Contaminant Chemistry	
Hydrogeology	Mobile, Dissolved (degrades/ volatilizes)	Mobile, Dissolved
Homogeneous, single layer	1[a]	1-2
Homogeneous, multiple layers	1	1-2
Heterogeneous, single layer	2	2
Heterogeneous, multiple layers	2	2
Fractured	3	3

[a] Relative ease of cleanup, where 1 is easiest and 4 is most difficult.
SOURCE: NRC (1994).

rely on moving fluid through the subsurface (e.g., injection of surfactants, oxidants, or carbon sources). Heterogeneities can make these technologies less effective due to bypass and/or limited contaminant contact time.

Complexity is also directly tied to the contaminants present at hazardous waste sites, which can vary widely and include organics, metals, explosives, and radionuclides. Some of the most challenging to remediate are dense nonaqueous phase liquids (DNAPLs), including chlorinated solvents. In general, different types of contaminants require different types of treatment and perhaps different remedial approaches altogether. Thus, the more types of contaminants found at a site, the more complex the site. Additionally, some contaminants are more resistant to natural biodegradation processes than others.

NRC (1994) provided a matrix that outlined the difficulty of groundwater remediation on a scale of 1 to 4 (with 4 representing the most difficult to remediate) as a function of hydrogeology and contaminant chemistry, including contaminant distribution in the subsurface (see Table 1-1). Ratings of 3 and 4 in Table 1-1 represent "complex sites" and include

- Sites having contamination in fractured media,
- Dissolved plumes extending more than 1000 m down-gradient of a source,
- Sites impacted by radioactive contaminants,
- Sites with DNAPL impacts extending to depths of 100 ft or greater, and
- Sites with residual NAPL that has diffused into fine-grained units.

Note that Table 1-1 does not factor in some of the topics discussed above (such as the size of the release and regulatory expectations) that can contribute to complexity.

Strongly Sorbed, Dissolved (degrades/ volatilizes)	Strongly Sorbed, Dissolved	Separate Phase LNAPL	Separate Phase DNAPL
2	2-3	2-3	3
2	2-3	2-3	3
3	3	3	4
3	3	3	4
3	3	4	4

BOX 1-2
Brief Overview of U.S. Cleanup Programs and Regulatory Terms Found in this Report

CERCLA: The CERCLA program (established in 1980 and also known as Superfund) locates, investigates, and cleans up the most problematic hazardous waste sites throughout the United States. At private sector sites, the EPA may perform the cleanup with federal funds and seek cost reimbursement from the responsible party or may issue orders or enter a judicially enforceable consent decree and oversee the implementation of long-term cleanups, short-term cleanups ("removal actions"), and other responses. At federal CERCLA sites, the federal party is primarily responsible for cleanup.

RCRA Corrective Action: RCRA is the primary federal statute regulating how wastes (solid and hazardous wastes) must be managed at facilities that treat, store, or dispose of hazardous wastes to avoid potential threats to human health and the environment. However, RCRA also provides corrective action order authority that governs the cleanup of solid waste management units at RCRA permitted facilities (including federal facilities). It is similar to CERCLA, but is primarily implemented by the states. EPA's policy is that the RCRA and CERCLA remedial programs should operate consistently and result in similar environmental solutions when faced with similar circumstances.

UST: The Underground Storage Tank program, which is part of RCRA, governs the cleanup of the nation's large numbers of leaking underground tanks. The sites are individually smaller in scope than a typical site regulated under CERCLA or RCRA corrective action. The UST program focuses on removing products (petroleum or industrial or dry cleaning chemicals) that have leaked out of the tanks, removal of soil, cleanup of the groundwater, and replacement of the tanks.

Brownfields: Brownfields are defined as real properties, the expansion, redevelopment, or reuse of which may be complicated by the presence or potential presence of a hazardous substance, pollutant, or contaminant. EPA's Brownfields program provides funds and technical assistance to states, communities, and

and Liability Act or CERCLA), which authorized broad federal authority to respond directly to the release of hazardous substances that endanger public health or the environment, in addition to taxing the chemical and petroleum industries to establish the Superfund Trust Fund. The NPL of the most contaminated sites was established under CERCLA.

Not long after CERCLA was enacted, it became clear that additional measures would be needed to combat the nation's burden of contaminated sites. In 1984, Congress amended RCRA (via the Hazardous and Solid

other stakeholders in economic redevelopment to work together to assess, safely clean up, and sustainably reuse Brownfields.

Federal Facilities Programs: A number of separate programs exist to address hazardous waste remediation on federal facilities. These include DoD's Military Munitions Response Program, the Installation Restoration Program, which addresses active bases, the Base Realignment and Closure facilities, and Formerly Used Defense Sites. The Department of Energy's Environmental Management program is another example (see Chapter 2). All such sites are variously regulated under CERCLA, RCRA, UST, or state regulations.

Maximum Contaminant Level Goals (MCLGs) are the health-based drinking water concentrations set in EPA's Safe Drinking Water program at a one-in-one million lifetime risk level for carcinogens and, for noncarcinogenic effects, at a concentration at which no adverse health effect are likely from long-term exposure. MCLGs are not enforceable under the Safe Drinking Water Act.

Maximum Contaminant Levels (MCLs) are the legally enforceable drinking water concentration limits for U.S. public drinking water supplies (i.e., supplies to more than 25 people). They are based on a balancing of the residual risk from ingesting the water, the feasibility of treatment to remove the chemical, the detection limit, and the costs to water suppliers. MCLs are enforced under the Safe Drinking Water Act.

Applicable or Relevant and Appropriate Requirements (ARARs) include two separate types of requirements. *Applicable Requirements* are any federal or duly promulgated state standard, requirement, criterion, or limitation under any other federal environmental law that would legally apply to a site. *Relevant and Appropriate Requirements* are any Federal or duly promulgated state standard, requirement, criterion, or limitation under any other federal environmental law that addresses problems or situations similar to the conditions at a site and that is "well suited" to a site. MCLs promulgated under the Safe Drinking Water Act are considered to be ARARs for sites regulated under CERCLA because of the potential for people to ingest the groundwater derived from a contaminated aquifer.

Waste Amendments) to implement more stringent standards for hazardous waste management, to impose restrictions that curbed the practice of land disposal of untreated hazardous waste, and to add authority for EPA and the states to remediate contamination on active RCRA permitted facilities. In 1986, the Superfund Amendments and Reauthorization Act (SARA) amended CERCLA to stress the importance of permanent or innovative solutions, incorporate a more rigorous process to define the goals of remediation that EPA has proposed in its regulations, provide new enforcement authorities and settlement tools, increase state involvement in CERCLA activities, and increase focus on the human health impacts of hazardous waste sites. SARA also established the Defense Environmental Restoration Program and its regulatory underpinnings.

The early years of both programs' implementation were marked by site studies rather than actual remediation. In 1988, EPA released interim measures for RCRA to allow action to be taken sooner to prevent exposure to contamination, and the agency began focusing on completing remedy construction, in particular pump-and-treat practices for groundwater containment and remediation. As more remediation began, it became clear that reaching drinking water standards such as maximum contaminant levels (MCLs), which were the applicable or relevant and appropriate requirements (ARARs) for many sites, was not always feasible, especially at sites with complicated hydrogeology and/or recalcitrant contaminants (see Box 1-2 for definitions of these terms). Thus, during the 1990s EPA continued to revisit and revise its policies for groundwater restoration. For those sites where restoration is impracticable for the foreseeable future given site conditions and the limitations of technologies, the agency created the Technical Impracticability (TI) Waiver (EPA, 1993). As specified in SARA, the TI Waiver was one of six waiver options that allowed for alternative remedial goals other than ARARs in specified portions of a site. For groundwater TI waivers, this required the designation of a "TI Zone" in which a specific ARAR (e.g., an MCL) would be waived. Outside of this zone, the original ARARs still need to be met.

By 1999, the CERCLA program was increasingly finding success in achieving remedy construction milestones on many of the less complex sites. Nonetheless, a 2001 report from Resources for the Future (Probst and Konisky, 2001) stated that most complex sites still had contamination in place at levels above those allowing for unlimited use and unrestricted exposure. Despite these findings, the dedicated taxes supporting the CERCLA program expired in 1995 and have not been reinstated, such that the trust fund was depleted in 2003 (although appropriations to the program continue). Other programs have fared better, such as the Brownfields program (which allows voluntary remediation of sites to promote the redevelopment and reclamation of properties where hazardous substances had been detected or are potentially present) to which $250 million per year was authorized in 2002. RCRA's UST program received additional support from the 2005 Energy Policy Act. In 2009, the American Recovery and Reinvestment Act boosted funding for all remediation-related programs at EPA by $800 million and for other federal remediation programs by $5 billion[3] (EPA, 2011b).

Today, EPA directives on groundwater remedies continue to evolve. In June 2009, the Office of Solid Waste and Emergency Response compiled all existing EPA groundwater policies into one singular directive (EPA, 2009). It reported that CERCLA action is only needed where groundwater contamination exceeds drinking water standards. The directive identified the role of institutional controls, which are non-engineered instruments such as administrative and legal controls that help minimize the potential for human exposure to contamination and/or protect the integrity of the remedy, and determined they are generally not to be the sole basis for a remedy. Classification of groundwater (i.e., whether an aquifer is a current or potential source of drinking water) is to be conducted only by EPA unless there is a state regulatory requirement to do so. And finally the directive acknowledged that EPA policy on point of compliance is to restore groundwater to the maximum extent practicable for beneficial reuse (see also Box 3-2). The report noted that in selecting remedial goals EPA is to consider an array of criteria, including drinking water standards, site-specific risk assessment, and land use.

THE LIFE CYCLE OF A CONTAMINATED SITE

The process for remediation of contaminated sites, from discovery to closure, was first documented in the National Contingency Plan (40 CFR 300 et seq.) in 1980 to reflect the needs of CERCLA. Other regulatory programs provide similar remedial guidance for active sites, including those with underground storage tanks. The Departments of Energy and Defense have developed their own processes that mirror the remedial process found within CERCLA, but using different terminology, while the states implement the federal laws over which they have primacy as well as state programs that encompass additional contaminated sites.

The life-cycle components of the various federal and state remedial programs are similar to one another and listed in Table 1-2 along with approximate time frames for their completion. Following discovery of contamination, a site must be characterized to determine the nature and extent of the contamination, a process that can extend for years into the future for some sites. One of the most important components of the site characterization step is the creation of an accurate conceptual site model (discussed at length in NRC, 2005a). If chemicals of concern are found to exceed certain regulatory limits, and/or a risk characterization indicates that unacceptable

---

[3] http://www.recovery.gov/Transparency/fundingoverview/Pages/contractsgrantsloans-details.aspx#EnergyEnvironment.

TABLE 1-2 Components and Approximate Time Frames in the Life Cycle of a Remediation Program

Activity	**Timing and Key Decisions →** *(shading indicates intensity of activity; timing arrow below indicates a representative time frame for complex sites)* *0 5 10 20 years+*
Site Characterization	
Risk Characterization	
Interim Response	
Establish Objectives	
Remedy Selection	
Remedy Design, Implement, Monitor & Assess Performance	
Future Management Decisions	
No Further Action -or- Long-term Management	

conditions exist, then several activities are possible. Interim responses may be necessary to reduce immediate threats. Once these are in place, remedial action objectives are set, and then remedial alternatives are evaluated and a remedy selected that will meet those objectives within a "reasonable"[4] time

[4] The definition of "reasonable" has been debated for many years at EPA and in state regulatory agencies. There are no statutory or regulatory definitions of this term in the context of soil and groundwater cleanup. EPA explicitly adopted no single definition for all sites because a "timeframe of 100 years may be reasonable for some sites and excessively long for others"

frame. Once the remedy has been designed and installed, monitoring of the impacted media and performance assessment of the remedial technology commence. Information from the monitoring program is used to inform future decision making, including the decision to continue remediation or transition to more passive management. From here, actions can lead to either site closure (including no further action required) or to long-term management. If residual contamination persists at levels above those allowing for unlimited use and unrestricted exposure, engineering and/or institutional controls will be needed. For example, institutional controls like deed restrictions are often necessary for long-term management at sites where physical or hydraulic containment of the contamination is a component of the final solution. Whether long-term management sites will ever attain contamination levels below those allowing for unlimited use and unrestricted exposure is often uncertain; at many sites, perpetual management may be necessary, particularly those with recalcitrant contaminants.

In practice, the process of moving a site from investigation to closure has been much more complex than implied in Table 1-2, and virtually all phases of remediation take more time and resources than originally contemplated. At NPL sites the time lapse from discovery to remedy implementation can exceed two decades. For example, two sites at Letterkenny Army Depot were listed on the NPL in 1987 and 1989, but as of 2011 neither had reached the point of having a final remedy selected (although interim actions have been taken to reduce risk including provision of alternative water supplies). There are numerous reasons for the long time lags between site discovery and closure, including the fact that remedial systems often require modification during implementation due to uncertainties in technology performance. Limited and shrinking resources (particularly at the state level) have also increased the time period between site discovery and eventual remediation.

Some states have proposed changes to their remediation programs in order to expedite moving sites through the system. For example, in 2009 New Jersey created a Licensed Site Remediation Professional (LSRP) Program to address a backlog of relatively simple sites that were not yet closed. Modeled after a similar program in Massachusetts, the New Jersey program transfers responsibility for remediation from the state Department of Environmental Protection (NJDEP) to private contractors licensed by the state in order to reduce the backlog of cases that need to be reviewed and approved by NJDEP. As of July 2010, a total of 392 LSRPs had been

---

(EPA, 1996). Because "reasonable" includes not just scientific judgments, but also values, risk tolerances, and preferences for discounting effects on future generations, definitions can vary by individual (Weitzman, 2001). The Committee, therefore, does not provide its own definition of "reasonable."

licensed within New Jersey, presumably allowing the NJDEP staff to dedicate its resources to the high priority, complex cases and manage cases more efficiently (NJDEP, 2011). Similarly, in California the State Water Resources Control Board has begun to allow closure of thousands of "low-threat" USTs even when groundwater contaminant concentrations exceed MCLs in some portion of the site (SWRCB, 2012). Sites are eligible if remediation has been attempted, the dissolved plume is shrinking, and the groundwater has no future as a drinking water source. California's Regional Water Quality Control Board (RWQCB) in Region 2 (San Francisco Bay area) has attempted to put forth a similar policy for low-threat chlorinated solvent sites (CA Region 2 RWQCB, 2009). Both California policies reflect the belief that at certain sites with low long-term risks to human health or the environment, closure could be granted despite some contaminant levels exceeding regulatory limits. Whether this approach for closure of "low-risk" of "low-threat" sites will be adopted by other regulatory agencies responsible for groundwater remediation is uncertain.

At sites regulated under CERCLA, the desired goal of the remedial process is to reach site closure as defined by unlimited use and unrestricted exposure (a goal which may or may not be practical to attain for decades). For non-CERCLA sites, site closure is often accompanied by a designation of "no further action." Within each of the major federal programs addressing subsurface contamination (CERCLA, RCRA, and RCRA UST) some proportion of the site population has reached this final stage. However, as mentioned before, a no-further-action designation does not necessarily mean that the site is contaminant-free. Indeed, many sites closed under the UST program have residual contamination left in place, some at levels above those allowing for unlimited use and unrestricted exposure. In the case of Superfund, an NPL delisting does not necessarily have to be based on the attainment of MCLs if the human health and environmental risk of the remaining contamination is minimal, groundwater migration is controlled, and remediation is technically impracticable (see Chapter 2). Sites that have residual contamination and require long-term management result in continued remediation costs and liability for the responsible parties or, in the case of "orphan" sites,[5] cost to taxpayers.

[5] Orphan sites are those private (thus, not military) Superfund facilities for which no viable potentially responsible party has been identified. These are transferred to state agencies for further management ten years after reaching the *construction completion* milestone (see Chapter 2).

THE REMEDIATION CHALLENGE

Over the past two decades, the NRC has published several reports on the technical, economic, institutional, and policy challenges arising from contamination of the nation's subsurface resources, with a particular focus on whether or not groundwater restoration is feasible or practicable (Box 1-3). Each of the NRC studies has, in one form or another, recognized that in almost all cases, complete restoration of contaminated groundwater is difficult, and in a substantial fraction of contaminated sites, not likely to be achieved in less than 100 years. The most difficult sites to remediate are characterized by their large size, heterogeneous hydrogeology, and/or multiple (and recalcitrant) contaminants. As suggested in Figure 1-1, sites contaminated with dense nonaqueous phase liquids (DNAPLs) like trichloroethene (TCE) and tetrachloroethene are particularly challenging to restore because of their complex contaminant distribution in the subsurface. At most complex sites, contamination will persist in the groundwater for a long time at levels above those allowing for unlimited use and unrestricted exposure. This reality, combined with the need to use the affected groundwater in some cases, has led to a considerable debate about the relative costs and benefits of remediating the sources of groundwater contamination as opposed to pathway interruption (e.g., vapor mitigation and wellhead treatment in the contaminant plumes).

Figure 1-2 shows four possible trajectories of post-remediation dissolved plume behavior at sites causing groundwater impacts. The first trajectory assumes no remedial action, such that the state of the plume remains as is and the regulatory goal at the receptor is never reached until the source naturally depletes. The second trajectory represents ineffective remediation where, after remediation stops, the dissolved plume returns to the original state or to one with a bigger footprint and higher concentrations resulting from source mass redistribution during the remediation attempt (e.g., the DNAPL pools were mobilized during remediation). The third trajectory shows a partially effective remedial action, but one in which the system will not reach an acceptable state for a very long time (e.g., because of matrix mass rebound after the removal of a DNAPL source that results in long-term plume persistence). In this situation, the question of whether to continue active remediation versus some more passive management like containment becomes paramount. The fourth trajectory, which might be called the best practicably achievable trajectory, represents a case where the remediation has resulted in a post-remediation dissolved plume where the remediation goals are achievable. Whether this trajectory can achieve remedial goals in a reasonable length of time is not known and depends on the scale of the x axis. Our ability to predict these trajectories for complex sites is highly uncertain, because of imprecise knowledge of source zone mass

BOX 1-3
Select NRC Studies Relevant to Groundwater Remediation at Sites with Persistent Contamination

The following five NRC reports have particular relevance to this report, as they address the feasibility of subsurface remediation from various perspectives:

Alternatives for Groundwater Cleanup[a] (NRC, 1994) reviewed extensive data from 77 pump-and-treat sites and found that ease of remediation depended on the nature of the contamination present and the site hydrogeology. Only two of 77 sites were rated as easy to clean up, and only eight of the 77 sites reached remedial goals, like obtaining MCLs in groundwater. The report suggested that an infeasibility fee be charged to potentially responsible parties (PRPs) to further research and development of new technologies to remediate such sites.

Groundwater and Soil Cleanup: Improving Management of Persistent Contaminants (NRC, 1999) provided a comprehensive review of groundwater and soil remediation technologies, focusing on three classes of contaminants that have proven very difficult to treat once released to the subsurface: metals, radionuclides, and DNAPLs, such as chlorinated solvents. The report concluded that "removing all sources of groundwater contamination, particularly DNAPLs, will be technically impracticable at many Department of Energy sites, and long-term containment systems will be necessary for these sites."

Natural Attenuation for Groundwater Remediation (NRC, 2000) focused on monitored natural attenuation (MNA) and considered when and where MNA will work. Prompted by the increasing use of MNA as a remedy at hazardous waste sites (from less than 5 percent of Records of Decision in 1985 to more than 25 percent in 1995), it evaluated the likelihood of success of MNA for many contaminant classes. The report found that the likelihood of MNA success for most compounds is low, despite the increase in its use at Superfund facilities. None of the 14 protocols reviewed in the report was completely adequate in its treatment of the important scientific and technological, implementation, and community concerns inherent to MNA. Thus, EPA was advised to provide new guidance on protocols.

and its distribution (sometimes referred to as "source zone architecture"[6]) and due to the diversity of opinions on the anticipated cost, effectiveness, and robustness of various remediation technologies.

[6] Source zone architecture refers to the distribution of DNAPL as either residual saturation (immobile ganglia and blobs) in more permeable media or as pools on tops of low-permeability layers. Residual DNAPL has a higher surface area, which provides greater exposure to flowing groundwater, contributing significantly to downgradient contaminant mass flux. In contrast, pools usually contain more DNAPL mass but have lower surface area exposed to clean groundwater and a correspondingly lower contribution to mass flux. See Figure 1-1.

Environmental Cleanup of Navy Facilities: Adaptive Site Management (NRC, 2003) developed the concept of adaptive site management (ASM) to deal with sites where remedial goals have not been reached after some significant amount of time operating the remedy (the so-called asymptote effect). The hallmark of ASM is doing things while a remedy is ongoing that will inform the process if the remedy fails. The report describes several management decision points at which new information from parallel activities could be incorporated to allow site remedies to be reconsidered over time.

Contaminants in the Subsurface (NRC, 2005a) responded to another trend in hazardous waste remediation—the use of aggressive source removal. Source removal via such technologies as *in situ* chemical oxidation, thermal treatment, and surfactant-enhanced flushing was often attempted without a clear understanding of whether those actions would in fact remove mass or lead to substantial changes in contaminant concentration in groundwater. The report defined five hydrogeologic settings, based on the degree of heterogeneity and permeability found in subsurface soils. In addition, it created a table for each source remediation technology discussing the extent to which that technology could meet five different goals in each of the five hydrogeologic settings. The goals included mass removal, concentration reduction, mass flux reduction, reduction of source migration potential, and a change in toxicity. The report concluded that available data from field studies do not demonstrate what effect source remediation is likely to have on water quality.

---

[a] Although sometimes used synonymously, there is an important difference between the terms remediation and cleanup. Remediation is the "removal of pollutants or contaminants from environmental media such as soil, groundwater, sediment, or surface water for the general protection of human health and the environment" (http://sis.nlm.nih.gov/enviro/iupacglossary/glossaryr.html); it does not imply removal or destruction of all contaminants. Cleanup is the restoration of the affected site to a condition allowing for UU/UE which generally implies meeting drinking water standards in the case of contaminated groundwater. This report primarily uses the term remediation to avoid confusion.

Key Challenges for Subsurface Remediation at DoD Facilities

The DoD has invested over $30 billion to address contamination of the soil and groundwater at military bases in the United States and abroad (OUSD, 2011). Under the Installation Restoration Program, many individual sites have been closed with no further action required. However, at complex sites characterized by multiple contaminant sources, large past releases of chemicals, or highly complex geologic environments, meeting the DoD's ambitious programmatic goals for remedy in place/response complete seems unlikely and site closure almost an impossibility. The recent

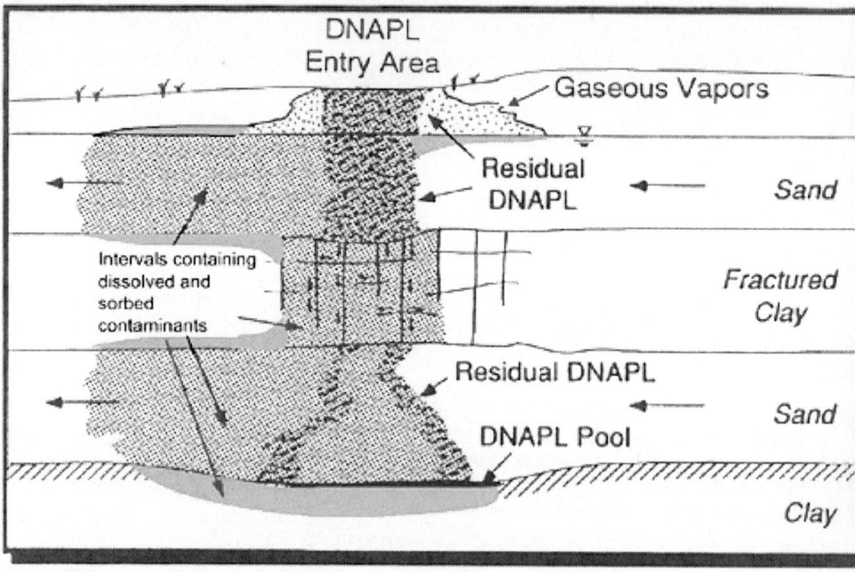

FIGURE 1-1 Hypothetical DNAPL release site. In addition to residual and pooled DNAPL sources, the figure depicts vapor-phase contaminants in the unsaturated zone and a plume of dissolved and sorbed contamination in the saturated zone downgradient of the DNAPL. Note that the residual DNAPL is more likely to occur in sparse pools and fingers, rather than in the massive bodies inferred in the picture. SOURCE: NRC (2005a); adapted from Cohen et al. (1993).

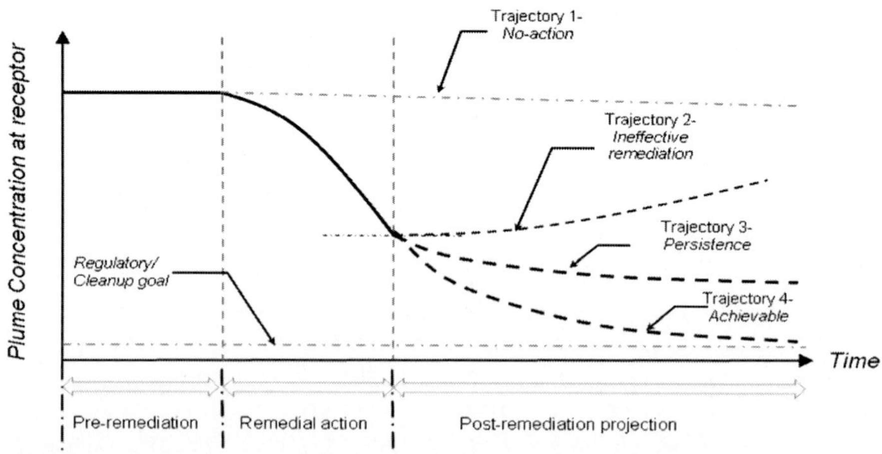

FIGURE 1-2 Schematic of possible post-remediation trajectories for plume behavior. The y axis could be any decision variable used to measure the remedial objective (e.g., the contaminant concentration at a point of compliance).

policy memorandum from the Air Force (Yonkers, 2011) regarding the new milestone of *accelerated site completion* does not appear to clarify or simplify military remediation requirements.

An example of the array of challenges faced by the DoD is provided by the Anniston Army Depot, where groundwater is contaminated with chlorinated solvents (as much as 27 million pounds of TCE [ATSDR, 2008]) and inorganic compounds. TCE and other contaminants are thought to be migrating vertically and horizontally from the source areas, affecting groundwater downgradient of the base including the potable water supply to the City of Anniston, Alabama. The interim Record of Decision called for a groundwater extraction and treatment system, which has resulted in the removal of TCE in extracted water to levels below drinking water standards. Because the treatment system is not significantly reducing the extent or mobility of the groundwater contaminants in the subsurface, the current interim remedy is considered "not protective." Therefore, additional efforts have been made to remove greater quantities of TCE from the subsurface, and no end is in sight. Modeling studies suggest that the time to reach the TCE MCL in the groundwater beneath the source areas ranges from 1,200 to 10,000 years, and that partial source removal will shorten those times to 830–7,900 years (Tetra Tech, 2011). Although Anniston is a strong candidate for a TI wavier, DoD officials have struggled to convince regulators of the need for alternative remedial objectives (at this and other complex military sites).

In part, the delays and transaction costs experienced at complex sites have led to the use of alternative contracting mechanisms for site remediation within the DoD, including performance-based contracting. In some cases, this has involved requesting guaranteed fixed-price proposals to achieve certain milestones within specified schedule deadlines. The intent of these contracting procedures is to accelerate remediation and reduce the overall life-cycle costs (Army, 2010). Anecdotal stories suggest that this process has indeed accelerated transition of sites to the status of remedy in place, but not to site closure.

It appears that future liabilities for the DoD are unknown because of the uncertain time frames to achieve remedial action objectives at the more complex sites. It is probable that these sites will require significantly longer remediation times than mandated, and thus, continued financial demands for monitoring, maintenance, and reporting. In addition, the tension between remedial strategies involving long-term containment compared to contaminant removal from the subsurface will likely continue, with a lack of efficient protocols that could potentially reduce overall life-cycle costs. Finally, consistent with DoD goals of achieving a greater level of environmental sustainability in all environmental programs (DoD, 2009), increased incorporation of sustainability metrics in remedial decision making appears likely.

STATEMENT OF TASK AND REPORT ROADMAP

Although technologies capable of removing substantial amounts of contaminants from groundwater have evolved significantly over the last 40 years, our ability to predict remediation performance, and its associated groundwater quality improvement, with adequate certainty is limited. Additional questions must be answered before management of sites can proceed in a way that is protective in an era of limited financial resources. The following questions guided the work of this NRC committee.

1. **Size of the Problem**
   At how many sites does residual contamination remain such that site closure is not yet possible? At what percentage of these sites does residual contamination in groundwater threaten public water systems?
2. **Current Capabilities to Remove Contamination**
   What is technically feasible in terms of removing a certain percentage of the total contaminant mass? What percent removal would be needed to reach unrestricted use or to be able to extract and treat groundwater for potable reuse? What should be the definition of "to the extent practicable" when discussing contaminant mass removal?
3. **Correlating Source Removal with Risks**
   How can progress of source remediation be measured to best correlate with site-specific risks? Recognizing the long-term nature of many problems, what near-term endpoints for remediation might be established? Are there regulatory barriers that make it impossible to close sites even when the site-specific risk is negligible and can they be overcome?
4. **The Future of Treatment Technologies**
   The intractable nature of subsurface contamination suggests the need to discourage future contaminant releases, encourage the use of innovative and multiple technologies, modify remedies when new information becomes available, and clean up sites sustainably. What progress has been made in these areas and what additional research is needed?
5. **Better Decision Making**
   Can adaptive site management lead to better decisions about how to spend limited resources while taking into consideration the concerns of stakeholders? Should life-cycle assessment become a standard component of the decision process? How can a greater understanding of the limited current (but not necessarily future) potential to restore groundwater be communicated to the public?

Although the focus of the study was on military sites, particularly those of the U.S. Army, the conclusions and recommendations are relevant to both public and private hazardous waste sites.

The study was intended to focus on those recalcitrant contaminants occurring most frequently at the most complex sites, in particular organic compounds present as DNAPLs. In addition, groundwater cleanup, as opposed to soil remediation, poses the greatest remediation challenge and was thus the primary focus of this study. Other topics relevant to the nation's subsurface remediation efforts that are not reviewed here include the impacts of agricultural activities on groundwater quality, abandoned mine sites, and impacts from municipal and solid waste landfills. Finally, although Department of Energy sites also illustrate the challenges of recalcitrant contamination requiring long-term management, because a number of NRC reports have reviewed sites with radioactive contaminants (NRC, 2005b, 2007, 2009) they are not discussed further here.

The questions in the statement of task are addressed variously throughout the report. Thus, Chapter 2 attempts to bound the size of the problem (first task item), including federal sites under the jurisdiction of EPA (CERCLA, RCRA, and UST programs), the military, the Department of Energy, and state remediation programs. For all programs, the Committee sought information on the total number of sites, the costs expended to date and to clean up remaining sites, and the number of sites affecting a drinking water supply. Chapter 2 (and Appendix C) also discusses sites that have been "closed" and characterized as successes to illustrate the point that many "closed" sites are still contaminated (though they are protective of human health and the environment).

Chapter 3 discusses elements primarily from the third task item but also from the second and fourth. With regards to the third task item, it outlines common remedial objectives (stemming from regulatory programs) including the use of MCLs and other risk-based objectives. It demonstrates the flexibility inherent in CERCLA for defining measurable remedial objectives that protect human health and the environment and prevent the spread of contamination, in the most cost-efficient way. It also discusses a suite of alternative remedial objectives that could be considered for sites slated for long-term management and the barriers that prevent more frequent use of these alternatives. The chapter introduces the concept of sustainability in remediation and its role as a remedial objective (from the fourth task item), and it provides the regulatory definition of "maximum extent practicable" (from the second task item).

Chapter 4 focuses on the current capabilities of technologies to remove or contain subsurface contamination (the second task item). For the major classes of removal technologies, including extraction, thermal, chemical, and biological technologies, as well as containment, the chapter updates the

NRC (2005a) report in addressing what is technically feasible in terms of removing contaminant mass. Case studies for the technologies are included both within the chapter and in Appendix B to illustrate the capabilities of existing technologies for removing mass from the subsurface. It should be noted that the percent contaminant removal that would be needed to reach unrestricted use, or to be able to extract and treat groundwater for potable reuse, can only be determined on a site-specific basis and is not addressed further in this report. It depends on knowing the amount of contamination present at a site as well as the removal capabilities of the chosen well-head treatment technologies.

Although not explicitly called for in the statement of task, the risks of leaving residual contamination in place in the subsurface are discussed comprehensively in Chapter 5. These include technological risks such as the failure of hydraulic containment or barrier technologies, or the inability of current treatment and containment systems to handle unregulated and unanticipated contaminants. Chapter 5 also discusses institutional issues that arise when contamination remains in place, such as economic and litigation risks like possible natural resource damage and trespass suits and the failure of institutional controls. The consequences of leaving contamination in place for water utilities and domestic wells are discussed.

Chapter 6 focuses on the future of treatment technologies (fourth task item). It provides a targeted discussion of those areas of technology development relevant to the problem of leaving contamination in place, but is not meant to be a comprehensive cataloging of remediation technologies (see Chapter 4). In addition to remediation technologies, it speaks to advances in our understanding of hydrogeology and contaminant transport pathways, improved diagnostics and new geophysical methods, and the use of sensors for monitoring long-term management. It should be noted that the report does not comprehensively discuss the need to discourage future contaminant releases, as significant progress has been made in this area. That is, it is now so expensive to manage contaminated sites that potentially responsible parties will go to great lengths to avoid causing groundwater contamination.

The report ends with a chapter on how better decision making can help manage sites with residual contamination (addressing the fifth task item, as well as the call for near-term endpoints in the third task item). This includes the introduction of several important decision points and a transition assessment to help move sites to one of three end states. The transition assessment is akin to the adaptive site management concept first developed in NRC (2003), but focuses specifically on complex sites where long-term management is likely needed. The chapter discusses the economic, risk assessment, and risk communication implications of this transition assessment. Life-cycle assessment is not discussed further because it goes beyond the issues presented by groundwater sites with residual contamination.

The Committee reached consensus on all conclusions and recommendations in the report except regarding a proposal for a public/private partnership that could be established to manage portfolios of sites in a manner similar to initiatives undertaken by private responsible parties (e.g., separate companies to manage legacy sites) or public agencies (e.g., Minnesota Pollution Control Agency's Closed Landfill Program). In these entities, liability and long-term responsibility for contaminated sites are transferred from the responsible party to a new entity. In the case of the Minnesota Pollution Control Agency, owners of sanitary landfills pay a fee to the program in exchange for transfer of all future liability and management costs. The Committee considered the concept of an industry/government/public organization that could be formed to assume management for a portfolio of sites, called the "environmental liability management organization (ELMO)." PRPs would pay ELMO to assume liability and site management, and the payment would cover expected damages and management costs for as long as the contamination remains above levels allowing for unlimited use and unrestricted exposure. The Committee could not agree on the details of such a proposed entity, but all members agreed that future consideration of such an organization could potentially provide a number of advantages to all parties, especially in the context of long-term management of sites.

Throughout the report are case studies of complex sites where it is most likely that contamination will remain in place after remedy operation. These sites are the most important to the Army in terms of being able to reach its 2014 goal of remedy in place/response complete and the updated goals of DoD, and in determining its future remediation liability. A list of the complex sites studied in depth by the Committee is found in Appendix B.

REFERENCES

Army. 2010. Use of Performance-Based Acquisition in the Army Environmental Cleanup Program.

ATSDR (Agency for Toxic Substance & Disease Registry). 2008. Follow-up Health Consultation: Anniston Army Depot. September 30, 2008.

CA Region 2 RWQCB. 2009. Assessment Tool for Closure of Low-Threat Chlorinated Solvent Sites. Groundwater Committee of the California RWQCB San Francisco Bay Region.

Cohen, R. M., J. W. Mercer, and J. Matthews. 1993. DNAPL Site Evaluation. Boca Raton, FL: C. K. Smoley Books, CRC Press.

Conger, J. 2011. Memo: New Goals for DERP. July 18, 2011.

DoD (Department of Defense). 2009. Consideration of Green and Sustainable Remediation Practices in the Defense Environmental Restoration Program. Office of the Secretary of Defense. August 10.

DoD. 2012. Manual, Defense Environmental Restoration Program No. 4715.20. March 9. Washington, DC: DoD Under Secretary of Defense for Acquisition, Technology & Logistics. http://www.dtic.mil/whs/directives/corres/pdf/471520m.pdf.

EPA (U.S. Environmental Protection Agency). 1993. Guidance for Evaluating Technical Impracticability of Ground-Water Restoration. Directive 9234.2-25. Washington, DC: EPA OSWER.

EPA. 1996. Presumptive Response Strategy and Ex-Situ Treatment Technologies for Contaminated Ground Water at CERCLA Sites. OSWER Directive 9283.1-12. EPA 540/R-96/023. Washington, DC: EPA OSWER.

EPA. 2001. Comprehensive Five-Year Review Guidance. EPA 540-R-01-007. OSWER Directive 9355.7-03B-P. Washington, DC: EPA OSWER.

EPA. 2004. Cleaning up the Nation's Waste Sites: Markets and Technology Trends. 2004 Edition.

EPA. 2009. Summary of Key Existing EPA CERCLA Policies for Groundwater Restoration. OSWER Directive 9823.1-33. Washington, DC: EPA OSWER.

EPA. 2010. Semiannual Report of UST Performance Measures, Mid-Fiscal Year 2010. Washington, DC: EPA Office of Underground Storage Tanks. http://www.epa.gov/oust/cat/ca_10_12.pdf.

EPA. 2011a. Facility Information: 2020 Corrective Action Universe. http://www.epa.gov/osw/hazard/correctiveaction/facility/index.htm#2020 [accessed November 8, 2011].

EPA. 2011b. American Recovery and Reinvestment Act, Quarterly Performance Report, Quarter 4, Cumulative Results as of September 30, 2011. http://www.epa.gov/recovery/pdfs/2011_Q4_Perf_Rpt.pdf.

EPA. 2012. National Priorities List. http://www.epa.gov/superfund/sites/npl/ [accessed August 17, 2012].

NJDEP (New Jersey Department of Environmental Protection). 2011. Site Remediation Reform Act. http://www.nj.gov/dep/srp/srra/ [accessed November 8, 2011].

NRC (National Research Council). 1994. Alternatives for Groundwater Cleanup. Washington, DC: National Academy Press.

NRC. 1999. Groundwater and Soil Cleanup: Improving Management of Persistent Contaminants. Washington, DC: National Academy Press.

NRC. 2000. Natural Attenuation for Groundwater Remediation. Washington, DC: National Academy Press.

NRC. 2003. Environmental Cleanup at Navy Facilities: Adaptive Site Management. Washington, DC: The National Academies Press.

NRC. 2005a. Contaminants in the Subsurface. Washington, DC: The National Academies Press.

NRC. 2005b. Improving the Characterization and Treatment of Radioactive Wastes for the Department of Energy's Accelerated Site Cleanup Program. Washington, DC: The National Academies Press.

NRC. 2007. Plans and Practices for Groundwater Protection at the Los Alamos National Laboratory. Washington, DC: The National Academies Press.

NRC. 2009. Advice on the Department of Energy's Cleanup Technology Roadmap: Gaps and Bridges. Washington, DC: The National Academies Press.

OUSD (Office of the Under Secretary of Defense for Acquisition, Technology and Logistics). 2011. FY2010 Annual Report to Congress.

Probst, K. N., and D. M. Konisky. 2001. Superfund's Future: What Will it Cost? Washington, DC: Resources for the Future.

SWRCB (State Water Resource Control Board). 2012. Low-Threat Underground Storage Tank Case Closure Policy (Effective August 17, 2012). Memo of August 24.

Tetra Tech. 2011. Focused Feasibility Study for Southeast Industrial Area OU1 of ANAD, Draft, May.

Weitzman, M. L. 2001. Gamma discounting. American Economic Review 91(1):260-271.

Yonkers, T. 2011. Policy for Refocusing the Air Force Environmental Restoration Program. February 24.

2

Magnitude of the Problem

This chapter focuses on the first topic of the Committee's statement of task: assessing the size of the groundwater contamination problem caused by residual subsurface contaminants. Specifically, in this chapter the Committee strives to answer three primary questions: (1) at how many sites does residual contamination remain such that site closure is not yet possible, (2) at what percentage of these sites does residual contamination in groundwater threaten public water systems,[1] and (3) what are the projected costs for reaching site closure or for long-term management? To answer these questions, the Committee gathered information on the major federal and state regulatory programs under which hazardous waste is cleaned up to determine the size and scope of these programs and relevant trends over time. The chapter also includes a discussion on "closed" sites (the meaning of which varies by program), because such sites may contain residual contamination at levels exceeding those allowing for unlimited use and unrestricted exposure (UU/UE).

NUMBER OF U.S. HAZARDOUS WASTE SITES

The Committee sought the following types of information to assess the magnitude of the nation's hazardous waste problem:

[1] The Safe Drinking Water Act defines public water systems as consisting of community water supply systems; transient, non-community water supply systems; and non-transient, non-community water supply systems—all of which can range in size from those that serve as few as 25 people to those that serve several million.

- Number of sites characterized by progress through the major phases of remediation from site discovery to site closure, as outlined in Table 1-1,
- Principal chemicals of concern, and
- Status of "closed" sites with respect to the potential presence of residual contamination.

At a national level, information was gathered from the U.S. Environmental Protection Agency (EPA) for sites that fall under the Comprehensive Environmental Response, Compensation, and Liability Act (CERCLA), Resource Conservation and Recovery Act (RCRA), or Underground Storage Tank (UST) programs using publicly available databases and via conversations with EPA program officers. Department of Defense (DoD) sites were explored with the aid of the online Annual Reports to Congress and via conversations with DoD staff. Information from the Department of Energy (DOE) and other federal agencies was collected from published literature. Another large group of sites includes those that fall under state purview, such as state Superfund, voluntary cleanup programs, Brownfields, and some dry cleaning sites. Information about such sites was gathered from a variety of sources, including state websites and databases, third-party websites, published literature, and conversations with state program managers.

The numbers in this chapter reflect the Committee's best efforts to compile available data on the magnitude of the problem, but there is significant uncertainty associated with some of the data. First, some of the reported data reflect detailed analyses (e.g., DoD, CERCLA, RCRA) while other data are only estimates. Second, there are differences in accounting across the programs that make it difficult to assess the magnitude of the hazardous waste problem on a consistent basis. In particular, CERCLA and RCRA's best available data are for facilities that could and often do contain many individual contaminated sites. To make matters even more confusing, the term "site" is used by the CERCLA and RCRA programs to mean an entire facility, while other programs use the term "site" to represent an individual contaminant release within a larger facility. *In this report the term "site" refers to an individual area of contamination within a facility; to avoid confusion, the term "Superfund site" is not used when referring to a facility on the Superfund list.* Finally, the statement of task requests information on the numbers of sites that have yet to reach "site closure"—a term that is defined differently by each of the large federal cleanup programs as well as by state agencies.

Considering these sources of uncertainty (estimates vs. real data, summing of facilities and individual sites, and the varying definitions of site closure), the overall total should be considered as a *rough* idea of the magnitude of the problem. Though it can be argued that there is limited utility in tallying the numbers of sites given these sources of uncertainty, this is done at the end of the chapter to provide the reader with an order-of-magnitude estimate of the size of the country's burden for cleanup of hazardous waste sites.

Department of Defense

The DoD environmental remediation program, measured by the number of facilities, is the largest such program in the United States, and perhaps the world. In budgetary terms, it is only exceeded by the U.S. Department of Energy's Environmental Management Program. The Installation Restoration Program (IRP), which addresses toxic and radioactive wastes as well as building demolition and debris removal, is responsible for 3,486 installations containing over 29,000 contaminated sites at active, Base Relocation and Closure (BRAC), and Formerly Used Defense Site (FUDS) properties (see Table 2-1). The Military Munitions Response Program, which focuses on unexploded ordnance and discarded military munitions, is beyond the scope of this report and is not discussed further here, although its future expenses are greater than those anticipated for the IRP. Additionally, DoD has responsibility for sites that are not included in the IRP totals, including 67 properties (primarily private waste disposal sites) in 31 states (OUSD, 2011). In total, the DoD has 141 installations that have been listed on the

TABLE 2-1 DoD Installation Restoration Program Installations, Sites, Expenses to Date, and Cost to Complete[a]

IRP	Number of Installations	Number of Sites	Costs Through FY10 (1000s)	Cost to Complete (1000s)
Active	1,622	21,528	$19,693,452	$7,230,071
Base Realignment and Closure(BRAC)	228	5,127	$8,085,265	$2,706,374
Formerly Used Defense Site (FUDS)	1,636	2,921	$3,136,362	$2,820,145
Total	3,486	29,576	$30,915,079[b]	$12,756,590

[a] According to the DERP Annual Report to Congress for FY 2010, the cost to complete (CTC) is derived from site-level funding information and can be impacted by prioritization, input from regulators and other stakeholders, the complexity of the cleanup, and the technologies that are available and chosen (DoD, 2012). The cost numbers are not adjusted for inflation.
[b] An additional $97.9 million was spent on remediation of sites not included in the Installation Restoration Program through 2010 (OUSD, 2011, p. E9-1).
SOURCE: OUSD (2011).

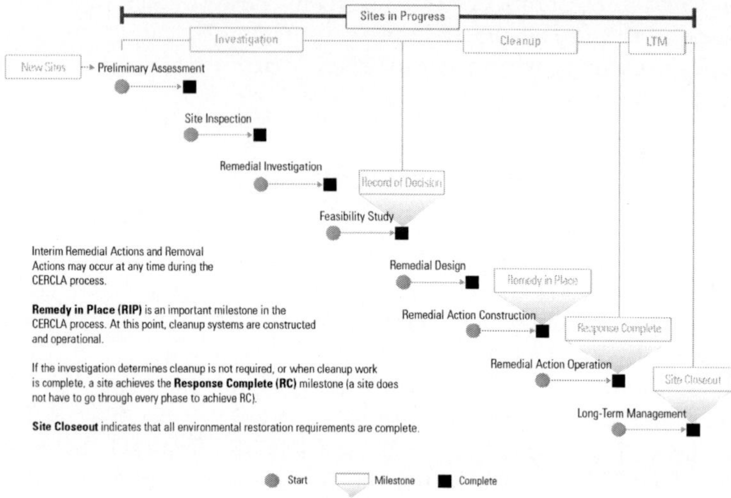

FIGURE 2-1 DoD CERCLA environmental restoration phases and milestones. SOURCE: Adapted from OUSD (2011).

National Priorities List (NPL) because they contain at least one very contaminated site, thus qualifying the entire installation for the NPL.

The phases and milestones used by the DoD to measure progress are shown in Figure 2-1. The DoD has established a performance goal for active and BRAC installations to achieve either the *remedy-in-place* or the *response-complete* milestone by 2014.[2] FUDS are supposed to achieve those milestones by 2020. As shown in Table 2-2, 79 percent of Installation Restoration Program sites have met that goal as of FY 2010. While impressive, these numbers should not be taken to imply that the remaining sites will be remediated at the same pace. This is because the bulk of the response-complete sites to date have been "low-hanging fruit," completed with little remediation activity. Indeed, at least 62 percent of the Installation Restoration Program sites that have achieved *response complete* (14,302 sites) did so without reporting a *remedy in place* (Deborah Morefield, DoD, personal communication, January 2011). Furthermore, in July 2011 DoD established more demanding goals based upon moving sites from *remedy in place* to *response complete* (Conger, 2011), such that success has been redefined within the agency to mean that 95 percent of Installation Restoration Program sites must achieve response complete by 2021.

The Defense Department's task is formidable because the remaining

[2] 2015 for Legacy BRAC sites.

site portfolio consists primarily of the largest and most complex sites, such as groundwater plumes containing difficult-to-remediate substances such as chlorinated solvents that can be present in the subsurface as dense nonaqueous phase liquids (DNAPLs). In the Committee's experience, these account for many of the 1,933 Installation Restoration Program sites where remedies are in place but which have not achieved response complete (Table 2-2), as well as many of the sites still undergoing study. DoD counts nearly 1,400 sites in the Long-Term Management phase, when the active response is complete, but where residual contamination remains above levels allowing for UU/UE. These sites may be subject to land use restrictions, periodic reviews, monitoring, and/or maintenance. Thus, the known number of DoD Installation Restoration Program sites with residual contamination in place is 4,329 (2,931 + 1,398). (Sites with a remedy in place or which are response complete are not included in this total because it would be impossible to know whether they contain residual contamination without considering each site.)

A snapshot of the DoD's contaminated sites is provided by a 2006 survey of occurrence data of hazardous contaminants at 440 installations for which the armed services had electronic records (Hunter, 2006). These installations accounted for about two-thirds of the total Installation Restoration Program's sites. The researchers reported that trichloroethene (TCE) has been found in groundwater at concentrations above the preliminary remediation goal at 69 percent of those installations. Another volatile solvent, tetrachloroethene (PCE), was found above its preliminary remediation goal at 57 percent of the 440 installations. Naphthalene, a key component of jet fuel, was found above its preliminary remediation goal at 48 percent of the installations. They also reported the widespread presence of toxic metals such as lead, arsenic, and nickel at high levels, but noted that most of those concentrations were consistent with naturally occurring background concentrations.

TABLE 2-2 DoD Installation Restoration Program Sites by Select Cleanup Phases or Milestones (see Figure 2-1)

IRP	Cleanup Planned or Under Way	Remedy in Place	Response Complete	Long-Term Management Under Way
Active	2,083	1,530	17,053	905
BRAC	529	396	4,065	403
FUDS	319	7	2,110	50
Total	2,931	1,933	23,228	1,398

NOTE: Remedy in Place is a subset of Cleanup Planned or Under Way and Long-Term Management Under Way is a subset of Response Complete.
SOURCE: OUSD (2011).

CERCLA

The CERCLA program was established to address hazardous substances at abandoned or uncontrolled hazardous waste sites. Through the CERCLA program, the EPA has developed the National Priorities List (NPL), which is periodically updated to reflect facilities with the highest priority hazardous waste sites.[3] The remedial actions at most nongovernmental CERCLA facilities are implemented by potentially responsible parties (PRPs) through legally enforceable administrative orders or settlement agreements, with EPA being the main agency responsible for enforcing the program. Where there are no viable nongovernmental PRPs, EPA performs the remediation pursuant to federal funds (i.e., the so-called Superfund, a term that has come to define the entire program). At governmental facilities, other federal agencies such as the DoD and DOE are responsible for cleaning up their sites in accordance with CERCLA requirements. States can also take the lead in determining remedial alternatives and contracting for the design and remediation of a site. Table 2-3 shows the phases of the CERCLA program, including the major milestones.

There are 1,723 facilities that have been on the NPL, including 59 that have been proposed by the EPA and are currently awaiting final agency action. Table 2-4 below shows a breakdown of these by status and milestone. As of June 2012, 359 of the 1,723 facilities have been "deleted" from the NPL, which means the EPA has determined that no further response is required to protect human health or the environment; 1,364 remain on the NPL. About 80 of those deleted facilities had contaminated groundwater and were evaluated more extensively by the Committee (see later section on closed sites and Appendix C). Facilities that have been deleted from the NPL are eligible for future Superfund-financed remedial action in the event of future conditions warranting the action. To provide some temporal perspective on these numbers, in 2004 there were 1,244 NPL facilities. At that time, 274 had been deleted from the NPL or referred for response to another authority.

Statistics from EPA (2004) illustrate the typical complexity of hazardous waste sites at facilities on the NPL. Volatile organic compounds (VOCs) are present at 78 percent of NPL facilities, metals at 77 percent, and semivolatile organic compounds (SVOCs) at 71 percent. All three contaminant groups are found at 52 percent of NPL facilities, and two of the groups at 76 percent of facilities (but not necessarily in the same matrix, i.e., soil, groundwater, sediment). In 1993, EPA (1993) reported that

[3] See http://www.epa.gov/superfund/programs/npl_hrs/nplon.htm for a description of how facilities are placed on the NPL. Note that CERCLA refers to facilities/installations as "sites" and smaller units within those facilities as "operable units"—terminology which is not used in this report unless an EPA CERCLA source is being cited, like Tables 2-3 and 2-4.

TABLE 2-3 Definition of CERCLA Milestones

PA/SI	*Preliminary Assessment/Site Inspection* Investigations of site conditions. If the release of hazardous substances requires immediate or short-term response actions, these are addressed under the Emergency Response program of CERCLA.
NPL Listing	*National Priorities List (NPL) Site Listing Process* A list of the most serious sites identified for possible long-term cleanup.
RI/FS	*Remedial Investigation/Feasibility Study* Determines the nature and extent of contamination. Assesses the treatability of site contamination and evaluates the potential performance and cost of treatment technologies.
ROD	*Records of Decision* Explains which cleanup alternatives will be used at a given NPL facility. When remedies exceed $25 million, they are reviewed by the National Remedy Review Board.
RD/RA	*Remedial Design/Remedial Action* Preparation and implementation of plans and specifications for applying site remedies. The bulk of the cleanup usually occurs during this phase.
Construction Completion	*Construction Completion* Identifies completion of physical cleanup construction, although this does not necessarily indicate whether final cleanup levels have been achieved.
Post Construction Completion	*Post Construction Completion* Ensures that CERCLA response actions provide for the long-term protection of human health and the environment. Included here are long-term response actions, operation and maintenance, institutional controls, five-year reviews, and remedy optimization.
NPL Deletion	*National Priorities List Deletion* Removes a site from the NPL once all response actions are complete and all cleanup goals have been achieved.
Reuse	*Site Reuse/Redevelopment* Information on how the CERCLA program is working with communities and other partners to return hazardous waste sites to safe and productive use without adversely affecting the remedy.

SOURCE: Adapted from http://www.epa.gov/superfund/cleanup/index.htm.

DNAPLs, which commonly include TCE and PCE, were observed directly in the subsurface at 44 of 712 NPL facilities examined. EPA (1993) also concluded that approximately 60 percent of NPL facilities at that time (1991) exhibited a medium-to-high likelihood of having DNAPL present as a source of subsurface contamination. Of the facilities on the NPL as of

TABLE 2-4 National Priority List Site Status

Status	Non-Federal	Federal	Total
Proposed Sites	55	4	59
Final Sites	1,147	158	1,305
Deleted Sites	344	15	359
Total	1,546	177	1,723
Milestones[a]	Non-Federal	Federal	Total
Partial Deletions[b]	40	17	57
Construction Completions[c]	1,053	70	1,123

[a] Sites that have achieved these milestones are included in one of the three NPL status categories (i.e., proposed, final, deleted).

[b] Partial deletion reflects the deletion from the NPL of specific operable units within a larger CERCLA facility. The EPA recognizes partial deletions to "communicate the completion of successful partial cleanups" and "help promote the economic redevelopment of Superfund sites" (60 FR 55466).

[c] "Construction completions" indicates completion of the physical construction of the remedy, although this does not necessarily indicate whether final remedial objectives have been achieved.

SOURCE: Modified from EPA's list of NPL Site Totals by Status and Milestone, as of June 1, 2012. http://www.epa.gov/superfund/sites/query/queryhtm/npltotal.htm

2004, 83 percent require remediation of groundwater, 78 percent soil, 32 percent sediment, and 11 percent sludge (EPA, 2004).

CERCLA uses additional metrics than those in Tables 2-3 and 2-4 to describe the program's progress. According to the Superfund National Accomplishments Summary Fiscal Year 2010 (http://www.epa.gov/superfund/accomp/numbers10.html), the program has *controlled potential or actual exposure risk to humans* at 1,338 NPL facilities and has *controlled the migration of contaminated groundwater* at 1,030 NPL facilities. At 66 NPL facilities all long-term protections necessary for anticipated use, including institutional controls, are in place and 475 facilities are classified as *ready for anticipated reuse*.

RCRA Corrective Action Program

Among other objectives, the Resource Conservation and Recovery Act (RCRA) governs the management of hazardous wastes at operating facilities that handle or handled hazardous waste. RCRA assigns the facility owners and operators the responsibility for corrective action, and it delegates oversight authority to the states (for those states that the EPA has authorized to implement the program). Because the RCRA program also

TABLE 2-5 Universe of RCRA Facilities

RCRA Milestone	Number of Facilities
CA 725 - Current Human Exposures Under Control	2,821
CA 750 - Groundwater Releases Controlled	2,465
CA 550 - Remedy Constructed	1,506
CA 900 - Corrective Action Performance Standards Attained (Controls Required or No Controls Necessary)	903
or	
CA 999 - Corrective Action Process Terminated[a]	

[a] CA 900 is the newer RCRA metric for corrective action complete. It is a voluntary reporting element, however, and not all EPA regions are using this metric at this time. CA 999 was used by some EPA regions in the past, but with differing definitions. This, too, was voluntary and has not been used for all facilities that meet its criteria. The cumulative number of CA 900 and CA 999 is 903.

SOURCE: Sara Rasmussen, EPA RCRA Office, personal communication, August 11, 2011 and September 7, 2011. CA denotes "corrective action."

governs waste generation and management, remediation to unlimited use and unrestricted exposure is not necessarily the focus as it is in CERCLA (although remediation under RCRA corrective action or CERCLA will substantively satisfy the requirements of both programs [EPA, 1996a]). Furthermore, RCRA remedies are not statutorily bound to comply with the nine criteria of the National Contingency Plan. Rather, EPA has emphasized the need to protect human health and the environment by dealing expeditiously with those sites that present the greatest risks.

Beginning in the late 1990s, the program emphasized achievement of two interim milestones: (1) the *human exposures environmental indicator* "ensures that people near a particular site are not exposed to unacceptable levels of contaminants," and (2) the *groundwater environmental indicator* "ensures that contaminated groundwater does not spread and further contaminate groundwater resources."[4] These indicators have now been satisfied at most of the highest-priority sites (see Table 2-5). Note that the points of compliance where cleanup objectives must be met at operating RCRA facilities may be defined by the property boundaries. The program has recently expanded its focus to include implementing more permanent solutions, and has created the milestone of *final remedy construction*, which is similar to the CERCLA milestone *construction complete*.

Although tens of thousands of waste handlers are potentially subject to RCRA, currently EPA has authority to impose corrective action on 3,747

[4] See also http://www.epa.gov/epawaste/hazard/correctiveaction/programs.htm.

RCRA hazardous waste facilities in the United States (deemed the "2020 Universe").[5] Federal facilities (primarily DoD or DOE) represent 5 percent of the 2020 Universe. The 2020 Universe contains a wide variety of facilities, including heavily contaminated properties yet to be cleaned up, others that have been cleaned up, and some that have not been fully investigated yet and may require little or no remediation. Multiple hazardous waste sites, designated as solid waste management units (SWMUs), may exist inside RCRA facilities, but numbers of SWMUs are not compiled by EPA headquarters.

Table 2-5 presents the national accomplishments and status of these facilities as of August 11, 2011. In terms of the number that have reached "closure," 903 RCRA facilities are categorized as either "Corrective Action Performance Standards Attained (Controls Required or No Controls Necessary) or "Corrective Action Process Terminated," leaving 2,844 needing additional remediation efforts.

Underground Storage Tank Program

In 1984, Congress recognized the unique and widespread problem posed by leaking underground storage tanks by adding Subtitle I to RCRA. This led to the creation of EPA's Office of Underground Storage Tanks (OUST) and the development and implementation of a regulatory program for UST systems. UST contaminants are typically light nonaqueous phase liquids (LNAPLs) such as petroleum hydrocarbons and fuel additives. Responsibility for the UST program has been delegated to the states (or even local oversight agencies such as a county or a water utility with basin management programs), which set specific cleanup standards and approve specific corrective action plans and the application of particular technologies at sites. This is true even for petroleum-only USTs on military bases, a few of which have hundreds of such tanks.

At the end of 2011, there were 590,104 active tanks in the UST program (EPA, 2011a). Active tanks are registered with the state subject to the Subtitle I regulations, but they do not necessarily have releases. Currently, there are 87,983 *leaking* tanks that have contaminated surrounding soil and groundwater, the so-called "backlog." The backlog number represents the cumulative number of confirmed releases (501,723) minus the cumulative number of completed cleanups (413,740). Since the mid-1990s the number of open releases has been declining, yet the pace at which the EPA cleans up the backlog has also slowed (EPA, 2009a). In a study of unaddressed confirmed releases from USTs in 14 states, EPA (2011b) reported that almost

[5] See http://www.epa.gov/osw/hazard/correctiveaction/ facility/index.htm#2020.

half the releases in the backlog are over 15 years old, and that 78 percent of the releases in the backlog have groundwater contamination.

Department of Energy

The DOE faces the task of cleaning up the legacy of environmental contamination from activities to develop nuclear weapons during World War II and the Cold War. Contaminants include short-lived and long-lived radioactive wastes, toxic substances such as chlorinated solvents, "mixed wastes" that include both toxic substances and radionuclides, and, at a handful of facilities, unexploded ordnance. Much like the military, a given DOE facility or installation will tend to have multiple sites where contaminants may have been spilled, disposed of, or abandoned that can be variously regulated by CERCLA, RCRA, or the UST program. The DOE Environmental Management program, established in 1989 to address several decades of nuclear weapons production, "is the largest in the world, originally involving two million acres at 107 sites in 35 states and some of the most dangerous materials known to man" (DOE, 2012a). Since 1989, DOE has also operated an office to develop scientific and technological advancements to meet environmental management challenges, called the Office of Engineering and Technology. In 2003, the Office of Legacy Management was established to focus on long-term care of legacy liabilities from former nuclear production areas following cleanup at each site.

Given that major DOE sites tend to be more challenging than typical DoD sites, it is not surprising that the scope of future remediation is substantial (NRC, 2009). Furthermore, because many DOE sites date back 50 years, contaminants have diffused into the subsurface matrix, considerably complicating remediation. Several previous NRC reports have summarized the nature and extent of contamination at DOE sites (for example, NRC, 1999). There are examples of success stories, such as the 2005 decommissioning of the Rocky Flats Site, arguably once the nation's most highly contaminated plutonium site.

DOE's Environmental Management has historically been responsible for restoration at 134 installations that have about 10,000 release sites, although 21 installations were transferred to the U.S. Army Corps of Engineers in 2004 and one installation was added in 2001 (EPA, 2004). EPA (2004) reported that DOE had completed active remediation at about half of its release sites, leaving about 5,000 sites where cleanup had not been completed. More recent reports suggest that about 7,000 individual release sites out of 10,645 historical release sites have been "completed," which means at least that a remedy is in place (DOE, 2011, pp. 52 ff), leaving approximately 3,650 sites remaining. In 2004, DOE estimated that almost all installations would require long-term stewardship (EPA, 2004).

Since 1998, the U.S. Army Corps of Engineers has managed the Formerly Used Sites Remedial Action Program (FUSRAP), established by DOE in 1974 to remediate nuclear weapons program sites formerly operated by the Manhattan Project and the Atomic Energy Commission. As of 2011, there are 24 active FUSRAP properties in ten states.

Other Federal Sites

Sites operated by civilian federal agencies include all federal agencies except for DOE and DoD. Federal agencies must comply with CERCLA and RCRA in the same manner as private parties and are liable for remediation at current or previously owned properties. As of April 1995, over 3,000 contaminated sites on 700 facilities, distributed among 17 non-DoD and non-DOE federal agencies, were potentially in need of remediation. The Department of Interior (DOI), Department of Agriculture (USDA), and National Aeronautics and Space Administration (NASA) together account for about 70 percent of the civilian federal facilities reported to EPA as potentially needing remediation (EPA, 2004).

EPA (2004) estimates that many more sites have not yet been reported, including an estimated 8,000 to 31,000 abandoned mine sites, most of which are on federal lands, although the fraction of these that are impacting groundwater quality is not reported. The Government Accountability Office (GAO) (2008) determined that there were at least 33,000 abandoned hardrock mine sites in the 12 western states and Alaska that had degraded the environment by contaminating surface water and groundwater or leaving arsenic-contaminated tailings piles.

State Sites

A broad spectrum of sites is managed by states, local jurisdictions, and private parties, and thus are not part of the CERCLA, RCRA, or UST programs. These types of sites can vary in size and complexity, ranging from sites similar to those at facilities listed on the NPL to small sites with low levels of contamination. A gross classification of such sites is (1) those covered under state programs that mandate remediation and (2) state voluntary cleanup programs and/or Brownfields sites.

The mandated programs, which are roughly patterned after the CERCLA program, generally include enforcement authority and state funds to finance the remediation of waste sites. Almost all states have such mandated hazardous waste programs, which generally include provisions for long-term remedial action, funding sources, enforcement authorities, staff to administer and oversee remediation, and efforts to ensure public participation (EPA, 2004). These sites are referred to as "state Superfund" sites in this report. For example, Georgia's state Superfund statute was enacted in 1992 and as of July 1, 2010, the list in Georgia had a total of 568 sites (Georgia Environmental Protection Division, 2010). It should be noted that a small and expensive part of the state Superfund caseload is likely to be the NPL facilities where no PRP was ever identified for which the responsibility for long-term operation and maintenance of the remedy is transferred from EPA to the state 10 years after a site remedy has been operational. These so-called orphan sites can create a substantial burden on state governments (see Box 2-1 for an example in Washington State).

Voluntary cleanup programs and Brownfields programs encourage private parties to remediate sites voluntarily rather than expend state resources on enforcement actions or remediation. Fifty states and territories have established voluntary cleanup programs, and 31 states have established separate Brownfields programs. States typically define Brownfields sites as industrial or commercial facilities that are abandoned or underutilized due to environmental contamination or fear of contamination. EPA (2004) postulated that only 10 to 15 percent of the estimated one-half to one million Brownfields sites have been identified.

Forty-one states have long-term stewardship programs for hazardous waste sites (EPA, 2004). The most common mechanisms used for long-term stewardship are educational materials, information systems such as signs, published notices, warnings about consumption of wildlife and fish, and government controls such as zoning. Scant funds have been committed to this effort (EPA, 2004). As of 2000, 40 states had a priority list or inventory of state sites (EPA, 2004), but the approach, definitions, and extent of these lists vary from state to state.

As of 2000, 23,000 state sites had been identified as needing further attention that had not yet been targeted for remediation (EPA, 2004). The same study estimated that 127,000 additional sites would be identified by 2030.

Dry Cleaner Sites

Active and particularly former dry cleaner sites present a unique problem in hazardous waste management because of their ubiquitous nature in urban settings, the carcinogenic contaminants used in the dry cleaning process (primarily the chlorinated solvent PCE, although other solvents have been used), and the potential for the contamination to reach receptors via the drinking water and indoor air (vapor intrusion) exposure pathways. Depending on the size and extent of contamination, dry cleaner sites may be remediated under one or more state or federal programs such as RCRA, CERCLA, or state mandated or voluntary programs discussed previously, and thus the total estimates of dry cleaner sites are not listed separately in

BOX 2-1
Wyckoff/Eagle Harbor Orphan Superfund Site, Washington State

The Wyckoff-Eagle Harbor Superfund site, located on the east side of Bainbridge Island, Washington, in central Puget Sound, was added to the NPL in 1987. The site is considered an "orphan" site because the previous owner is defunct with no insurance coverage available to address the legacy contamination. The site includes a former wood-treating facility and shipyard, and contaminated sediments in Eagle Harbor adjacent to these former facilities. The Wyckoff wood-treating facility operated on the site for 85 years, and these operations resulted in soil and groundwater contamination (including creosote, pentachlorophenol, and various polycyclic aromatic hydrocarbons). The shipyard contaminated the harbor sediments with organic compounds and heavy metals, including lead, copper, and mercury.

EPA has divided the site into four operable units (OUs) one of which included groundwater beneath the Former Process Area. A Record of Decision (ROD) for the groundwater OU was signed in 2000 with the preferred remedy being physical containment combined with a pump-and-treat system to reduce groundwater discharges to Puget Sound.

Because of concerns about long-term containment of the groundwater OU, steam technology was pilot tested to achieve mass removal from the subsurface. The pilot study was determined to be unsuccessful, partly because of improper operation of the technology. Nonetheless, a 2005 Engineering Evaluation of Remediation Scenarios for the site concluded that any source depletion technology would not likely be sufficient to reduce groundwater concentrations to the levels specified in the ROD. An Explanation of Significant Differences published in 2007 modified some details of the remedy, but containment remained the remedy for the groundwater OU. As of 2007, the on-site groundwater extraction system, which

Table 2-6. However, dry cleaner sites are discussed here because of the high prevalence of active and inactive dry cleaner sites across the nation, their frequent proximity to residential neighborhoods, the highly recalcitrant and toxic nature of the contaminants released, and the importance of the vapor intrusion pathway. Thirteen states[6] have legislation specific to dry cleaner sites including earmarked funds for site investigation and remediation. Cu-

[6] These states—Alabama, Connecticut, Florida, Illinois, Kansas, Minnesota, Missouri, North Carolina, Oregon, South Carolina, Tennessee, Texas, and Wisconsin—are members of the State Coalition for Remediation of Drycleaners (SCRD). California, Maryland, New Jersey, New York, and Virginia are also represented within the coalition as being active in the area of dry cleaner remediation, although they do not have dry cleaner-specific programs. Established in 1998, the Coalition's primary objectives are "to provide a forum for the exchange of information and the discussion of implementation issues related to established state dry cleaner programs; share information and lessons learned with states without dry cleaner-specific pro-

has been in operation since 1993, had removed approximately 100,000 gallons of nonaqueous phase liquid (NAPL) and treated over 475 million gallons of contaminated groundwater. It is estimated that the volume of NAPL in the subsurface of the Former Process Area is approximately 1.2 million gallons. Costs to date for the remedy are not readily available, but are estimated to be more than $130 million.

There is uncertainty about the Washington Department of Ecology's (WDOE) ability to effectively carry out long-term stewardship of the site consistent with the current ROD/Explanation of Significant Differences, given that the remedy must be maintained for hundreds of years. The two primary concerns include (1) the long-term environmental consequence of leaving large amounts of mobile contamination beneath the Former Process Area, given its sensitive location on the shores of Puget Sound, and (2) the financial burden that this action places on Washington State—an in-perpetuity and federally mandated obligation for the State to maintain active operation and maintenance of the remedy, including periodic rebuilding of the containment components such as the groundwater extraction system and perimeter sheet pile wall. The WDOE estimates that life-cycle costs are in excess of hundreds of millions of dollars.

As a result of these concerns, the WDOE has not yet entered into a long-term Superfund State Contract with the EPA for the long-term operations and maintenance for the soil and groundwater OUs. In 2010, WDOE undertook an assessment of alternatives that could potentially decrease or eliminate the need for long-term stewardship. No decision has yet been made regarding the implementation of a new remedy for the groundwater OU as of June 2012.

The lessons learned from this site are a significant concern nationwide, given the expected large number of orphan sites under the CERCLA program and the difficult financial conditions currently facing state governments, who will ultimately be responsible for these orphan sites.

mulative statistics of remediation for these states provides an illustration of the state of progress in remediating U.S. dry cleaner sites (SCRD, 2010a):

- 3,817 sites in dry cleaning programs,
- 2,177 sites where contamination assessment work has been initiated,
- 1,221 sites where contamination assessment work has been completed,
- 574 sites where remediation has been initiated,
- 205 sites where remediation has been completed, and
- 693 sites closed.

grams; and encourage the use of innovative technologies in dry cleaner remediation" (http://www.drycleancoalition.org). Approximately one-third of the nation's dry cleaners are located in states participating in the SCRD (EPA, 2004).

In 2004, there were an estimated 30,000 commercial, 325 industrial, and 100 coin-operated *active* dry cleaners in the United States (EPA, 2004). Despite their smaller numbers, industrial dry cleaners produce the majority of the estimated gallons of hazardous waste from these facilities (EPA, 2004). As of 2010, the number of dry cleaners has grown, with an estimated 36,000 active dry cleaner facilities in the United States—of which about 75 percent (27,000 dry cleaners) have soil and groundwater contamination (SCRD, 2010b).

In addition to active sites, dry cleaners that have moved or gone out of business—i.e., *inactive* sites—also have the potential for contamination. Unfortunately, significant uncertainty surrounds estimates of the number of inactive dry cleaner sites and the extent of contamination at these sites. Complicating factors include the fact that (1) older dry cleaners used solvents less efficiently than younger dry cleaners thus enhancing the amount of potential contamination and (2) dry cleaners that have moved or were in business for long amounts of time tend to employ different cleaning methods throughout their lifetime. EPA (2004) documented at least 9,000 inactive dry cleaner sites, although this number does not include data on dry cleaners that closed prior to 1960. There are no data on how many of these documented inactive dry cleaner sites may have been remediated over the years. EPA *estimated* that there could be as many as 90,000 inactive dry cleaner sites in the United States.

COST ESTIMATES

In addition to tracking the number of hazardous waste sites that have not yet reached closure, the Committee sought information on the cleanup costs expended to date and cost estimates for reaching closure (including estimates for remediation efforts and for long-term management, within the next 30 to 50 years) for each of the programs discussed in the previous section. This information was available for some of the programs but not all (as summarized in Table 2-6). Cost estimates to reach closure (i.e., where no further action is required) are notoriously uncertain and subject to change whenever new contamination is discovered, technology performance and its cost becomes better known, and regulatory perspectives or requirements change. Some cost estimates may be based on unrealistic expectations of remediation performance, particularly in situations with recalcitrant contaminants in complex geologic settings. Also, cost-to-complete estimates frequently underestimate the cost of long-term management. Thus, the Committee, based on its experience, has low confidence in the following cost projections.

Department of Defense

The Installation Restoration Program reports that it has spent approximately $31 billion through FY 2010, and estimates for "cost to complete" exceed $12 billion (Table 2-1). The program's cost to complete actually rose by more than $587 million between 2009 and 2010, despite an annual expenditure totaling nearly $1.3 billion (OUSD, 2011, p. E-9; DEPARC, 2010, p. C-1-1). DoD has collected almost $578 million from non-DoD parties as cost-sharing for IRP projects. The lion's share, over $548 million, has been Shell Oil's payment for remediation at the Rocky Mountain Arsenal in Colorado, where Shell produced pesticides after the Army stopped manufacturing chemical weapons (OUSD, 2011, p. D-6).

CERCLA

Implementation costs for the CERCLA program are difficult to obtain because most remedies are implemented by private, nongovernmental PRPs and generally there is no requirement for these PRPs to report actual implementation costs. PRPs have historically paid for 70 percent of costs associated with facilities on the NPL. EPA (2004) estimated that the cost for addressing the 456 facilities that have not begun remedial action is $16-$23 billion.[7] A more recent report from the GAO (2009) suggests that individual site remediation costs have increased over time (in constant dollars) because a higher percentage of the remaining NPL facilities are larger and more complex (i.e., "megasites") than those addressed in the past. Additionally, GAO (2009) found that the percentage of NPL facilities without responsible parties to fund cleanups may be increasing.

When no PRP can be identified, the cost for Superfund remediation is shared by the states and the Superfund Trust Fund. The Superfund Trust Fund has enjoyed a relatively stable budget—e.g., $1.25 billion, $1.27 billion, and $1.27 billion for FY 2009, 2010, and 2011,[8] respectively—although recent budget proposals seek to reduce these levels. States contribute as much as 50 percent of the construction and operation costs for certain CERCLA actions in their state. After ten years of remedial actions at such NPL facilities, states become fully responsible for continuing long-term remedial actions.

[7] This total is based on an average cost per operable unit of $1.4 million for RI/FS, $1.4 million for remedial design, $11.9 million for remedial action, and $10.3 million for long-term remedial action (EPA, 2004).

[8] See http://www.epa.gov/planandbudget/archive.html.

RCRA Corrective Action Program

EPA headquarters has no information on either costs expended or costs to closure because RCRA regulations do not require responsible RCRA parties to provide cost information (Sara Rasmussen, EPA, personal communication, February 24, 2010). In 2004, EPA estimated that remediation of the remaining RCRA sites will cost between $31 billion and $58 billion, or an average of $11.4 million per facility (EPA, 2004) (hence, the estimate of $11.4 × 2,844 = $32.4 billion in Table 2-6). It is unclear whether this cost estimate represents only capital costs for the remedy or also includes long-term management costs.

Underground Storage Tank Program

There is limited information available to determine costs already incurred in the UST program. EPA (2004) estimated that the cost to close all leaking UST (LUST) sites could reach $12-$19 billion or an average of $125,000 to remediate each release site (this includes site investigations, feasibility studies, and treatment/disposal of soil and groundwater). Based on this estimate of $125,000 per site, the Committee calculated that remediating the 87,983 backlogged releases would require $11 billion. The presence of the recalcitrant former fuel additive methyl tert-butyl ether (MTBE) and its daughter product and co-additive tert-butyl alcohol could increase the cost per site. Most UST cleanup costs are paid by property owners, state and local governments, and special trust funds based on dedicated taxes, such as fuel taxes.

Department of Energy

To gain an understanding of the DOE costs that would be comparable to other federal programs, the Committee reviewed the Department's FY 2011 report to Congress, which shows that DOE's anticipated cost to complete remediation of soil and groundwater contamination ranges from $17.3 to $20.9 billion. The program is dominated by a small number of mega-facilities, including Hanford (WA), Idaho National Labs, Savannah River (SC), Los Alamos National Labs (NM), and the Nevada Test Site. Given that the cost to complete soil and groundwater remediation at these five facilities alone ranges from $16.4 to $19.9 billion (DOE, 2011), the Committee believes that the DOE's anticipated cost-to-complete figure is likely an underestimate of the Agency's financial burden; the number does not include newly discovered releases or the cost of long-term management at all sites where waste remains in the subsurface.

Data on long-term stewardship costs, including the expense of operating and maintaining engineering controls, enforcing institutional controls, and monitoring, are not consolidated but are likely to be substantial and ongoing. The Office of Legacy Management, which is responsible for managing non-operational facilities once the Environmental Management program has completed its work, had a $38.8 million annual budget for FY 2012 for "long-term surveillance and maintenance" (DOE, 2012b). Stewardship costs for just the five facilities managed by the National Nuclear Security Administration (Lawrence Livermore National Laboratory, CA, Livermore's Site 300, Pantex, TX, Sandia National Laboratories, NM, and the Kansas City Plant, MO) total about $45 million per year (DOE, 2012c).

Through 2010, the FUSRAP program had spent $2.03 billion, and the annual budget normally ranges from $130 million to $140 million. No cost data are available on estimated costs to complete remedial actions for this program.

Other Federal Sites

EPA (2004) reports that there is a $15-$22 billion estimated cost to address at least 3,000 contaminated areas on 700 civilian federal facilities, based on estimates from various reports from DOI, USDA, and NASA.

States

EPA (2004) estimated that states and private parties together have spent about $1 billion per year on remediation, addressing about 5,000 sites annually under mandatory and voluntary state programs. If remediation were continued at this rate, 150,000 sites would be completed over 30 years, at a cost of approximately $30 billion (or $20,000 per site).

IMPACTS TO DRINKING WATER SUPPLIES

The Committee sought information both on the number of hazardous waste sites that impact a drinking water aquifer—that is, pose a substantial near-term risk to public water supply systems that use groundwater as a source. Unfortunately, program-specific information on water supply impacts was generally not available. Therefore, the Committee also sought other evidence related to the effects of hazardous waste disposal on the nation's drinking water aquifers.

Program-Specific Reports of Groundwater Impacts

Despite the existence of several NPL and DoD facilities that are known sources of contamination to public or domestic wells (e.g., the San Fernando and San Gabriel basins in Los Angeles County—Land et al., 2011), there is little aggregated information about the number of CERCLA, RCRA, DoD, DOE, UST, or other sites that directly impact drinking water supply systems. None of the programs reviewed in this chapter specifically compiles information on the number of sites currently adversely affecting a drinking water aquifer. However, the Committee was able to obtain information relevant to the groundwater impacts from some programs:

- **DoD.** The Army informed the Committee that public water supplies are threatened at 18 Army installations (Laurie Haines, U.S. Army Environmental Command, 2010, personal communication). Also, private drinking water wells are known to be affected at 23 installations. A preliminary assessment in 1997 showed that 29 Army installations may possibly overlie one or more sole source aquifers (based on simply comparing the general aquifer locations from EPA maps to Army installation locations). Each of the other armed services is also responsible for groundwater contamination that has affected drinking water supplies. Some of the best known are Camp Lejeune Marine Corps Base (NC), Otis Air National Guard Base (MA), and the Bethpage Naval Weapons Industrial Reserve Plant (NY) (see Appendix B).

- **CERCLA.** Each individual remedial investigation/feasibility study (RI/FS) and Record of Decision (ROD) should state whether a drinking water aquifer is affected, although this information has not been compiled. Canter and Sabatini (1994) reviewed the RODs for 450 facilities on the NPL. Their investigation revealed that 49 of the RODs (11 percent) indicated that contamination of public water supply systems had occurred. "A significant number" of RODs also noted potential threats to public supply wells. Additionally, the authors note that undeveloped aquifers have also been contaminated, which prevents or limits the unrestricted use (i.e., without treatment) of these resources as a future water supply.

  The EPA also compiles information about remedies implemented within Superfund. EPA (2007) reported that out of 1,072 facilities that have a groundwater remedy, 106 specifically have a water supply remedy, by which we inferred direct treatment of the water to allow potable use or switching to an alternative water supply. This suggests that 10 percent of NPL facilities adversely affect or significantly threaten drinking water supply systems. This estimate is further bolstered by EPA (2010b), which reports that of the 311 decision documents from FY 2005–2008 with "groundwater other" remedies, 8 percent of these (26) include water supply remedies.

- **RCRA.** Of the 1,968 highest priority RCRA Corrective Action facilities, EPA (2008) reported that there is "unacceptable migration of contaminated groundwater" at 77 facilities. Also, 17,042 drinking water aquifers have a RCRA facility within five miles (Roger Anzzolin, EPA, personal communication, 2010), but without additional information, it is impossible to know if these facilities are actually affecting the water sources.

- **UST.** In 2000, 35 states reported USTs as the number one threat to groundwater quality (and thus indirectly to drinking water) (EPA, 2000). However, more specific information on the number of leaking USTs currently impacting a drinking water aquifer is not available.

Other Evidence That Hazardous Waste Sites Affect Water Supplies

The U.S. Geological Survey (USGS) has compiled large data sets over the past 20 years regarding the prevalence of VOCs in waters derived from domestic (private) and public wells. VOCs include solvents, trihalomethanes (some of which are solvents [e.g., chloroform], but may also arise from chlorination of drinking water), refrigerants, organic synthesis compounds (e.g., vinyl chloride), gasoline hydrocarbons, fumigants, and gasoline oxygenates. Because many (but not all) of these compounds may arise from hazardous waste sites, the USGS studies provide further insight into the extent to which anthropogenic activities contaminate groundwater supplies (although it should be remembered that it was not the goal of these studies to uniquely identify the source of the contamination). The following paragraphs do not discuss metals and other inorganic groundwater contaminants described in the USGS studies, because of the many other possible natural sources for these constituents.

Zogorski et al. (2006) summarized the presence of VOCs in groundwater, private domestic wells, and public supply wells from sampling sites throughout the United States. Using a threshold level of 0.2 µg/L—much lower than current EPA drinking water standards for individual VOCs (see Table 3-1)—14 percent of domestic wells and 26 percent of public wells had one or more VOCs present. The detection frequencies of individual VOCs in domestic wells were two to ten times higher when a threshold of 0.02 µg/L was used (see Figures 2-2 and 2-3). In public supply wells, PCE was detected above the 0.2 µg/L threshold in 5.3 percent of the samples and TCE in 4.3 percent of the samples. The total percentage of public supply wells with either PCE or TCE (or both) above the 0.2 µg/L threshold is 7.3

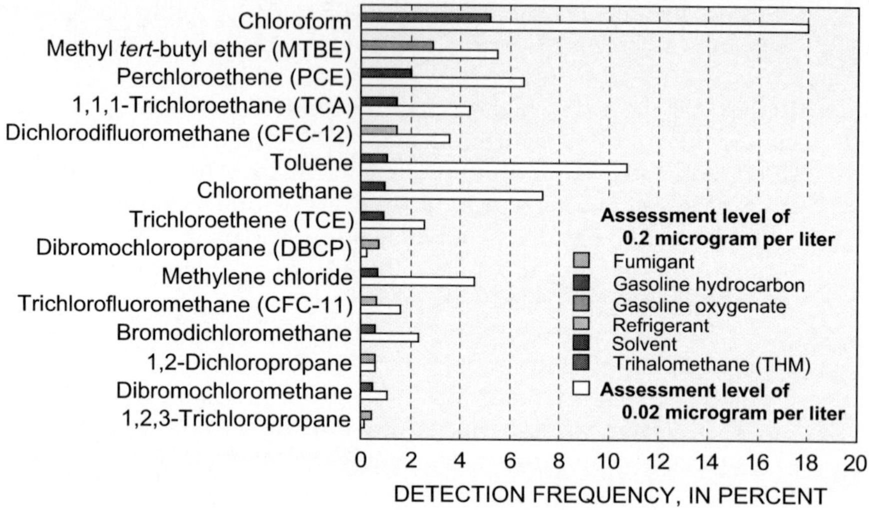

FIGURE 2-2 Detection frequencies in domestic well samples for 15 most frequently detected VOCs at levels of 0.2 and 0.02 μg/L.
SOURCE: Zogorski et al. (2006) with illustration provided by USGS National Water Quality Assessment program.

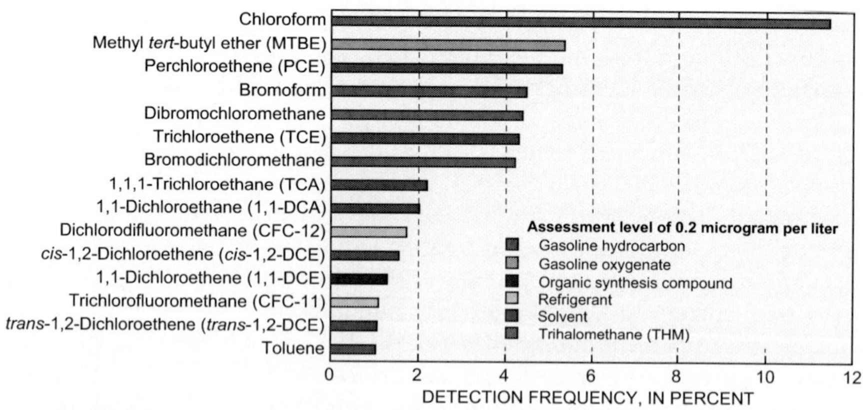

FIGURE 2-3 The 15 most frequently detected VOCs in public supply wells. SOURCE: Zogorski et al. (2006) with illustration provided by USGS National Water Quality Assessment program.

percent. The following conclusions were drawn in the Zogorski et al. study: (1) public wells are more vulnerable to contamination than private domestic wells, (2) public wells had higher concentrations of VOCs (50 percent of public wells had total VOC concentrations > 1 μg/L), and (3) public wells were more likely to have mixtures of VOCs than domestic wells. These effects were attributed, by the authors, to the larger withdrawal rates and closer proximity to urban areas of many public supply wells.

Further analysis of domestic wells by DeSimone et al. (2009) showed that organic contaminants were detected in 60 percent of 2,100 sampled wells. Wells were sampled in 48 states in parts of 30 regionally extensive aquifers used for water supply. Aquifers were randomly selected for sampling and there was no prior knowledge of contamination. Seventeen VOCs were detected in more than 1 percent of wells at concentrations greater than 0.02 μg/L (see Figure 2-4 below, VOCs are in black). TCE was detected above the maximum comtaminant level (MCL) of 5 μg/L in 0.1 percent of wells. PCE was detected above the MCL of 5 μg/L in 0.05 percent of wells.

Rowe et al. (2007) compiled data for 2,400 domestic wells sampled from 1985 until 2002. Sixty-five percent of domestic wells had a VOC detection 0.02 μg/L or greater (31 percent had a single VOC, 34 percent had more than one VOC). The top five VOCs detected were chloroform (25.6 percent), toluene (17.9 percent), 1,2,4-trimethylbenzene (15.2 percent), PCE (11 percent), and chloromethane (9.7 percent). PCE, TCE, and chloromethane were the compounds with the largest fraction of samples at 0.1 × MCL or greater. The presence of a LUST site within 1 km of the sampled well strongly correlated with MTBE detections, and the presence of an RCRA site (as determined by the EPA Envirofacts database) within 1 km of the well strongly correlated with the detections of PCE, TCE, and 1,1,1-TCA.

Toccalino and Hopple (2010) and Toccalino et al. (2010) focused on 932 public supply wells across the United States. The public wells sampled in this study represent less than 1 percent of all groundwater that feeds the nation's public water systems. The samples, however, were widely distributed nationally and were randomly selected to represent typical aquifer conditions. Overall, 60 percent of public wells contained one or more VOCs at a concentration of ≥ 0.02 μg/L, and 35 percent of public wells contained one or more VOCs at a concentration of ≥ 0.2 μg/L. The percentages are higher than those reported by Zogorski et al. (2006), but this study focused on a larger suite of VOCs (85 vs. 55 compounds). Overall detection frequencies for individual compounds included 23 percent for PCE, 15 percent for TCE, 14 percent for MTBE, and 12 percent for 1,1,1-TCA (see Figure 2-5). PCE and TCE exceeded the MCL in approximately 1 percent of the public wells sampled. About 70 percent of VOC detections were from sand and gravel aquifers. Public wells in sand and gravel aquifers more often

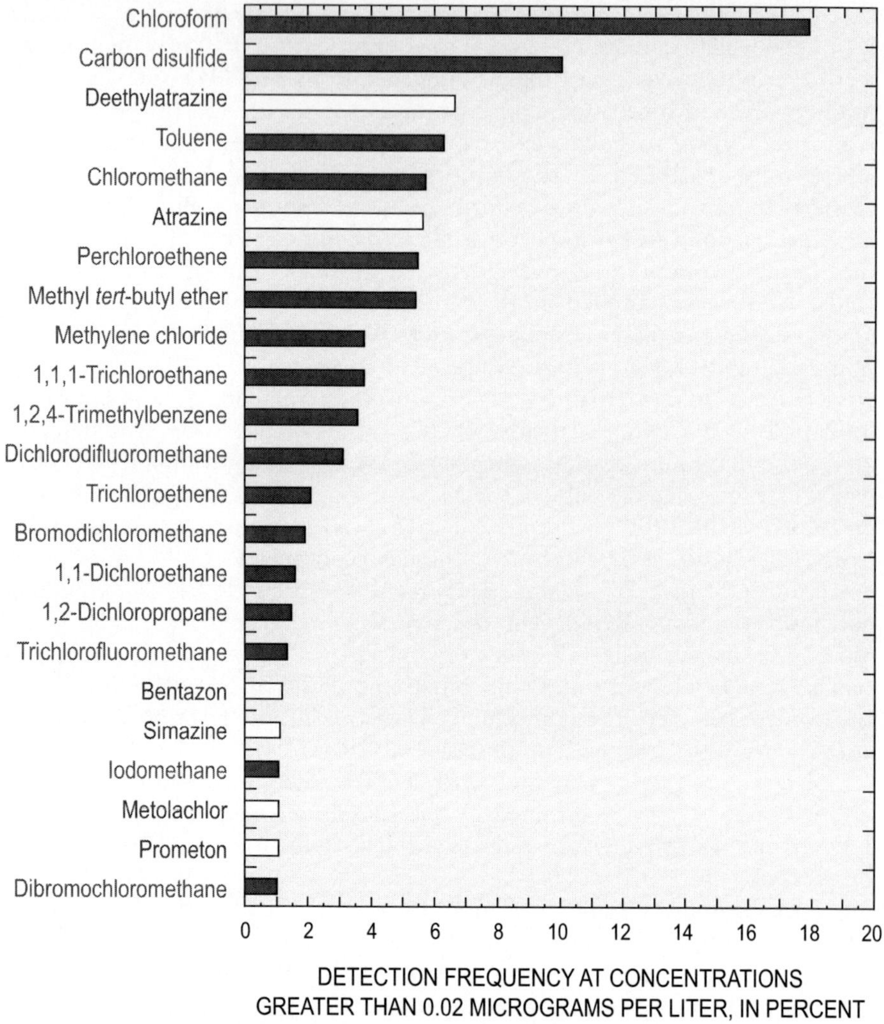

FIGURE 2-4 VOCs (in black) and pesticides (in white) detected in more than 1 percent of domestic wells at a level of 0.02 μg/L.
SOURCE: DeSimone et al. (2009).

withdraw water from shallower unconfined aquifers than from deeper confined aquifers. Thus, VOCs were detected more frequently in samples from unconfined aquifers than from confined aquifers, highlighting the vulnerability of shallow unconfined aquifers. Overall, the detection frequencies of some VOCs were 2-fold to 6-fold greater in public wells than domestic

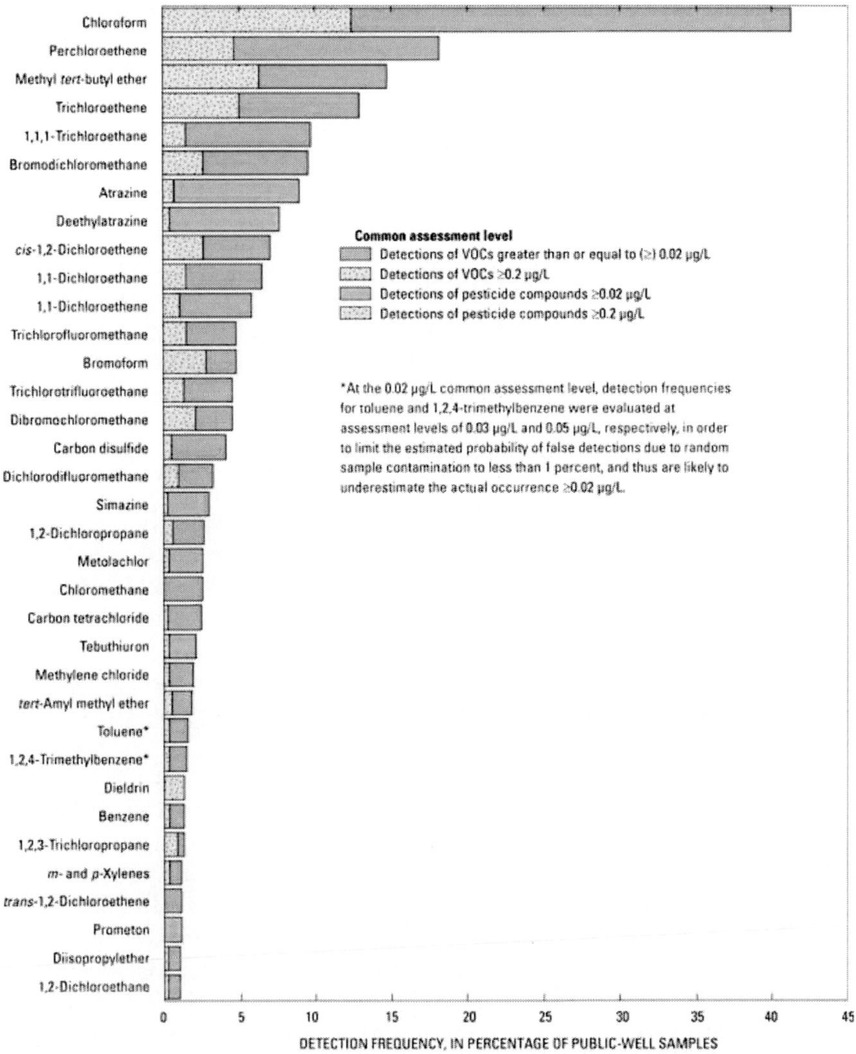

FIGURE 2-5 VOCs and pesticides with detection frequencies of 1 percent or greater at assessment levels of 0.02 µg/L in public wells in samples collected from 1993–2007.
SOURCE: Toccalino and Hopple (2010) and Toccalino et al. (2010)

wells—again, likely because of proximity of public wells to developed areas and higher pumping rates used for public wells versus domestic wells.

Overall, the USGS studies show that there is widespread, very low level contamination of private and public wells by VOCs, with a reasonable estimate being 60 to 65 percent of public wells having detectable VOCs. According to the data sets of Toccalino and Hopple (2010) and Toccalino et al. (2010), approximately 1 percent of sampled public wells have levels of VOCs above MCLs. Thus, water from these wells requires additional treatment to remove the contaminants before it is provided as drinking water to the public. EPA (2009b) compiled over 309,000 groundwater measurements of PCE and TCE from raw water samples at over 46,000 groundwater-derived public water supplies in 45 states. Compared to the USGS data, this report gives a lower percentage of water supplies being contaminated: TCE concentration exceeded its MCL in 0.34 percent of the raw water samples from groundwater-derived drinking water supply systems.

There are other potential sources of VOCs in groundwater beyond hazardous waste sites. For example, chloroform is a solvent but also a disinfection byproduct, so groundwater sources impacted by chlorinated water (e.g., via aquifer storage/recharge, leaking sewer pipes) would be expected to show chloroform detections. Another correlation seen in the USGS data is that domestic and public wells in urban areas are more likely to have VOC detections that those in rural areas. This finding is not unexpected given the much higher level of industrial practices in urban areas that can result in releases of these chemicals to the subsurface.

Another way to estimate the number of public water supplies affected by contaminated groundwater is to consider the number of water supply systems that specifically seek to remove organic contaminants. The EPA Community Water System Survey (EPA, 2002) reports that 2.3 to 2.6 percent of systems relying solely on groundwater have "organic contaminant removal" as a treatment goal. For systems that use both surface water and groundwater, 10.3 to 10.5 percent have this as a treatment goal. While it is possible that this range (2 to 10 percent) may be the fraction of water supplies impacted by groundwater contamination, this is at best only a rough estimate and highly uncertain. A water utility could (or may be forced to) use an alternative water supply, rather than treat a contaminated source, which would make this a lower estimate. On the other hand, the category "organic contaminants" includes pesticides, which may come from nonpoint sources rather than contaminated sites, meaning this range could be an overestimate.

In summary, it appears that the following conclusions about the contamination of private and public groundwater systems can be drawn: (1) there is VOC contamination of many private and public wells (upwards of 65 percent) in the United States, but at levels well below MCLs; the origin of this contamination is uncertain and the proportion caused by releases from hazardous waste sites is unknown; (2) approximately one in ten NPL facilities is impacting or significantly threatening a drinking water supply system relying on groundwater, requiring wellhead treatment or the use of alternative water sources; and (3) public wells are more susceptible to contamination than private wells, due their higher likelihood of being in urban areas and their higher pumping rates and hydraulic capture zones.

THE PARADOX OF "CLOSED" SITES

In considering the size of the nation's hazardous waste problem, one question that has arisen is the definition of the term "closure" as it relates to these sites. Does a closed site mean no residual contamination above regulatory limits or is the definition flexible depending on the risk environment in which the regulatory decision to close a site is made? Indeed, there is confusion about the definition of "site closure"—not only to the public, the regulated community, and between regulatory agencies, but even within EPA's own closure guidance (EPA, 2011c). For example, EPA (2011c) on page 1-2 states that "site completion typically occurs when it is determined that no further response is required at the site, all cleanup levels have been achieved, and the site is deemed protective of human health and the environment." It goes on to say that "site completion signifies the end of all response actions at a site" and "it is anticipated that no further Superfund response is necessary to protect human health and the environment." However, it then states on page 4-3 that "operation and maintenance are not defined as a response action by the NCP, and may continue after site completion and deletion." Furthermore, the guidance states that the final closure must explain whether a five-year review is appropriate. However, a five-year review is only required when contaminants are left in place above UU/UE levels, such as the drinking water standards. It is no wonder that stakeholders are confused by the site closure metric, as operation and maintenance of a remedy may continue for many decades after "closure."

To better understand the status of "closed" sites and whether these sites could in fact demand future resources for monitoring, reporting, or additional remediation, the Committee reviewed an Interstate Technology and Regulatory Council (ITRC)[9] survey of "closed" underground storage tanks, EPA cleanup success stories, and 80 facilities delisted from the NPL

[9] The ITRC, which consists of states, federal agencies, industry, and other stakeholders, "develops guidance documents and training courses to meet the needs of both regulators and environmental consultants, and it works with state representatives to ensure that ITRC products and services have maximum impact among state environmental agencies and technology users" (http://www.irtcweb.org/aboutIRTC.asp).

where groundwater was contaminated. This review revealed that these sites vary widely in the extent to which they contain contaminant levels that are actually higher than MCLs or other levels that would allow for unlimited use and unrestricted exposure. The Committee found that there was no publicly available mechanism for tracking these sites subsequent to closure, nor do the federal programs maintain a central repository of information about their closed sites (except for NPL-delisted facilities). Thus, little quantitative data or information are available to assess such sites.

It is clear that the definition of site closure varies from program to program, such that a site closed under one program would not necessarily be closed under another, even for the same type of waste site. Perhaps the most prominent example of this is the way that the states have defined site closure for underground storage tanks. Cleanup goals for tanks have often been expressed as removal of contaminants "to the maximum extent practicable," which, as discussed earlier in the context of UST remediation, can be interpreted many different ways—from no interpretation at all to a maximum allowable LNAPL thickness in a monitoring well (e.g., sheen or 1/8-inch thickness). The ITRC's recent survey of state UST programs (ITRC, 2009) revealed that many states rely solely on best professional judgment of maximum extent practicable (which would obviously vary from site to site within the state), while a few others are starting to consider site-specific risk. Still other states close USTs when contaminant levels are no longer "detectable."

The potential for misunderstanding in the labeling of sites as "successes" is illustrated by an EPA (2009c) review of 13 DNAPL sites—some CERCLA, one RCRA, and some state sites. These sites were chosen because they are examples of where source reduction has contributed to a site meeting remedial objectives (such as groundwater MCLs). However, closer inspection of the 13 sites by this Committee revealed that five of the sites reported only soil contamination and thus the Committee could not determine if they were examples of the more intractable problems found at groundwater sites. Of the remaining eight sites with contaminated groundwater, EPA's report states that only three sites were "able to achieve MCLs onsite" although two others achieved MCLs at an offsite point of compliance [see EPA (2009c), Table D-1 in Appendix D].

This Committee conducted a more in-depth analysis of 80 Superfund facilities (identified by EPA personnel) that had groundwater contamination that were eventually deleted from the NPL. For each of the 80, the Committee analyzed five-year review reports, site closure documents, RODs, and fact sheets produced by EPA; the full analysis can be found in Appendix C. Sixty percent were industrial facilities, 22 percent were landfills, and the rest were potable well fields, military bases, or other facility types. As would be expected of complex Superfund facilities, almost all of the 80 had groundwater contaminated by VOCs, SVOCs, metals, or some combination thereof.

The Committee first determined what the remedial action objectives were for each delisted NPL facility. Of the 80, 45 had remedial objectives that specified a contaminant concentration goal for groundwater, either MCLs or some other level. For seven, the stated objectives involved some other specific metric (such as prevention of contaminated groundwater migration, exposure prevention, etc.). Finally, 28 had no explicitly stated objective other than the goal of "protecting human health and the environment." This broad goal statement was most typical of NPL facilities delisted early in the program; indeed, for many of these early delisted facilities a later ROD amendment, consent decree, or five-year review report appears to establish that there *were* numeric concentration goals for groundwater. For the Committee's subsequent analysis (see below), for any facility where groundwater contaminant concentrations were compared to MCLs in five-year review reports, the facilities were categorized as either meeting or not meeting MCLs, even if this was not an original goal of the ROD.

The primary objective of the Committee's analysis was to determine the extent to which the 80 delisted facilities had actually met MCLs in groundwater. According to information that could be easily gleaned from EPA's CERCLIS database, 37 of the 80 reported achieving MCLs prior to deleting the facility from the NPL (see Figure 2-6). Of this subgroup, 14 achieved MCLs after some length of time operating an active remedy (like pump and

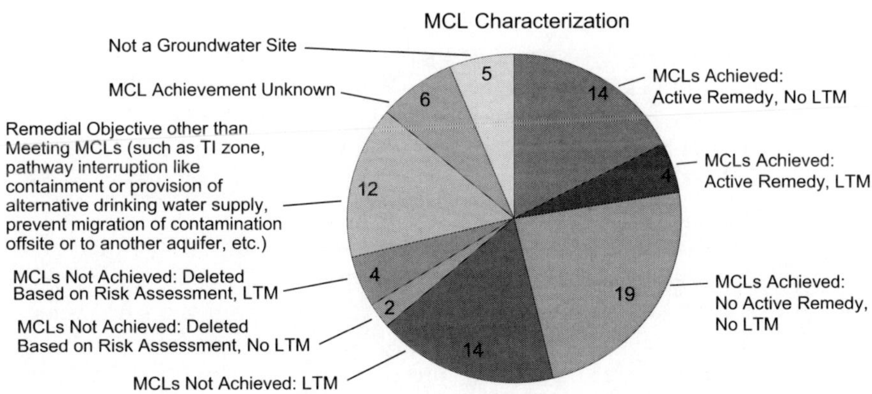

FIGURE 2-6 Pie chart of 80 groundwater facilities delisted from the NPL categorized by whether they reached MCLs and whether long-term monitoring is in place. LTM = long-term monitoring.

treat or thermal treatment) and no longer require long-term monitoring. Four facilities deleted from the NPL have achieved MCLs and still have long-term monitoring in place. At 19 the MCLs were achieved without an active remedy (thus, no long-term monitoring is in place), suggesting that either there was no substantial groundwater contamination when the facility was added to the NPL or that natural attenuation occurred during the RI/FS process to significantly reduce contaminant concentrations.

More interesting are the 20 facilities with contaminated groundwater that were deleted from the NPL where MCLs have not been met (as of August 2011 and as related in readily accessible EPA documents). Fourteen of these have been shown to have contaminant concentrations that are trending downward, and thus must continue to do five-year reviews. Six were deleted after a site-specific risk assessment demonstrated that the risks were below an acceptable threshold, even if contaminant concentrations were above MCLs, and four of the six must do long-term monitoring.

Twelve of the 80 were delisted after successfully installing containment or another protective remedy and thus could not be considered as having met or not met MCLs, because that was not the goal of the remedy. For example, at Schofield Barracks in Hawaii, the Army was able to delist the facility after providing an alternative source of water to local residents and determining that the contamination present in the subsurface was no longer presenting a human health risk (see Box 2-2). Because contamination remains in place, the facility must undergo five-year reviews in perpetuity, but this facility is anecdotally referred to as "closed." Also included in this category are facilities that were granted a Technical Impracticability waiver for some portion of the facility (at which MCLs are waived). Thus, it would be impossible to consider the sites as having achieved MCLs or not.

For six facilities there was insufficient information in the documentation available from EPA to determine if MCLs were met or not. Presumably, these six could have been binned into one of the other categories if further information had been sought from EPA regional offices.

Finally, five facilities did not appear to have ever had groundwater contamination.

The Committee cautions that there is some amount of uncertainty associated with this analysis due to the uneven and sparse nature of the documentation available on delisted NPL facilities from the EPA website. In particular, frequently found statements such as "a site is meeting health-based standards" were difficult to interpret as having met MCLs or not. The documents for a given facility were not necessarily consistent with one another, especially with respect to the statement of remedial goals. For the purposes of the analysis, the most recent documents were weighted more heavily. Despite these uncertainties, only half of groundwater-contaminated facilities deleted from the NPL, which are considered success stories for

BOX 2-2
Schofield Barracks, HI, Case Study

This site is an example of a delisted NPL facility at which restoration was considered not practical and that will require long-term management and monitoring. Schofield Barracks is a U.S. Army post located in the City and County of Honolulu and in the Wahiawa District of the island of Oahu, Hawaii. Established in 1908, the 17,725-acre facility served as a major support facility during World War II and is the largest Army base outside the contiguous United States.

The hydrogeology at Schofield is complex, including a highly fractured basalt aquifer that causes extreme heterogeneity on a local scale. Depth to groundwater is 500–600 feet from the surface. Contaminated sites include a former landfill on 35 acres that contains solid, domestic waste; industrial waste from vehicle equipment and maintenance, solvents, and sewage sludge; medical waste; explosives (both ordnance and unexploded); and construction and demolition waste from various military installations. Contaminants detected at levels above MCLs in the groundwater system beneath the landfill were TCE and carbon tetrachloride (CCl_4), antimony, and manganese. Other chlorinated VOCs such as PCE were detected at low levels (less than MCLs). The precise source for these contaminants in the groundwater remains unidentified.

In 1985, high levels of TCE (as much as 100 ppb) were found to be contaminating wells that supplied water to about 25,000 people living at Schofield Barracks, which was the catalyst for the site being listed on the NPL (EPA, 2010a; U.S. Army Environmental Command, 2007). As a result, there was a temporary switch from well water to city and county water supplies. In 1986, an air stripping treatment unit was established to treat water from the four existing production wells to reduce concentrations of TCE in the drinking water used at the base. Public drinking water wells that serve 55,000 people are located within three miles of the base, but they do not appear to have been affected by the contamination.

The Army divided the site into four Operable Units (OU2 is the groundwater plume and OU4 is the former landfill), for which a ROD was signed in 1996 (EPA, 1996b). Because of the difficult hydrogeologic conditions and the inability to conclusively locate the source of contamination, the Army applied for and received a Technical Impracticability waiver for the site. Treatment for the drinking water wells has maintained an average concentration of TCE below 5 µg/L since air strippers were installed in 1986. The installation was delisted from the NPL in 2000. The Army is conducting the five-year reviews, the second of which was completed in 2007. Site inspection shows that the remedies (for both contaminated groundwater and the landfill) are functioning properly (U.S. Army Environmental Command, 2007).

site closure, have actually achieved MCLs. Of course, at *all* of the deleted facilities, human health and the environment are currently protected. What is also clear from this analysis is that many site-specific, pragmatic factors come into play when decisions are made on the future of the facility (i.e., no further action or some kind of long-term management).

As discussed further in Chapter 3, the cleanup goals of the federal programs range from preventing or minimizing exposure, to meeting engineering milestones (such as remedy selection, design completion, completing construction, completing the active remedy), to attaining the ultimate goal of achieving UU/UE conditions at a site. The military's primary goal is to achieve *remedy in place* or *response complete* at its sites by 2014, with little mention of site closure or attaining unrestricted use of the site.

All of these issues suggest that there can be no generalizations about the condition of sites referred to as "closed," particularly assumptions that they are "clean," meaning available for unlimited use and unrestricted exposure. Indeed, the experience of the Committee in researching "closed sites" suggests that many of them contain contaminant levels above those allowing for unlimited use and unrestricted exposure, even in those situations where there is "no further action" required. Rather, site closure may simply mark the beginning of a long-term operation and maintenance phase involving oversight of institutional controls. Furthermore, it is clear that states are not tracking their caseload at the level of detail needed to ensure that risks are being controlled subsequent to "site closure." Thus, reports of cleanup success should be viewed with caution.

CONCLUSIONS AND RECOMMENDATIONS

The Committee's rough estimate of the number of sites remaining to be addressed and their associated future costs is presented in Table 2-6, which lists the latest available information on the number of facilities (for CERCLA and RCRA) and contaminated sites (for the other programs) that have not yet reached closure, and the estimated costs to remediate the remaining sites. The Committee used these data to estimate the total number of complex sites with residual contamination, as described below.

At least 126,000 sites across the country have been documented that have residual contamination at levels preventing them from reaching closure. This number is likely to be an underestimate of the extent of contamination in the United States for a number of reasons. First, for some programs data are available only for contaminated facilities rather than individual sites; for example, RCRA officials declined to provide an average number of solid waste management units per facility, noting that it ranged from 1 to "scores." CERCLA facilities frequently contain more than one individual release site. The total does not include DoD sites that have reached *remedy in place* or *response complete*, although some such sites may indeed contain residual contamination. Finally, the total does not include sites that likely exist but have not yet been identified, such as dry cleaners or small chemical-intensive businesses (e.g., electroplating, furniture refinishing) that

TABLE 2-6 Rough Estimate of the Total Number of Currently Known Facilities or Contaminated Sites That Have Not Reached Closure and Estimated Costs to Complete

Program/Agency	Number of Contaminated Facilities	Number of Contaminated Sites	Estimated Cost to Complete[a]
DoD		4,329	$12.8 billion
CERCLA	1,364		$16–23 billion
RCRA	2,844		$32.4 billion
UST		87,983	$11 billion
DOE		3,650	$17.3–20.9 billion
Other Federal Sites		> 3,000	$15–22 billion
State Sites		>23,000	$5 billion[b]
Total	>126,000		**$110–127 billion**[c]

NOTE: Munitions were excluded from the DoD numbers, but some munitions are found under RCRA.

[a]Cost figures are undiscounted 2010 dollars. The Committee's cost-to-complete estimate is lower than EPA (2004) because some activities were excluded by the Committee (e.g., MMRP).

[b]For State sites, assumed $20K/site.

[c]Data presented as a range to reflect ranges presented in the original data sets. However, many programs simply provided a single estimate.

have not been investigated for possible contamination. There is overlap between some of the categories (e.g., some sites are counted under both the CERCLA and DoD or DOE categories), but in the Committee's opinion this overlap is not significant enough to dismiss the conclusion that the total number of 126,000 is an underestimate. If more accurate numbers were desired, consistent information would need to be collected on the number of contaminated sites across the various programs.

No information is available on the total number of sites with contamination in place above levels allowing for unlimited use and unrestricted exposure, although the total is certainly greater than the number of sites tallied in Table 2-6. For the CERCLA program, many facilities have been delisted with contamination remaining in place at levels above unlimited use and unrestricted exposure (as much as half according to the Committee's analysis of 80 delisted NPL facilities with groundwater contamination). Depending on state closure requirements, USTs are often closed with contamination remaining due to the biodegradability of petroleum hydrocarbons. Most of the DOE sites, including those labeled as "completed," contain recalcitrant contamination that in some cases could take hundreds of years to reach UU/UE levels.

A small percentage (about 12,000 or less than 10 percent) of the 126,000 sites are estimated by the Committee to be complex from a hydrogeologic and contaminant perspective. This total represents the sum of the remaining DoD, CERCLA, RCRA, and DOE sites and facilities, based on the assumption that many of the simpler sites in these programs have already been dealt with. Although the complexity of the typical RCRA facility can be debated, there are undoubtedly some UST, state, and other federal sites with complex hydrogeologic conditions or contaminants that were not included. This estimate is admittedly uncertain and based largely on the Committee's experience with a wide range of hazardous waste sites. Data on the complexity of sites has not been tallied by any of the programs, and can only be gathered accurately through site-specific data from a random sampling of sites.

Approximately 10 percent of CERCLA facilities affect or significantly threaten public water supply systems, but similar information from other programs is largely unavailable. Surveys of groundwater quality report that 0.34 to 1 percent of raw water samples from wells used for drinking water (including public supply and private wells) contain mean VOC concentrations greater than the MCL, although there are no data linking these MCL exceedances to specific hazardous waste sites. The percentage of drinking water wells with samples containing low-level VOC concentrations is likely to be higher for areas in close proximity to contaminated sites, for urban rather than rural areas, and in shallow unconfined sandy aquifers.

Information on cleanup costs incurred to date and estimates of future costs, as shown in Table 2-6, are highly uncertain. Despite this uncertainty, the estimated "cost to complete" of $110-$127 billion is likely an underestimate of future liabilities. Remaining sites include some of the most difficult to remediate sites, for which the effectiveness of planned remediation remains uncertain given their complex site conditions. Furthermore, many of the estimated costs (e.g., the CERCLA figure) do not fully consider the cost of long-term management of sites that will have contamination remaining in place at high levels for the foreseeable future.

The nomenclature for the phases of site cleanup and cleanup progress are inconsistent between federal agencies, between the states and federal government, and in the private sector. Partly because of these inconsistencies, members of the public and other stakeholders can and have confused the concept of "site closure" and NPL deletion with achieving UU/UE goals for the site, such that no further monitoring or oversight is needed. In fact, many sites thought of as "closed" and considered "successes" will require oversight and funding for decades and in some cases hundreds of years to remain protective. CERCLA and other programs have reduced public health risk from groundwater contamination by preventing unacceptable exposures in water or air, but not necessarily by reducing contamination to levels allowing for unlimited use and unrestricted exposure for every contaminant throughout the affected aquifers.

More consistent and transparent terminology that simply and clearly differentiates the discrete phases of remediation and facilitates logical tracking of progress would improve communication with the public. Improvements in terminology among state and federal regulators and PRPs are particularly important in the later stages of remediation. For example, once a remedy has been implemented and operated for some time, classifying the site as a "long-term management site," rather than deleting it from the NPL or classifying it as "closed," would more accurately communicate its status. Sites that attain contaminant concentrations consistent with unlimited use and unrestricted exposure could be classified as "unrestricted-use sites." These classifications would directly reflect progress toward the goals of most state and federal groundwater cleanup programs.

REFERENCES

Canter, L. W., and D. A. Sabatini. 1994. Contamination of public ground water supplies by Superfund sites. International Journal of Environmental Studies 46(1):35-57.

Conger, J. 2011. Memo: New goals for DERP. July 18, 2011.

DEPARC. 2010. Fiscal Year 2009 Defense Environmental Programs Annual Report to Congress, April. Page C-1-1.

DeSimone, L. A., P. A. Hamilton, and R. J. Gilliom. 2009. Quality of water from domestic wells in principal aquifers of the United States, 1991–2004. Overview of major findings. U.S. Geological Survey Circular 1332, 48 pp.

DoD (Department of Defense). 2012. Manual Number 4715.20. Defense Environmental Restoration Program (DERP) Management.

DOE (Department of Energy). 2011. FY2011 Budget Vol. 5 (pages 60-71).

DOE. 2012a. FY2012 Budget Vol. 5 (page 12).

DOE. 2012b. FY2012 Budget Vol. 2 (page 74).

DOE. 2012c. FY2012 Budget Vol. 1 (page 282).

EPA (U.S. Environmental Protection Agency). 1993. Evaluation of the Likelihood of DNAPL Presence at NPL Sites: National Results. EPA 540R-93-073. Washington, DC: EPA Office of Solid Waste and Emergency Response.

EPA. 1996a. Memorandum from Steven A. Herman, Assistant Administrator, Office of Enforcement and Compliance Assurance, and Elliot Laws, Assistant Administrator for Solid Waste and Emergency Response, to EPA Regions. Re: Coordination between RCRA Corrective Action and Closure and CERCLA Site Activities at 2 (September 24, 1996).

EPA. 1996b. EPA Superfund Record of Decision for Schofield Barracks.

EPA. 2000. National Water Quality Inventory: 2000 Report. http://www.epa.gov/305b/2000report/.

EPA. 2002. Community Water System Survey 2000. EPA 815-R-02-005B. Washington, DC: EPA Office of Water.

EPA. 2004. Cleaning up the Nation's Waste Sites: Markets and Technology Trends. Washington, DC: EPA Office of Solid Waste and Emergency Response.

EPA. 2007. Technologies for Site Cleanup: Annual Status Report (Twelfth Edition). EPA-542-R-07-012. Washington, DC: EPA Office of Solid Waste and Emergency Response.

EPA. 2008. Facilities on the RCRA 2008 GPRA Corrective Action Baseline. http://www.epa.gov/epawaste/hazard/correctiveaction/pdfs/base08fc.pdf. Accessed August 16, 2012.

EPA. 2009a. Semi-Annual Report of UST Performance Measures End of Fiscal Year 2009. http://www.epa.gov/oust/cat/ca_09_34.pdf.

EPA. 2009b. The Analysis of Regulated Contaminant Occurrence Data from Public Water Systems in Support of the Second Six-Year Review of National Primary Drinking Water Regulations. EPA 815-B-09-006. Washington, DC: EPA.

EPA. 2009c. DNAPL Remediation: Selected Projects Where Regulatory Closure Goals Have Been Achieved. EPA 542/R-09/008. Washington, DC: EPA Office of Solid Waste and Emergency Response.

EPA. 2010a. Region 9 Online Site Description. http://yosemite.epa.gov/r9/sfund/r9sfdocw.nsf/7508188dd3c99a2a8825742600743735/1818fba2414310dc88257007005e9452!OpenDocument#descr. Accessed August 2010.

EPA. 2010b. Superfund Remedy Report, Thirteen Edition. EPA 542-R-10-004. Washington, DC: EPA Office of Solid Waste and Emergency Response.

EPA. 2011a. Semi-Annual Report of UST Performance Measures End of Fiscal Year 2011. http://www.epa.gov/oust/cat/ca_11_34.pdf.

EPA. 2011b. The National LUST Cleanup Backlog: A Study of Opportunities. Washington, DC: EPA OUST.

EPA. 2011c. Close Out Procedures for National Priorities List Sites at 1-2. OSWER Directive 9320.222. http://epa.gov/superfund/programs/npl_hrs/closeout/pdf/2011guidance.pdf.

GAO (Government Accountability Office). 2008. Hardrock Mining. Information on Abandoned Mines and Valuc and Coverage of Financial Assurances on BLM Land. GAO-08-574T. Washington, DC: GAO.

GAO. 2009. Superfund: Litigation Has Decreased and EPA Needs Better Information on Site Cleanup and Cost Issues to Estimate Future Program Funding Requirements. GAO-09-656. Washington, DC: GAO.

Georgia Environmental Protection Division. 2010. Hazardous Site Inventory Introductory Information. http://www.georgiaepd.org/Files_PDF/gaenviron/hazwaste/intro.pdf.

Hunter, P. 2006. DoD-Wide Occurrence of Emerging COCs in Groundwater & Soil. SERDP-ESTCP Partners Symposium 2006, Washington, DC, November.

ITRC (Interstate Technology and Regulatory Council). 2009. Evaluating LNAPL Remedial Technologies for Achieving Project Goals. Washington, DC: ITRC.

Land, M., J. T. Kulongoski, and K. Belitz. 2011. Status of Groundwater Quality in the San Fernando–San Gabriel Study Unit, 2005. California GAMA Priority Basin Project. Scientific Investigations Report 2011–5206. U.S. Geological Survey.

NRC (National Research Council). 1999. Groundwater and Soil Cleanup. Washington, DC: National Academy Press.

NRC. 2009. Advice on the Department of Energy's Cleanup Technology Roadmap. Washington, DC: The National Academies Press.

OUSD (Office of the Under Secretary of Defense for Acquisition, Technology and Logistics). 2011. FY10 Defense Environmental Programs Annual Report to Congress: Final Report, July 2011.

Rowe, B. L., P. L. Toccalino, M. J. Moran, J. S. Zogorski, and C. V. Price. 2007. Occurrence and potential human-health relevance of volatile organic compounds in drinking water from domestic wells in the United States. Environmental Health Perspectives 115(11):1539-1546.

SCRD (State Coalition for the Remediation of Dry Cleaners). 2010a. May 2010 newsletter. http://www.drycleancoalition.org/download/news0510.pdf.

SCRD. 2010b. December 2010 newsletter. http://drycleancoalition.org/download/news1210.pdf.

Toccalino, P. L., and J. A. Hopple. 2010. The Quality of Our Nation's Waters—Quality of Water from Public-Supply Wells in the United States, 1993–2007—Overview of major findings. U S. Geological Survey Circular 1346, 58 pp.

Toccalino, P. L., J. E. Norman, and K. J. Hitt. 2010. Quality of Source Water from Public-Supply Wells in the United States, 1993–2007. U.S. Geological Survey Scientific Investigations Report 2010-5024, 206 pp.

U.S. Army Environmental Command. 2007. Second Five-Year Review Report for Operable Units 2 and 4, Schofield Army Barracks, Sites 12 and 19.

Zogorski, J. S., J. M. Carter, T. Ivahnenko, W. W. Lapham, M. J. Moran, B. L. Rowe, P. J. Suillace, and P. L. Toccalino. 2006. The Quality of Our Nation's Waters—Volatile Organic Compounds in the Nation's Ground Water and Drinking-Water Supply Wells. U.S. Geological Survey Circular 1292, 101 pp.

3

Remedial Objectives, Remedy Selection, and Site Closure

The issue of setting remedial objectives touches upon every aspect and phase of soil and groundwater cleanup, but none perhaps as important as defining the conditions for "site closure." Whether a site can be "closed" depends largely on whether remediation has met its stated objectives, usually stated as "remedial action objectives." Such determinations can be very difficult to make when objectives are stated in such ill-defined terms as removal of mass "to the maximum extent practicable." More importantly, there are debates at hazardous waste sites across the country about whether or not to alter long-standing cleanup objectives when they are unobtainable in a reasonable time frame. For example, the state of California is closing a large number of petroleum underground storage tank sites that are deemed to present a low threat to the public, despite the affected groundwater not meeting cleanup objectives (California State Water Quality Control Board, 2010; Doyle et al., 2012). In other words, some residual contamination remains in the subsurface, but this residual contamination is deemed not to pose unacceptable future risks to human health and the environment. Other states have pursued similar pragmatic approaches to low-risk sites where the residual contaminants are known to biodegrade over time, as is the case for most petroleum-based chemicals of concern (e.g., benzene, naphthalene). Many of these efforts appear to be in response to the slow pace of cleanup of contaminated groundwater; the inability of many technologies to meet drinking water-based cleanup goals in a reasonable period of time, particularly at sites with dense nonaqueous phase liquids (DNAPLs) and complicated hydrogeology like fractured rock; and the limited resources available to fund site remediation.

This chapter focuses on the remedial objectives dictated by the common regulatory frameworks under which groundwater cleanup generally occurs. It first describes the phases of cleanup for the primary federal programs and their milestones, the gaining of which is often used as a metric of progress and ultimately success. The Comprehensive Environmental Response, Compensation, and Liability Act (CERCLA) and the Resource Conservation and Recovery Act (RCRA) guidance outline criteria for setting remedial objectives and points of compliance, and for selecting remedies to meet them. The chapter closes with a discussion of alternative strategies to address the current limitations on achieving groundwater restoration, such as CERCLA Technical Impracticability waivers for some portion of the site. This includes sustainability concepts that have become relevant to decision making regarding remedy selection and modification in the past few years.

The topic of setting cleanup objectives has a long history and was a significant component of the debates in the 1980s during the passage of the Superfund Amendments and Reauthorization Act (SARA) in 1986 and the establishment of the ARAR process in Section 121 of SARA. Several National Research Council (NRC) reports (1994, 2005) have provided insights and recommendations on improving the process of establishing objectives for groundwater cleanup. The DoD has also provided recommendations for setting objectives through reports published through the Environmental Security Technology Certification Program (e.g., Sale and Newell, 2011). Recently the Interstate Technology and Regulatory Council (ITRC) provided a comprehensive guidance document on setting objectives for remediation at DNAPL sites (ITRC, 2011). All these efforts have informed this overview of the objective setting process, which considers how that process might evolve in light of advances in our understanding of technical limitations to aquifer restoration.

THE CLEANUP PROCESS AND ASSOCIATED OBJECTIVES

The current regulatory framework for remediation of hazardous waste sites evolved from a complex collection of federal, state, tribal, and even local statutes, regulations, and policies. CERCLA and RCRA are the two federal programs that govern most subsurface cleanup efforts, and most state programs are similar to or even authorized under these federal models.

CERCLA

CERCLA provides federal authority for cleanup of sites with hazardous substances, usually excluding petroleum-only sites. At sites with no viable responsible party, EPA can fund remedial activities from the Superfund—a special account initially funded by a tax on petroleum and chemical companies, but presently derived from general tax revenues. However, at a majority of sites, the response is funded by private parties, either through a legally binding agreement to perform the remedy (e.g., an Administrative Order of Consent) or by reimbursing EPA for its remedial costs. At federal facilities cleanup is funded by the agency responsible for releasing contamination.

Initial Phases

A site regulated through CERCLA generally progresses through the Preliminary Assessment/Site Inspection, listing on the National Priorities List (NPL), site investigation (Remedial Investigation), remedial alternative assessment (Feasibility Study), remedy selection (Record of Decision), remediation implementation (remedial design followed by construction), and long-term monitoring and institutional controls until the site media concentrations are at or below unrestricted use levels (see Table 2-3). If there is an immediate threat to human health or the environment ("imminent and substantial endangerment"), the Preliminary Assessment/Site Inspection may trigger an interim emergency response.

The Remedial Investigation consists of detailed site characterization, while the Feasibility Study incorporates the evaluation of remedial alternatives that might meet remedial action objectives. The Remedial Investigation and Feasibility Study may be conducted concurrently, and, in any case, they influence each other. The Remedial Investigation generally includes a human health risk assessment and the determination of site-specific remedial action objectives. The Feasibility Study develops a series of remedial alternatives that describe the placement, timing, and remedial technology for cleanup activities, and it includes a detailed comparison of these alternatives with respect to criteria established under CERCLA regulations (see below).

Setting of Cleanup Goals and Selection of Remedies

CERCLA's overarching groundwater remediation goal is to restore groundwater to its "beneficial use" "wherever practicable" (EPA, 2009a). A common beneficial use of groundwater, if conditions are appropriate, is that it be a source of drinking water. In addition, the groundwater plume "should not be allowed to migrate and further contaminate the aquifer or other media (e.g., vapor intrusion into buildings; sediment; surface water; or wetland)" (EPA, 2009a).

The alternative remedial strategies in the Feasability Study are evaluated based on a balancing of the nine criteria of the National Oil and Hazardous Substances Pollution Contingency Plan, usually called the National Contingency Plan (EPA, 1990):

1. Overall protection of human health and the environment (a threshold criterion that must be met by the chosen remedy)
2. Compliance with applicable or relevant and appropriate requirements (ARARs) (also a threshold criterion)
3. Long-term effectiveness and permanence (a balancing criterion)
4. Reduction of toxicity, mobility, or volume (a balancing criterion)
5. Short-term effectiveness (a balancing criterion)
6. Implementability (a balancing criterion)
7. Cost (a balancing criterion)
8. State acceptance (modifying criterion that is considered but not required to be met or balanced)
9. Community acceptance (modifying criterion)

Threshold Criteria. The first two criteria, called *threshold* criteria, must be met by the chosen remedy. The criterion "protective of human health" is sometimes embodied in a quantitative risk assessment and has been interpreted as having a calculated excess lifetime cancer risk between 10^{-6} and 10^{-4} or a hazard index < 1.0.[1] "Protective of the environment" is less clearly defined.

At most Superfund facilities with groundwater contamination, federal and state drinking water standards (such as maximum contaminant levels, MCLs, and non-zero maximum contaminant level goals) are established as ARARs and hence the groundwater cleanup goals. The designation of a drinking water standard as an ARAR is often independent of whether the particular groundwater is, in fact, currently used as a source of drinking water or is likely to be so used in the future, as long as it is *capable* of being used as a source of drinking water.

There is considerable variability in how EPA and the states consider groundwater as a potential source of drinking water. EPA has defined groundwater as not capable of being used as a source of drinking water if (1) the available quantity is too low (e.g., less than 150 gallons per day can be extracted), (2) the groundwater quality is unacceptable (e.g., greater than 10,000 ppm total dissolved solids, TDS), (3) background levels of metals or radioactivity are too high, or (4) the groundwater is already contaminated by manmade chemicals (EPA, 1986, cited in EPA, 2009a). California, on the other hand, establishes the TDS criteria at less than 3,000 ppm to define a "potential" source of drinking water. And in Florida, cleanup target levels

[1] The hazard index (HI) is "the sum of more than one hazard quotient for multiple substances and/or multiple exposure pathways. The HI is calculated separately for chronic, subchronic, and shorter-duration exposures." The hazard quotient is "the ratio of an exposure level to a substance to a toxicity value selected for the risk assessment for that substance (e.g., LOAEL or NOAEL)" http://www.epa.gov/oswer/riskassessment/glossary.htm.

for groundwater of low yield and/or poor quality can be ten times higher than the drinking water standard (see Florida Administrative Code Chapter 62-520 Ground Water Classes, Standards, and Exemptions). Some states designate *all* groundwater as a current or future source of drinking water (GAO, 2011). Although EPA generally defers to state or local groundwater classifications on these issues (EPA, 2009a), EPA policy recognizes that less stringent cleanup levels may be appropriate for groundwater that is not a current or reasonably expected future source of drinking water (GAO, 2011).

In addition to federal ARARs, states may propose requirements as state ARARs, subject to EPA acceptance. There is considerable variability between federal and some state ARARs, even for the same chemicals or situation, as described in Box 3-1. Table 3-1 demonstrates that the MCL for an individual compound can range over more than an order of magnitude, with some states being much more stringent than EPA. There are multiple reasons for these differences including differences in risk targets, different interpretations of technical feasibility, and different interpretations of toxicological findings.

Another example of variability among EPA and the states concerns the point of compliance. EPA has long directed that the point of compliance monitoring of the final cleanup levels for contaminated groundwater can apply "at and beyond the edge of the waste management area when waste is left in place" (EPA, 1988a, 1990, 1991a). (Note that the drinking water standard in this situation still defines whether the groundwater within the source area may be subject to unrestricted use.) At landfills the application of this policy is relatively straightforward, while at sites where DNAPL has migrated from the original area of release the application of this strategy may be more uncertain.[2] On the other hand, some states require that all points within a contaminated aquifer meet the state ARAR. All this variability can lead to different remedial objectives, different decisions about the chosen remedy, and different long-term outcomes.

Although the most commonly used ARAR, it is noteworthy that MCLs are not based on consideration of the vapor intrusion pathway, suggesting that there can be limitations to relying on ARARs based solely on drinking water ingestion in making decisions regarding remediation of groundwater contamination. Vapor intrusion is discussed further in Chapters 5 and 6.

Balancing Criteria. On a case-by-case basis, the remedy selection criteria (particularly the balancing criteria) are "balanced in a risk management

[2] DNAPL may migrate within the area of waste management. At some CERCLA sites, the edge of the waste management area has been "flexibly applied," while at others the edge of the waste management area has been "rigorously applied."

BOX 3-1
State/Federal Differences in Goals for Groundwater Restoration

The differences between state and federal goals for groundwater restoration often hinge on the present and expected future use of the groundwater in question. However, even if the defined use of the groundwater is for drinking, there can still be differences in the actual numeric goals. This is because states have the option of developing their own, more restrictive MCLs that will replace the EPA's MCL as the enforceable limit. Examples for different chemicals are given in Table 3-1, which provides a sense of the potential magnitude of state/federal differences but is not meant to be comprehensive.

In some cases, the difference between the federal MCL and the state MCL is more than an order of magnitude. For example, the federal drinking water limit for *cis*-1,2-dichloroethene (*cis*-1,2-DCE) is 70 ppb (1 ppb = 1 μg/L), whereas the California standard is 6 ppb. Both values are based on non-cancer liver toxicity in animals, with the differences mainly due to varying interpretations of toxicological findings. As another example, the federal MCL for carbon tetrachloride is 5 ppb, whereas the California standard is 0.5 ppb. Both carbon tetrachloride standards had similar conclusions regarding liver cancer in rodents as the critical endpoint. The differences for carbon tetrachloride are related to measurement feasibility and determination of the practical quantitation limit, rather than to differences in the underlying risk assessment (CalEPA, 2000).

In some cases, there are chemicals for which there are state standards but no federal standards. One example is perchlorate, where the Massachusetts standard is 2 ppb and the California standard is 6 ppb. Although both states chose the same toxicological study as the basis for establishing these limits, Massachusetts adopted a more conservative approach, both with respect to interpretation of the underlying human exposure study by Greer and coworkers (Zewdie et al., 2010), as well as with application of uncertainty factors to derive the non-cancer toxicity criterion (i.e., the reference dose or RfD). In addition, Massachusetts applied different assumptions regarding drinking water intake and other sources of perchlorate. Although the calculated health-based value for Massachusetts was 0.49 ppb, the state chose 2 ppb for risk management purposes to minimize compliance issues. In contrast, the California health-based value of 6 ppb is the same as the standard.

The reasons for differences in drinking water limits are varied and include the application of different toxicity studies to establish underlying health-based values, differences in application of uncertainty factors, variations in selection of exposure assumptions, and differences in risk management considerations. In some cases, the differences reflect the date when a standard was set, and does not always incorporate the new information that has become available for the more recent standard.[1] State/federal differences in drinking water limits may result in different levels of cost effectiveness and health protectiveness of remedial decisions across sites, as well as present risk communication challenges.

---

[1] Due to lack of consideration of technical feasibility, advisory values can lower than mandated values, but they are not mandatory.

judgment as to which alternative provides the most appropriate solution for the site" (EPA, 1990). Under CERCLA, there is a preference for a permanent solution; indeed, EPA "expects to use treatment to address the principal threats[3] posed by a site, wherever practicable" (EPA, 1996a). However, there is "nothing in CERCLA §121 to suggest that selecting permanent remedies is more important than selecting cost-effective remedies" (*Ohio v. EPA*, 997 F.2d 1520, 1533, D.C. Cir. 1993). Rather, the emphasis on permanent solutions and treatment is balanced by the co-equal mandate that remedies be cost-effective through the addition of the wording *to the maximum extent practicable* (EPA, 1996a) (see Box 3-2). EPA believes that "certain source materials are generally addressed best through treatment because of technical uncertainties regarding the long-term reliability of containment of these materials, and/or the consequences of exposure should a release occur," while other source materials generally can be reliably contained (EPA, 1996a).

An issue discussed in Chapter 7 but introduced here is that of the discount rate and its role in remedy selection in addressing one of the nine NPL criteria, namely cost effectiveness. During the feasibility study, cost estimates are developed for each remedial option to identify their relative cost effectiveness. Once costs are identified and quantified for each remedial option, they are discounted to a present value to adjust for differing annual costs across options. For example, some remedies may have large costs in the near future and other remedies may have large costs in the distant future; discounting is a mechanism to compare the costs of remedial options using a common dollar metric. The logic for discounting is that if firms were able to invest these funds they would earn a positive rate of return in the future, which means that expenditures in the present have a higher cost than expenditures in the future.

Currently, the annual cost of each option in EPA feasibility studies for private parties is discounted to present values using a presumptive value of 7 percent, which EPA argues reflects the long-term return to private capital in the United States (OMB, 2003; EPA and USACE, 2000; EPA, 2010a). Discount rates from Appendix C of OMB Circular A-94 (OMB, 2012), which currently are significantly lower than 7 percent, are generally used for all federal facilities.

Under the current approach to discounting, options with costs in the distant future will have lower present values than options with front-loaded costs. For example, with the discount rate of 7 percent, $1 next year is worth about 94¢ today and $1 in 50 years is worth about 3¢ when dis-

[3] In addition to drum wastes and other similar source material, principal threats are where the toxicity and mobility of the source material combine to present an ingestion risk of greater than 10^{-3} (EPA, 1991c).

TABLE 3-1 Examples of State versus Federal Maximum Contaminant Levels

Name	Tetrachloroethene (PCE)	Trichlorethene (TCE)	*cis*-1,2-Dichloroethene (*cis*-1,2-DCE)	1,2,3-Trichloro-propane (1,2,3-TCP)
U.S. EPA	5 ppb	5 ppb	70 ppb	n/a
California	5 ppb	5 ppb	6 ppb (state MCL)	n/a
Florida	3 ppb (state MCL)	3 ppb (state MCL)	70 ppb	n/a
Massachusetts	5 ppb	5 ppb	70 ppb	n/a
New Jersey	1 ppb (state MCL)	1 ppb (state MCL)	70 ppb	n/a
New York	5 ppb	5 ppb	5 ppb (state MCL)	5 ppb (state MCL)

[a]EPA interim advisory level for perchlorate is 15 ppb.

[b]The Massachusetts MCL "is directed at the sensitive subgroups of pregnant women, infants, children up to the age of 12, and individuals with hypothyroidism. They should not consume drinking water containing concentrations of perchlorate exceeding 2 ppb. MassDEP [Massachusetts Department of Environmental Protection] recommends that no one consume

counted to the present. Thus, a cost-efficiency determination tends to favor selection of options that have larger costs in the future and lower near-term costs. Pump and treat, in particular, is an option that discounting favors because the remedy might operate for decades and the present-value calculation indicates the costs of this operation beyond 50 years is $0. A lower discount rate, such as the 3 percent social rate for public projects, would increase the present value of $1 in 50 years to 23¢ today, but it is still likely

Carbon Tetrachloride	Perchlorate	Source	Internet URL
5 ppb	n/a[a]	National Primary Drinking Water Regulations	http://www.epa.gov/safewater/contaminants/index.htm#listmcl
0.5 ppb (state MCL)	6 ppb (state MCL)	State Code of Regulations (Chapter 15, Title 22, Articles 4 and 5.5)	http://www.cdph.ca.gov/certlic/drinkingwater/Documents/Lawbook/dw-regulations-01-01-2009.pdf
3 ppb (state MCL)	n/a	State Code of Regulations (Chapter 62-550)	http://www.dep.state.fl.us/legal/Rules/drinkingwater/62-550.pdf
5 ppb	2 ppb (state MCL)[b]	2008 Standards and Guidelines for Contaminant in Mass. Drinking Water	http://www.mass.gov/dep/water/drinking/standards/dwstand.htm
2 ppb (state MCL)	n/a	State Code of Regulations (N.J.A.C. 7:10)	http://www.state.nj.us/dep/watersupply/sdwarule.pdf
5 ppb	n/a	State Code of Regulations (Part 5, Subpart 5-1)	http://www.health.state.ny.us/environmental/water/drinking/part5/tables.htm

water containing perchlorate concentration greater than 18 ppb" (http://www.mass.gov/dep/water/drinking/standards/dwstand.htm).
SOURCE: Modified, with permission, from Julie Blue, Cadmus Group, Inc. (2009).

that the alternative with higher future costs would be selected over options with high costs in the near future.

Most economists agree that discounting is necessary, because to not discount would overlook the differential time paths of costs across remedy options. There is a long-standing debate over what discount rate is appropriate for use in environmental cases where the costs may be intergenerational. While it is beyond the Committee's charge to opine on the appropriate discount rate, discounting should be considered very carefully

BOX 3-2
Guidance on Definition and Application of "Maximum Extent Practicable"

The Committee was charged with answering the question: what should be the definition of "to the extent practicable" when discussing contaminant mass removal. Terms like "maximum extent practicable (MEP)," "to the extent practical," "practicability," etc., are routinely heard when discussing what can be achieved during groundwater remediation. For example, EPA groundwater remediation guidance, which applies to all EPA non-UST cleanup programs, repeatedly states that EPA's goal is to attain drinking water standards "wherever practicable." The UST regulations 40 CFR 280.64, which apply only to light nonaqueous phase liquid (LNAPL), requires removal of free product "to the maximum extent practicable" as determined by the implementing agency at sites where free product is present. These terms are not defined explicitly or quantitatively in the federal or state statutes, regulations, or settlements and administrative orders that dictate remediation requirements for soil and groundwater. That is, statements as explicit as "70% reduction in concentration" or "removal of mobile DNAPL" are not provided as definitions of "maximum extent practicable."

The main statutory reference to the term "maximum extent practicable" is found in CERCLA in reference to practicability during remedy selection, where practicability reflects a balancing of the nine criteria specified in the NCP (EPA, 2009a, p. 4, footnote 9). EPA guidance states that CERCLA's emphasis on permanent solutions and treatment should be balanced by "the co-equal mandate for remedies to be cost-effective" through the addition of the wording "to the maximum extent practicable" (EPA, 1996a). EPA considers cost to be relevant to technical impracticability because that term is "ultimately limited by cost," although EPA policy is that cost should generally play a subordinate role in a technical impracticability determination unless compliance would be "inordinately costly" (EPA, 1996a).

For this limited use of the term "maximum extent practicable," an explicit definition is already available. EPA has concluded that treatment is *not practicable* when

in the weighing of alternatives along with the other four National Contingency Plan (NCP) balancing criteria listed above. Specifically for projects whose duration exceeds 30 years, EPA and the Army Corps of Engineers (2000) recommend that the present value analysis include a "no discounting" scenario to demonstrate (for comparison purposes only) the impact of the discount rate on the total present value cost of the remedy and the relative amounts of future annual expenditures.

Modifying Criteria. Normally the lead agency evaluates a number of remedial alternatives against the first seven criteria and presents that evaluation, designating a preferred alternative to the public (i.e., community

(1) "treatment technologies are not technically feasible or are not available within a reasonable time frame;" (2) "the extraordinary volume of materials or complexity of the site may make implementation of the treatment technologies impracticable;" (3) "implementation of a treatment-based remedy would result in greater overall risk to human health and the environment due to risks posed to workers, the surrounding community, or impacted ecosystems during implementation (to the degree that these risks cannot be otherwise addressed through implementation measures);" or (4) "implementation of the treatment technology would have severe effects across environmental media" (EPA, 1997a). As an example of the second item above, the use of containment as a presumptive remedy for municipal landfills (EPA, 1997b) means that removal of waste from source areas in those situations can be interpreted as generally not practicable. This case-by-case application of the concept of practicability has been upheld in several court cases [*State of Ohio v. U.S. Env'l't Prot. Agency*, 997 F.2d. at 1532 and *U.S. v. Ottati & Goss, Inc.*, 900 F.2d 429 (1st Cir. 1990) (opinion by now Supreme Court Justice Breyer)]. Thus, as long as the remedy is chosen in accordance with the NCP and is performing in accordance with reasonable environmental engineering practices, that is the end of decision making with respect to what is *practicable* for remedy selection.

The term "maximum extent practicable" is often used informally as a measure of remediation progress even though it has no regulatory bearing in that context. In Chapter 7, the Committee suggests that remedies at complex sites be regularly assessed to determine whether they are being implemented in a manner consistent with good environmental engineering practice and their resulting performance. If a remedy reaches a point where continuing expenditures bring little or no reduction of risk prior to attaining drinking water standards, the Committee recommends that there should be a reevaluation of the future approach to cleaning up the site (called a Transition Assessment). When this point is reached, the chosen active remedy can be said, de facto, to have been operated to the "maximum extent practicable."

stakeholders) in the form of a Proposed Plan. With regard to the two final, *modifying* criteria, neither the state nor the community have the legal authority to "veto" a remedy. The provision does mean that the lead agency must engage in a formal community involvement process and, at each NPL facility, provide a technical assistance grant to one eligible nongovernmental organization to hire an independent technical consultant to advise the community. EPA recognizes about 70 Community Advisory Groups at NPL facilities across the country. From 1988 to 2010, 323 technical advisory grants have been awarded (205 providing $50,000 or less and 15 providing a total of more than $250,000) (Catalogue of Federal Domestic Assistance, 2011). Following the public comment period, the lead agency selects a remedy and memorializes it in a Record of Decision.

After Remedy Selection

Following remedy selection decision, the remedy is designed, constructed, and operated. Once an active remedy is operating properly and successfully, it is considered to have met the EPA *Construction Complete* milestone. Operation and maintenance continue as long as an active remedy is needed to be protective. Optimization evaluations and five-year reviews are performed if chemical concentrations remain above unrestricted use levels in groundwater, soil, soil vapor, and other media (EPA, 2001a). As described in greater detail in Chapter 7, at these later stages monitoring data may be gathered, the remedy may be adjusted, and institutional controls (designed to minimize the potential for human exposure to residual contamination and/or protect the integrity of the remedy) are imposed. According to the NCP, institutional controls are supposed to supplement, not substitute for, active remediation "unless such active measures are determined not to be practicable, based on the balancing of trade-offs among alternatives that is conducted during remedy selection" [40 CFR § 300.430(a)(iii)(D)].

RCRA Corrective Action

Congress enacted RCRA in 1976 to regulate, by permit, the treatment, storage, and disposal of hazardous wastes. In 1984 it amended the law to regulate cleanup at facilities with RCRA permits (40 CFR section 264.101). Though RCRA is a federal law, most RCRA implementation is conducted by the states and territories. Today 43 states and territories have been delegated primacy over their RCRA Corrective Action programs. Therefore, there is more variation in RCRA oversight than under EPA's CERCLA program.

The RCRA remedy selection process and criteria are generally similar to the CERCLA process (EPA, 1996b, 1997a, 2011a). Implementation of corrective action can vary from site to site (and state and state) but it invariably begins with an evaluation of site conditions through an RCRA facility assessment conducted by either EPA or the authorized state. Similar to the Preliminary Assessment/Site Inspection phase of CERCLA, this involves examination of the facility's solid waste management units to determine if a release occurred or if the potential for a future release exists. Interim action to stop the spread of contamination or provide an alternate source of drinking water may be required during this stage. Additional information can be necessary to support interim actions and can be obtained by the site owner through an RCRA Facility Investigation. This investigation involves sampling and modeling to determine the nature and extent of contamination, the site hydrogeology, and the source zone architecture, similar to the Remedial Investigation process under CERCLA. If it is determined that corrective action is required, the site owner will conduct a corrective measures study. Not unlike the feasibility study in CERCLA, a corrective measures study evaluates and selects the remedy and is conducted by the facility owner with oversight from the EPA or the state.

The RCRA program recommends that corrective action be based on risk (EPA, 1997c). EPA's RCRA guidance specifies that cleanup levels be set at federal drinking water standards (where they exist) or be based on a residential drinking water exposure scenario where groundwater is currently used or may be reasonably expected to be used as a source of drinking water (EPA, 2004). RCRA regulations define the point of compliance as the "vertical surface located at the hydraulically down gradient limit of the waste management area that extends down to the uppermost aquifer underlying the regulated units" (EPA, 2004), which conceptually is the boundary of the waste disposal or other management area at the RCRA facility. The exact location is determined on a site-by-site basis.

The two primary RCRA milestones include the *human exposures environmental indicator* and the *groundwater environmental indicator* (see Chapter 2). The objectives that are frequently called for in site-specific agreements between owners and operators of treatment, storage, and disposal facilities and regulatory authorities are typically defined in terms of concentrations of particular contaminants as measured at the boundaries of given units of real property.

Public participation is a part of the corrective measures selection process, but community acceptance (the ninth NCP criterion) is not a statutory requirement for RCRA sites. While in many cases regulators may have established a robust community involvement process, in general this is less extensive than at sites regulated under CERCLA. For example, there are funding sources, such as Technical Assistance Grants, available for CERCLA public involvement that do not exist for RCRA, and regional Superfund programs have Community Involvement Coordinators.

While RCRA permits do not have a statutory requirement for five-year reviews, periodic reviews may be built into RCRA permits. EPA views RCRA permits as "living documents that can be modified to allow facilities to implement technological improvements, comply with new environmental standards, respond to changing waste streams, and generally improve waste management practices" (EPA, 2011b).

As part of RCRA, UST cleanup is also overseen by state and territories or their subjurisdictions. Of interest for this chapter is that the definition of UST "closure," which is a major goal of UST programs, varies significantly from state to state. According to the ITRC 2009 report, historic cleanup goals for LNAPLs have been to remove them "to the maximum extent practicable (MEP)," although some states provide no interpretation of MEP and others specify a maximum allowable amount of LNAPL in a monitoring well (e.g., no visible sheen or 1/8-inch thickness). Some state statutes include "LNAPL thickness-in-a-well requirements" and definitions for when LNAPL remediation efforts may be discontinued. Some states may be bound by statute to remove all LNAPL based on a law or policy stipulating nondegradation of waters.

Federal Facilities

Current and former federal facilities are subject to the same environmental cleanup laws as other properties (see Section 120 of CERCLA), but there are differences. For example, in 1986 Congress established the Defense Environmental Restoration Program (as part of the Superfund Amendments and Reauthorization Act, SARA, 1986), requiring the Defense Department to fund its own cleanups. Other federal agencies are similarly liable for the remediation of their properties.

In general, the Defense Department manages most of its facilities under CERCLA, whether or not they have been listed on the NPL. A major reason for this is that in 1987 President Reagan assigned *lead agency* status to federal responsible parties. At NPL sites, the lead agency is supposed to negotiate a Federal Facilities Agreement with EPA and its state counterparts. These agreements define the scope and timing of the cleanup, and they establish a dispute resolution mechanism whereby the EPA administrator is ultimately responsible for resolving differences between regulators and responsible parties. Federal responsible parties are responsible for conducting five-year reviews under CERCLA, but EPA must approve the finding of protectiveness.

The major federal responsible party agencies, the Departments of Defense and Energy, maintain robust community involvement programs, even at facilities that are not on the NPL. Currently the Defense Department sponsors 191 site-specific Restoration Advisory Boards covering 218 installations (DoD, 2010), and DOE hosts similar bodies at most of it major sites.

A fraction of contaminated federal facilities are regulated under RCRA. In 1992, Congress amended the law to make explicit that states have the legal authority to enforce RCRA cleanup requirements at federal facilities. In the Committee's opinion, this may be a major reason that federal agencies prefer CERCLA, where they will maintain lead agency status, even though RCRA provides greater flexibility in establishing remedial objectives and points of compliance.

Federal facilities that are being transferred to non-federal ownership are subject to additional oversight under CERCLA Section 120(h). In most cases, remedies must be in place and operating properly and successfully before a parcel can be transferred (EPA, 2010b), although groundwater concentrations need not meet drinking water standards prior to transfer. EPA and state regulators must issue a Finding of Suitability for Transfer, providing EPA with authority over federal cleanup at closing military bases and other properties, even if they are not on the NPL (DoD, 1994). There are also provisions for Leasing and Early Transfer, in which non-federal entities may use or take ownership of property before cleanup has been completed (EPA and DoD, 2005; DOE, 1998). In general, this means that regulators must approve of remedies if a transfer is to occur. However, properties that were transferred before the 1986 Superfund Amendments, such as the Defense Department's Formerly Used Defense Sites and the former Atomic Energy Commission's Formerly Used Site Remedial Action Program sites, are subject to CERCLA as managed by the Army Corps of Engineers. They are regulated only by the states and territories unless they are placed on the NPL, which gives EPA regulatory oversight as well.

Lessons Learned

The process outlined above for CERCLA and its counterparts occurs in a straightforward way at only relatively small or simple sites. In reality, the remedial action process is much more complex and nonlinear, particular for the type of sites that are the focus of this study. The process at a particular site can also be more flexible than implied in the description above. The Committee's combined experience provides the following general observations about how cleanup can deviate from the idealized RCRA and CERCLA models. First, a significant amount of cleanup can be implemented through interim and emergency responses. Second, the study phase is often protracted, for several reasons. And third, at many complex sites attaining drinking water standards throughout the contaminated groundwater zone is difficult and unlikely for many decades, which can complicate the latter stages of remediation.

Interim and Emergency Responses

At most complex sites, actual cleanup activity begins long before the selection of a final remedy. First and foremost, easily accessible source materials can be and are quickly removed, such as piles of drums on the ground surface, leaking lagoons, and surface pits. Sites with surface contamination are typically fenced to prevent easy access. Second, measures are taken to interrupt exposure pathways. For example, in the San Gabriel Valley, California, wellhead treatment was provided to ensure that the public water supply, which derives from contaminated groundwater, meets health standards.[4] At the Hopewell Precision Superfund facility in Hopewell Junction, New York, impacted homes were provided with water filtration and vapor mitigation systems.[5] People whose private wells were contaminated with perchlorate from the Olin plume in San Martin, California, were provided with bottled water (California Regional Water Quality Control Board, Central Coast Region, 2003). Containment remedies are often applied at the earliest stages of site response to prevent the spread of contamination. For example, at the MEW Superfund Study Area in Mountain View, California, responsible parties quickly installed slurry walls around the known source areas on their properties, and they found and plugged abandoned agricultural wells that served as vertical conduits for contamination to move between aquifers (EPA, 2009b).

Regulators and responsible parties often agree to conduct source removal or containment long before the full extent of contamination is even mapped. At the CTS Asheville site in North Carolina, EPA conducted soil vapor extraction as an emergency response prior to a remedial investigation (EPA, 2010c). At the MEW site responsible parties removed contaminated soil, conducted soil vapor extraction, and installed localized groundwater extraction and treatment systems long before the development of a regional remediation strategy (EPA, 2009b).

The lesson learned from existing case studies and the experience of the Committee is that in geographic locations where there are numerous separate sources affecting the same aquifer, a regional remediation strategy that addresses sources in a variety of federal and state programs (e.g., CERCLA, RCRA, and UST) early in the process and with the involvement of all stakeholders can allow for Interim and Emergency Responses to be implemented in a more effective manner. It is also consistent with EPA environmental justice program's efforts to use "an integrated One EPA presence" to engage communities in the Agency's work to protect human health and the environment (EPA, 2011c).

Protracted Study

Robust, reliable site characterization is essential to effective site cleanup. Without it, remedies may fail to address significant problems or may even spread contamination. For a number of reasons, investigation of a complex site is always protracted. First, it is inherently difficult to characterize groundwater contamination and develop an accurate conceptual model at complex sites. Once sampling schedules are established, it can be

---

[4] See http://yosemite.epa.gov/r9/sfund/r9sfdocw.nsf/3dec8ba3252368428825742600743733/538dd2f968eac4fb88257007005e9460!OpenDocument.

[5] See http://www.epa.gov/region02/superfund/npl/hopewell.

difficult to change them. This is especially true at sites where assessment is an exercise in routine data gathering, rather than an attempt to improve the understanding of site conditions. In addition, at virtually all sites sampling results in new discoveries that may change the sampling strategy. Second, the nature of the study process is adversarial (i.e., where the work and funding come from the responsible party, but the final decision about moving forward rests, as it must, with the regulators). Third, the process of having a large number of government experts (both state and federal) review different portions of the responsible party's submissions adds time. During the Remedial Investigation/Feasibility Study phase multiple documents are created, including specialized studies. Actual cleanup, of course, cannot proceed until regulators review responsible party documents, the responsible parties respond to regulator comments, and all outstanding issues are resolved. A lack of adequate staffing in state and federal agencies aggravates this situation (e.g., Sweeney, 2010). Finally, the interpretation of study data is nontrivial and often the subject of disputes between EPA and the potentially responsible parties.

In recent years, agencies have emphasized the establishment of data quality objectives to be certain that the quality of samples will be high enough to answer key questions about and to test hypotheses of the conceptual model. EPA's Data Quality Objective Process (EPA, 2006) discusses how to "clarify study objectives, define the appropriate type of data [to collect], and specify tolerable levels of potential decision errors that will be used as the basis for establishing the quality and quantity of data needed to support decisions." Nonetheless, in the Committee's experience there is still a strong tendency to collect too much information for fear of missing a key data point, leading to protracted study at these complex sites.

There have been initiatives, such as the Air Combat Command's "Streamlined Oversight" project (U.S. Air Force, 1995), piloted at Langley Air Force Base, Virginia, in which regulators and responsible parties have formed partnerships to jointly solve problems, eliminating much of the back-and-forth shuffle of documents, but those programs remain the exception rather than the rule.

The Limits of Aquifer Restoration

As shown in many previous reports (EPA, 2003; NRC, 1994, 1997, 2003, 2005), at complex groundwater contamination sites (particularly those with low solubility or strongly adsorbed contaminants), conventional and alternative remediation technologies have not been capable of reducing contaminant concentrations (particularly in the source area) to drinking water standards quickly. Because the history of groundwater cleanup is still relatively recent, in that few sites with remedies have been operating for more than 25 years, the time to achieve restoration cannot be easily predicted based on empirical observations, but it likely extends for decades. As a practical matter, at both Superfund and RCRA sites a variety of strategies are being used, which recognize that drinking water standards are unlikely to be attained within source areas. These methods include the use of monitoring compliance points outside the source area, use of containment zones for petroleum and low-risk solvent sites (by the California Water Resources Board and the Regional Water Quality Control Board-San Francisco Bay Area, respectively), Texas' Municipal Settings Designation, Florida's Natural Attenuation Default Source Concentration, and EPA's Technical Impracticability (TI) waivers, among others.

Because of the diversity of chemicals and conditions at sites, the limits of existing technologies, and the inevitable lack of agreement on the proper balance between the nine criteria of the NCP, there is no precise formula or clear trigger for determining when restoring an aquifer to drinking water standards is practicable or what is a reasonable remediation time frame in which to accomplish this, and these debates are likely to continue. Rather, general remedy selection principles (laid out in many EPA guidance documents and described in this chapter) should be applied to the specific conditions at a site to determine the remedy. The remedial alternatives should be reviewed to determine the timeframe, the cost, and the practicality of reducing the concentrations in groundwater to drinking water standards. This requires the transparent exchange of technical and cost information between regulators and responsible parties. Other implications of the limits of aquifer restoration are discussed more fully in Chapter 7.

THE FUTURE OF CLEANUP OBJECTIVES

The Committee assumes that drinking water standards will remain the long-term goal of groundwater remedies for the foreseeable future. Drinking water standards define unlimited groundwater use and unrestricted exposure, and until they are met, five-year reviews (at sites regulated under CERCLA) and institutional controls are needed. Despite these requirements, the Committee believes that new approaches to setting cleanup goals should be considered, to the extent permitted by law. These include giving more attention to site-specific risks, setting alternative concentration limits, seeking TI waivers, reclassifying groundwater, and considering sustainability, as discussed below.

A More Central Role for Risk Assessment

During the RI phase of CERCLA, information is collected that can be used to conduct human health and ecological risk assessments, usually following EPA's Risk Assessment Guidance for Superfund (RAGS) (EPA, 1989, 1991b). Site-specific risk assessment integrates information on the physical conditions at the site, the nature and extent of contamination, the toxicological and physicochemical characteristics of the contaminants, the current and future land use conditions, and the dose-response relationship between projected exposure levels and potential toxic effects. The end result is a numerical value of potential additional risk to the hypothetical receptor from the contaminant source under present conditions (i.e., the "no-action" scenario), along with a discussion of the attendant uncertainty. The calculated risk values are typically compared to the range of acceptable risk defined by EPA or by state regulations (often 10^{-6} to 10^{-4} for carcinogenic compounds). If the risk estimate is greater than the acceptable target risk level, target cleanup level objectives are identified for the site using the assumptions developed in the risk assessment related to potential levels of exposure.

In the Committee's experience, EPA and state drinking water standards usually drive groundwater cleanup rather than the results of site-specific risk assessment. This can lead to responsible parties, regulators, and the public having an incomplete understanding of risk-related issues, including the plausibility of the scenarios that are driving decision making, the likely site risks at the present and in the future, and site risks reduced to date. Hamilton and Viscusi (1999) provided several examples, taken from Superfund, of the importance to risk estimates of assumptions regarding the selected scenarios (e.g., future on-site resident as compared to present-day off-site residents). These authors also provided estimates of both individual and population risks, demonstrating that at some sites population risks, reflected in the number of estimated cancer cases, can be small, often well below 1. Rarely is such information provided as part of the typical risk assessment process. Moreover, exposure pathways, such as inhalation of vapors from off-gassing during showering or inhalation of chemicals from vapor intrusion, are often not reflected in ARARs for groundwater, especially MCLs. Failing to consider these pathways could yield over- or underestimates of risks.

More Comprehensive Consideration of Time

The risk-based methods typically used at contaminated sites evaluate carcinogenic and non-cancer risk to a hypothetical individual over the course of the person's lifetime. These methods do not factor in the changes in concentrations or exposure over the lifetime of a contaminant source. For example, the future potential risk is calculated over a 30-year timeframe[6]

[6] Thirty years is the typical exposure duration in the baseline risk assessment under RAGS.

based on the reasonable maximum exposure concentration determined in the remedy-selection risk assessment, even if the chemical concentration decreases going forward without any active remedy. In this example, the risk reduction predicted as a result of the selected remedy (and, therefore, the cost effectiveness and practicality of that remedy) would be overestimated since some of the risk reduction would have occurred even in the absence of active remediation. The only example known to the Committee of where time has been considered more explicitly is EPA's guidance on remediation of polychlorinated biphenyl sites. This guidance recommends that the calculated risk consider concentration decreases over time from volatilization (e.g., a 72 percent reduction in concentration over 30 years) and biodegradation (e.g., a half-life of 50 years) (EPA, 1990).

Similarly, there is also no formal framework for considering the impact on risk of concentrations ceasing to decline once a remedy has been in place for an extended period (see Chapter 7 for more discussion). This is not an uncommon occurrence at complex sites, where contaminant concentrations may reach an asymptote beyond which there is very little, if any, further decline in concentration despite continued operation of an active remedy. In such situations, the reduction in potential risk has also plateaued (i.e., risk reduction ceases) and is not achieving its full extent as predicted in the Record of Decision.

The Committee believes that more formal consideration of the time element in risk assessment (i.e., by linking predicted changes in concentrations with and without a remedy to changes in risk, assuming that drinking water is a complete exposure pathway) can be important in understanding the cost effectiveness of a remedy. In addition, once a remedy is implemented, understanding the risk–concentration response function over time will provide risk managers and the community with a more complete understanding of the changing risk profile and, if restoration is not practical, facilitate decision making regarding the need for long-term management. These issues are discussed further in Chapter 7.

Population Impacts and Risks to Remediation Workers and Surrounding Residents

Population risk is commonly represented as the number of cases of disease or fatalities in a particular exposure setting. For cancer risks, this might be presented as the hypothetical estimated number of cancer cases associated with a particular exposure scenario. The value is calculated based on population size and average risk level.[7] Estimation of population

[7] We note this is a simplification of the calculation and factors such as the age distribution of a population may also be considered.

risk over the lifetime of a hazardous waste site is rarely if ever conducted because there is no regulatory requirement to do so, nor is there a currently prescribed regulatory context for considering the results of such an evaluation. Because of the relatively small population size affected by a given Superfund facility, the total number of cancer cases associated with contaminant exposure is likely to be small (often less than 0.1 cancer cases per site according to Hamilton and Viscusi, 1999). In contrast, consideration of population risk is an important component of many federal rules in other settings, including other environmental exposures as well as occupational and pharmaceutical exposures. For example, in the setting of the arsenic MCL, EPA considered population size in reducing arsenic in drinking water from 50 μg/l to the options of 3, 5, or 10 μg/L. Benefits included the number of lung and bladder cancer cases averted in the potentially exposed population and the subsequent impact on reduction in costs of morbidity and mortality (EPA, 2001b).

Another risk component to remedial decision making related to population size that is infrequently quantified in any formal analytical way is that of short-term risks created as part of remedial activities. For every remedial alternative, there may be short-term risks to workers during implementation of the remedy (e.g., due to excavation of large volumes of waste and contaminated soil at landfills and/or treatment facilities), short-term risks of injury to local residents and populations along the transportation route due to traffic accidents during transportation of such wastes, and long-term risks to local residents who live near the redisposal site (e.g., Greenberg and Beck, 2011). Despite the existence of these risks, few remedy selection decisions consider them in a quantitative way (Leigh and Hoskin, 2000).

Population risks and risks from physical injuries to remediation workers (including on-site injuries) have been quantified in the site remediation context using the metric of years of potential life lost (YPLL) (Cohen et al., 1997). YPLL, which is used in public health decision making, considers the number of fatalities resulting from particular activities and the age of the individuals experiencing the fatalities. For some of the hypothetical case studies in the analysis by Cohen et al. (1997), the increase in YPLL to remedial workers was greater than the reduction in YPLL for the population at the site. Use of this concept in decision making, particularly when remediation extends over long periods, could be useful in selecting among remedial alternatives to find those with the largest overall benefit to public health (e.g., by reducing YPLL to the greatest extent practicable). An example of the practical application of the YPLL concept in environmental decision making (albeit not in the context of groundwater cleanup) is provided by Frost et al. (2002) who concluded that, in the analogous context of selecting among different strategies for reducing drinking water arsenic in Albuquerque, NM, the drinking water treatment approach of coagulation and microfiltration yielded greater public health benefit (i.e., fewer YPLL) than other drinking water treatment approaches. Note that the Committee is not suggesting that population-size considerations should be used to select remedies that are less health protective—i.e., remedies should result in post-remedy risks within the target risk range for all populations.

Uncertainty and Variability in Risk Analyses

At complex sites, risk-based methods could be employed to more fully understand the nature of existing risk and expectations regarding future risks reduced under different scenarios through more explicit uncertainty analysis. The results of uncertainty analysis could help identify areas where additional data collection may be beneficial and provide risk managers and communities with a greater understanding of the implications of specific decisions.

In the context of risk assessment, *variability* refers to the natural variation that occurs across space and time in a population—including differences in exposure point concentrations, intake assumptions (e.g., breathing rates), and pharmacokinetic differences among individuals as a function of age, genetics, or other factors. For a given contaminated site, such differences yield a distribution of risks across the population of present and future residents living near a site. Use of probabilistic risk analysis methods, in which a distribution of risks is presented, provides a more complete understanding of variability.

Uncertainty refers more generally to a lack of knowledge about specific parameters. For example, the shape of the dose-response curve for a carcinogen at low doses (e.g., whether is it linear, nonlinear, sublinear) is often uncertain. Reasons for uncertainty may include a lack of experimental data in the dose range of interest or lack of understanding of the mode of action for carcinogenesis. In the case of groundwater contamination, an important source of uncertainty in risk assessment is often the choice of use scenarios for contaminated groundwater. If the water is presumed to be used for drinking water, then the relevant pathways would include ingestion, inhalation from off-gassing during showering and vapor intrusion, and dermal uptake. If the water is not used (and not reasonably expected to be used in the future) for drinking water, then only inhalation from vapor intrusion would constitute a complete exposure pathway.

Tools are available for formally characterizing uncertainty in risk assessment and could be applied more frequently at contaminated sites. For example, expert judgment could be used, based on assumptions regarding demographics and types of water usage in an area, to formally elicit determinations as to the likelihood of certain use scenarios. Sensitivity analyses could be conducted comparing different use scenarios, incorporating tem-

TABLE 3-2 Examples of Uncertainty and Variability in Risk Assessment for Contaminated Groundwater

Parameter	Source of Uncertainty	Source of Variability
Carcinogenicity of chlorinated solvent	Lack of toxicological understanding in low-dose region	Differences in metabolizing ability across individuals, resulting in differences in susceptibility to toxicity
Sampling for compliance with cleanup targets	Differences in detection limits as a function of changing technologies	Differences in concentrations across time and space
Groundwater use	Changes in water use patterns in the future can affect the plausibility of the use scenario	High variability in water ingestion rates
Exposed population	Changing demographics	Age distribution of population, which can affect water consumption patterns

porality into the analysis. More sophisticated modeling software tools are becoming available to conduct analyses of variability and uncertainty. For example, the Analytica program (Mansfield et al., 2009) provides general mathematical modeling language to develop uncertainty models, which can then be combined with probabilistic modeling. EPA has employed Analytica to combine information on variability in exposure along with uncertainty in dose-response function to evaluate the benefits and costs of air quality regulations. While such analyses can be complex and time consuming, they are likely to be worthwhile for certain situations at recalcitrant sites, and informative in the context of making decisions regarding the need for alternative endpoints. Table 3-2 presents some sources of uncertainty and variability in risk assessment for contaminated groundwater.

Additional Strategies for Goal Setting

Many strategies have been developed and accepted by regulators to acknowledge site complexity and inherent technical and cost barriers to achieving drinking water standards, yet provide a path forward that reduces risk and retains the ability to determine when unrestricted use is appropriate. Examples include applying for and being granted a TI waiver, groundwater reclassification, applying EPA's flexible guidance on determining if a requirement is an ARAR (including applying the exceptions, exemptions, and variances associated with the federal or state requirement), use of compliance monitoring points outside the presumed source area, and use of alternate concentration limits, among others (ESTCP, 2011; EPA, 1996c, 2009a).

There are multiple benefits of using these additional strategies, each of which is provided for in EPA guidance (EPA, 1996c, 2009e, 2011d). These strategies can meet most regulatory requirements, establish common expectations, protect public health through exposure control, provide a pathway toward meeting the DoD milestones of *remedy in place* and *response complete*, manage remedial project risks, and potentially use resources more efficiently. The challenges are also significant, and include regulatory reluctance to adopt such additional strategies because of (a) scientific disagreements on the fate of chemicals or technological performance; (b) disagreement about what is a reasonable time frame or what is cost effective; (c) community concerns; and (d) uncertainty about the ability to control long-term risks. Whether these alternative strategies can still be protective while leading to reductions in life-cycle costs is difficult to quantify, but intuitively cost savings seem likely.

ARAR Waivers

Under CERCLA, the selection of an ARAR requires a careful application of site-specific facts to the site of interest. A requirement under other environmental laws may be either *applicable*[8] (i.e., it would apply, but for this being a Superfund facility) or *relevant and appropriate* (i.e., it addresses problems or situations similar to the conditions at the site and is a requirement that is "well suited" to the site) [see 40 CFR §300.5 and 40 CFR §300.400(g)]. To determine if a requirement is well suited, one must assess the nature of the substances at the site; the physical, chemical, and microbial characteristics at the site; the circumstances surrounding the release; and the ability of the action to address the release. Thus, an ARAR must be determined on a case-by-case basis and the analysis may provide substantial flexibility.

ARARs, even if applicable, may also be waived, e.g., TI waivers. There is no rigid definition of what constitutes technical impracticability (EPA, 1993; AEC, 2004). Eighty-five TI waivers have been issued for ground-

---

[8] "Applicable requirements" are those "cleanup standards, standards of control, and other substantive environmental requirements, criteria, or limitations promulgated under federal environmental or state environmental or facility sitting laws that specifically address a hazardous substance, pollutant, contaminant, remedial action, location, or other circumstance found at a CERCLA site." 40 C.F.R. § 300.5; EPA, Draft ARARs Guidance at pp. 1-10. There should be a one-to-one correspondence between the requirement and the circumstances at the site. Id. "Applicability" implies that the remedial action or the circumstances at the site satisfies all of the jurisdictional prerequisites of a requirement. Id. at 1-10.

water through February 2012, based on a variety of site-specific factors. These factors, summarized in ESTCP (2011), include (1) complex geologic features, (2) confirmed presence of DNAPLs or other recalcitrant contamination, (3) a combination of the above, (4) excessive cost, (5) physical limitations due to surface structures, and (6) perceived technical limitations of remediation technologies.

Another less common ARAR waiver is the Greater Risk ARAR waiver, which applies if activities taken to meet an ARAR would cause greater harm (like remobilizing DNAPL or dewatering wetlands) than waiving the ARAR; an example is Onondaga Lake, which has elemental mercury as a DNAPL (ESTCP, 2011). The other strategies discussed below have been used much less frequently than waiving an ARAR due to technical impracticability.

Alternate Concentration Limits

Alternate concentration limits (ACLs), which apply at CERCLA and RCRA sites, allow the use of a remediation goal in groundwater that is protective of surface water into which contaminated groundwater discharges, rather than the drinking water standard. The basic concept is that when the groundwater plume enters surface waters, the remedial goal should be consistent with the permitted discharge program governing point source discharges into surface water, as regulated under the Clean Water Act. EPA (2005) clarified ACL policy at sites regulated under CERCLA by identifying a number of considerations. For example, one has to consider whether all plumes discharge to surface water (e.g., a deeper aquifer might not), whether there are potential degradation products between the source and the points of entry (e.g., trichloroethene degrades to vinyl chloride), and whether groundwater can be restored to beneficial use within a reasonable timeframe.

One example where an ACL was adopted is the Naval Surface Warfare Center Ammunition Burning Grounds in Crane, Indiana (ESTCP, 2011). Downgradient of the explosives-contaminated site are several springs that discharge into a nearby creek, which serves as a public water supply 11 miles downstream. Rather than setting the 3 µg/L drinking water standard for the chemical explosive RDX as the site remediation goal, an ACL for RDX of 140 parts per billion (ppb = µg/L) at the spring was set. It was based on ensuring that Indiana Water Quality Standard of 240 ppb would be achieved in the non-potable surface water and the 3 parts per billion RDX would be met at the public water supply. Note that because the ACL is less stringent than the contaminant level that would allow for unlimited use and unrestricted exposure, five-year reviews continue at this site.

Groundwater Management and Reclassification of Groundwater

Groundwater management or containment zones and reclassification of groundwater uses are similar to TI zones, in that they refer to a volume in the saturated zone that is allowed to exceed water quality standards, but the rationale may differ. Most regulators, as a matter of policy, have designated the goal for most groundwater as attaining the highest beneficial use (i.e., use as drinking water), even where the natural or background quality is relatively poor. At some sites, regulators have explicitly recognized that the groundwater in a particular area is unlikely to be used for drinking water now and in the future. A variety of terms have been developed for the affected area in such circumstances, such as plume management zone (Texas), groundwater management zone (Delaware, Illinois, New Hampshire), and containment zone (California Regional Water Quality Control Board – Region 2). An example is the Joliet Army Ammunition Plant, where explosives-contaminated groundwater has a cleanup timeframe of up to 340 years (ESTCP, 2011). There are three groundwater management zones at Joliet, for which the remedial objectives are higher concentrations than elsewhere. Contamination within the groundwater management zones is being addressed through a number of approaches including deed restrictions and continued groundwater and surface water monitoring.

Groundwater reclassification refers to changing the beneficial use of an aquifer such that it is no longer considered a potential source of potable water. At the Altus Air Force Base in Oklahoma, where DNAPL is likely present in fractured bedrock, the groundwater was reclassified to Class III, based primarily on the presence of elevated TDS in the aquifer. This classification will not allow the groundwater to be used for drinking water, although it does permit agricultural and industrial uses. The cleanup objective is to contain the plume, rather than restore it to maximum beneficial use, and the point of compliance is the base boundary (ESTCP, 2011). In New Jersey, there are groundwater classification exemption areas, which are used as an institutional control that provides for the protection of human health as long as the contaminant concentrations in the areas exceed the New Jersey groundwater quality standard.

***

The ESTCP (2011) report on alternative strategies for site management makes it clear that alternative remedial objectives are not used in many situations where they might apply, despite their attractiveness for dealing with complex sites. TI waivers have been approved by EPA, albeit at only a small percentage of NPL sites (3 percent) for which they could likely be used (ESTCP, 2011). (It should be emphasized that at Superfund facilities groundwater restoration remains the goal outside of the agreed-upon TI zone.)

A similar initiative directed by the ITRC suggests that many state regulators wish to see a change in the process for overseeing groundwater cleanup activities. ITRC established a committee to develop an integrated strategy for cleanup of groundwater sites impacted by DNAPLs. That committee's report (ITRC, 2011) includes recommendations on alternative procedures for setting objectives at these sites. For example, the report recommends that "setting of remedial objectives should be based on realistic assumptions and expectations"—a reference to the technical limitations on achieving MCLs. This is a clear indication of the desire to recognize technical limitations *before* a remedy is selected, equivalent to the use of a TI waiver prior to the completion of a Record of Decision. One of the most significant parts of the ITRC report is the emphasis on greater accountability in setting cleanup objectives. That is, it recommends that interim or functional objectives (see NRC, 2005, for an extensive discussion of "functional" versus "absolute" objectives) be established that can be observed within a 20-year timeframe in order to ensure that potentially responsible parties and engineering firms are held accountable, even where restoration remains the long-term goal. Timeframes beyond 20 years were felt to reduce the likelihood of holding parties accountable for remedial performance.

Despite these initiatives, there is still widespread reluctance by federal and state regulatory agencies to accept the concept of alternative remedial objectives. Reasons for this reluctance are not difficult to comprehend. Notwithstanding EPA's written guidance, some regulators may seek more aggressive remedies. It is also likely that some regulators inherently make the most protective decision on cleanup objectives and are reluctant to accept the need to revise objectives. For example, it appears that one of the factors that may make issuance of TI waivers difficult is that when such a waiver is granted, an alternative groundwater remediation goal is set for the TI zone in lieu of the unrestricted use level. Thus, the perception is that in order to grant such a waiver, one must "abandon" achieving the unrestricted use of the groundwater in some portion of the aquifer. Some state regulatory bodies argue that as a matter of policy, state non-degradation policies should also be used to require cleanup to drinking water standards of groundwater already degraded and maintain that the goal of restoration is paramount regardless of the technical or economic constraints. Other causes of this reluctance to use alternative strategies on the part of regulatory agencies include, in the Committee's experience, rotating project managers and lack of incentives to reach a compromise between the potentially responsible parties and the regulators. On the other hand, potentially responsible parties may be reluctant to accept an alternative remedial objective because of the transaction costs associated with the process, or because of future litigation risks should residual contamination persist (see Chapter 5).

Sustainability as a Cleanup Objective

The historic definition of sustainable is "[d]evelopment that meets the needs of the present without compromising the ability of future generations to meet their own needs" (Brundtland, 1987). According to the Bruntdland report (1987), the most "sustainable" policies address environmental, economic, and social aspects of a problem (the so-called triple bottom-line approach)—a definition much broader than that encompassed by the federal and state hazardous waste laws. If sustainability is to be a remedial goal, this broad policy definition needs to be translated into concrete direction on how to clean up a site "sustainably."

Incorporating sustainability concepts into remediation decision making is a developing, but still incomplete, practice at EPA and other agencies. EPA, DoD, the states, and others have "green" or sustainable remediation policies (DoD, 2009; Army Corps of Engineers, 2010; EPA, 2008; ITRC, 2011). All ten EPA Regions have adopted Clean and Green policies for contaminated sites, generally with green remediation goals including to minimize total energy use and to reduce, reuse, and recycle materials and wastes (EPA, 2011e). However, "green" remediation and even some of these agency guidance documents that use the word "sustainability" do not include all of the elements of sustainability found in the Brundtland report. For example, EPA's definition of green remediation is the "practice of considering all environmental effects of remedy implementation and incorporating options to minimize the environmental footprints of cleanup actions" (EPA, 2011e). This is narrower than the concept of sustainable remediation as "balance[ing] outcomes in terms of the environmental, social, and economic elements of sustainable development" (see Table 3-3 below and Bardos et al., 2011; NRC, 2011). In fact, some argue that sustainable decisions should consider community improvements, jobs, and quality of life, and the benefits to the surrounding community (NRC, 2011). Several examples of sustainable remediation that illustrate the range of concepts that can be incorporated are given in Box 3-3.

Each of the Sustainable Remedy Selection environmental factors listed in Table 3-3 (i.e., column 1), and some of the social and economic factors (columns 2 and 3), fit into the standard EPA and state remedy selection criteria. For example, impacts on human health and safety (a social factor), impacts on various environmental media and natural resources, and community involvement can be assessed under existing remedy selection schemes. However, ethical and equity considerations, indirect economic costs and benefits, and employment and capital gain (among others) are

TABLE 3-3 Sustainable Remedy Selection Factors

Environmental	Social	Economic
1. Impacts on air (including climate change) 2. Impacts on soil and ground condition 3. Impacts on groundwater and surface water 4. Impacts on ecology 5. Use of natural resources and waste	1. Impacts on human health and safety 2. Ethics and equality 3. Impacts on neighborhood and locality 4. Communities and community involvement 5. Uncertainty and evidence	1. Direct economic costs and benefits 2. Indirect economic costs and benefits 3. Employment and employment capital 4. Induced economic costs and benefits 5. Project lifespan and flexibility

SOURCE: Adapted, with permission, from CL:AIRE (2011).

not explicitly provided for in any cleanup statute or existing programs. Many of these broader societal factors could be taken into account at federal facilities if the government decided to expend its own funds, but they are likely to be difficult to include as enforceable requirements on private sector decision making without amendments to existing cleanup statutes.

Industry groups are currently driving sustainable remediation efforts. For example, approximately 87 percent of the largest companies in the Drugs and Biotechnology, Household and Personal Products, and Oil and Gas Operations sectors have environmental sustainability programs, according to a survey of the five largest U.S. companies in each of the 26 industrial sectors (Cowan et al., 2010). Most companies develop their own sustainability policies based on their sector, stakeholder interests, products or services, and business model. In the hazardous waste arena, the leader in sustainability is the Sustainable Remediation Forum (or SURF, http://www.sustainableremediation.org), which includes industry, government agencies, environmental groups, consultants, and academia. The SURF approach, described in greater detail below, advises that one "should balance the level of sustainability analysis in accordance with the budget and available resources" (Holland, 2011; Ellis and Hadley, 2009).

A Method for Estimating Sustainability

There are a variety of potential methods for including sustainability factors in selecting a remedy, but none are generally accepted and no U.S. regulatory agency has formally adopted a methodology. The SURF Framework (Holland, 2011) "provides a systematic, process-based, holistic approach for: (1) performing a tiered sustainability evaluation, (2) updating the conceptual site model (CSM) based on the results of the sustainability

BOX 3-3
Examples of Sustainability in Hazardous Waste Remediation

There are a number of clear examples of hazardous waste site remediation where sustainability is being taken into consideration in the remedy selection process.

One example is the Bell Landfill NPL site in northern Pennsylvania. Large trucks were previous used to carry landfill leachate to a wastewater treatment plant with the proper permit—a 640-km road trip. Chemical analysis of the leachate showed that the only remaining components were dissolved iron and manganese. Now, a spray irrigation system is used to distribute the leachate onto the landfill cap, which is covered with grass. As a result, the grass on the cap no longer dies during the summer, and the local unpaved roads are no longer impacted by the heavy truck traffic during wet weather. Changing how the leachate was disposed of also avoided the release of about 3,400 tons of CO_2.

At the Brevard, NC, polymer recycling site, off-spec films were previously disposed of in an industrial landfill that contains up to 80,000,000 pounds of PET. They are now being excavated, inspected, and shipped to China where the material is being recycled (the final use of the material is not known). Once the project is complete, the landfill will be converted into parkland and deeded to the State Forest. This is an example of resource recovery and recycling, leading to lower greenhouse gas emissions (which could be as much as 100,000 tons of CO_2). Note that the life-cycle assessment for this project included all of the impacts associated with shipping the materials to China.

Another example of sustainability in site remediation is at DuPont's Chambers Works site—a 146-acre landfill with about 10 million tons of waste. Three remediation options were evaluated: excavation, stabilization, and bioremediation. Qualitative consideration of a number of factors, including the amount of CO_2 produced, led to the choice of bioremediation. Using bioremediation instead of excavation was predicted to reduce potential emissions by over 2,500,000 tons of CO_2, avoid odor problems in the adjacent community, and avoid the need for round-the-clock intense lighting and heavy equipment operation, which would disturb nearby residents.

At a Naval Air Station Superfund facility in Weymouth, Massachusetts, EPA modified an excavation remedy to allow reuse of the soil as a subgrade fill layer rather than disposing of the soil offsite, which "significantly reduced energy consumption associated with truck trips for off-site disposal and importing common fill and allowed for the beneficial reuse of the excavated materials in a manner which is protective of human health and the environment." Emissions of regulated air pollutants were also reduced (EPA, 2010d).

The Reichhold Chemical Site is a former paint and coatings manufacturer located south of downtown Chicago. The site was redeveloped following RCRA clean closure that left no residual contamination on the site. Two large retail stores were opened on this formerly abandoned site, and 500 new inner city jobs were created. In addition to the obvious economic benefits, there is also the social benefit of having major retailers in the community; residents previously had to drive over 10 miles to find comparable services.

evaluation, (3) identifying and implementing sustainability impact measures, and (4) balancing sustainability and other considerations during the remediation decision-making process."

The SURF approach includes a series of separate toolkits (organized into tables) for the investigation, remedy selection, remedial design and construction, and operation and maintenance phases of site cleanup. For each phase, the team identifies parameters, objectives, metrics, and benefits and challenges to applying these metrics to each phase of the remediation (Butler et al., 2011). For example, the project team and stakeholders review which of the potential sustainability parameters (i.e., consumables, physical disturbances and disruptions, land stagnation, air impacts, water impacts, solid wastes, job creation, and remediation labor) are appropriate for consideration at a particular site (see Butler et al., 2011). For each of the relevant parameters, the team identifies the applicable objectives, the metrics for measuring the achievement of each objective, the benefits that are likely to be derived, and challenges of using this parameter for each remedy being considered for the site. The team considers these factors, benefits, and tradeoffs explicitly in the table. The results obtained during this exercise are balanced with project considerations to determine the most appropriate remedy.

Critical to the implementation of the SURF approach is the preferred future use of the site, including consideration of (a) local laws, ordinances, and deed restrictions; (b) the end use of the site and the likely future development around the site; (c) the capacity to establish and maintain necessary institutional controls; (d) potential liabilities and community needs; and (e) long-term technical and environmental issues (Holland, 2011).

Legal Basis for Considering Sustainability

As mentioned previously, sustainability criteria are not included explicitly in CERCLA or RCRA guidance on remedy selection or modification (e.g., the feasibility study guidance, EPA, 1988b). Consideration of social factors (such as jobs or the economic well-being of a community) is not traditionally within the statutory authority of environmental regulators and is particularly difficult to envision. For example, if consideration of the impact on job creation for each remedial alternative were required, the result could be that the most expensive remedy is chosen since it is likely to create more jobs. Similarly, if job creation is considered on a site-specific basis, it may be necessary to evaluate the net gain or loss of jobs caused by the devotion of a company's capital to remediation versus expanding their production or other economic activities.

Such dramatic changes in remedy selection criteria are more appropriately adopted by statute (i.e., create a tenth criterion and specify how social factors are to be weighed on a site-specific and remedy-specific basis). More detailed direction than can be found in SURF guidance will be necessary concerning how to balance social factors, economic factors, and environmental factors. Absent a statutory basis (either federal or state), regulators cannot require a more costly remedy than a remedy that is consistent with the current statute and regulations. Of course, potentially responsible parties including the military may decide voluntarily to implement a remedy that goes beyond what might be selected by application of the nine remedy selection factors, based on a general good neighbor policy or adoption of a policy such as sustainable development. There is greater incentive to use sustainability factors in remedy selection when the costs of the remedial alternatives are similar. However, a more sustainable remedy is not necessarily a less expensive one. Thus, it remains to be seen whether implementation of more sustainable remedial alternatives will be feasible at hazardous waste sites.

CONCLUSIONS AND RECOMMENDATIONS

At most hazardous waste sites in the United States, meeting drinking water standards is the long-term goal of remediation. Unfortunately, drinking water MCLs will not likely be met in many affected aquifers for decades, especially at complex sites. Fortunately, EPA's current remediation guidance provides flexibility within the remedy selection process in a number of ways, although there are legal and practical limits to this flexibility. The following conclusions and recommendations discuss the value of exploring goals and remedies based on site-specific risks, sustainability, and other factors.

By design (and necessity), the CERCLA process is flexible in (a) determining the beneficial uses of groundwater; (b) deciding whether a regulatory requirement is an ARAR at a site; (c) using site-specific risk assessment to help select the remedy; (d) using at least some sustainability factors to help select the remedy; (e) determining what is a reasonable timeframe to reach remedial goals; (f) choosing the point of compliance for monitoring; and (g) utilizing alternate concentration limits, among others. **These flexible approaches to setting remedial objectives and selecting remedies should be explored more fully by state and federal regulators, and EPA should take administrative steps to ensure that existing guidance is used in the appropriate circumstances.** Often the same level of protection can be attained for lower costs by exercising this flexibility.

To fully account for risks that may change over time, **risk assessment at contaminated groundwater sites should compare the risks from taking**

"no action" to the risks associated with the implementation of each remedial alternative over the life of the remedy. Risk assessment at complicated groundwater sites is often construed relatively narrowly, with an emphasis on risks from drinking water consumption and on the MCL. Risk assessments should include additional consideration of (a) short-term risks that are a consequence of remediation; (b) the change in residual risk over time; (c) the potential change in risk caused by future changes in land use; and (d) both individual and population risks.

Progress has been made in developing criteria and guidance concerning how to consider sustainability in remedy selection. However, in the absence of statutory changes, remedy selection at private sites regulated under CERCLA cannot consider the social factors, and may not include the other economic factors, that fall under the definition of sustainability. At federal facility sites, the federal government can choose, as a matter of policy, to embrace sustainability concepts more comprehensively. Similarly, private companies may adopt their own sustainable remediation policies in deciding which remedial alternatives to support at their sites. **New guidance is needed from EPA and DoD detailing how to consider sustainability in the remediation process to the extent supported by existing laws, including measures that regulators can take to provide incentives to companies to adopt more sustainable measures voluntarily.**

REFERENCES

AEC (U.S. Army Environmental Command). 2004. Technical Impracticability Assessments: Guidelines for Site Applicability and Implementation. Phase II Report. http://aec.army.mil/usaec/cleanup/techimprac.pdf.

Army Corps of Engineers. 2010. Decision Framework for Incorporation of Green and Sustainable Practices into Environmental Remediation Projects. Interim Guidance 10-01; March 5, 2010. http://www.environmental.usace.army.mil/pdf/IG%2010 01%2003_05_10%20 doc.pdf.

Bardos, P., B. Bone, R. Boyle, D. Ellis, F. Evans, N. D. Harries, and J. W. N. Smith. 2011. Applying sustainable development principles to contaminated land management using the SURF-UK Framework. Remediation 21(2):77–100.

Bruntland, G. 1987. Our Common Future: The World Commission on Environment and Development. Oxford: Oxford University Press. http://www.un-documents.net/wced-ocf.htm.

Butler, P., L. Larsen-Hallock, C. Glenn, and R. Armstead. 2011. Metrics for integrating sustainability evaluations into remediation projects. Remediation 21(3):81-87. http://www.sustainableremediation.org/library/guidance-tools-and-other-resources.

Cadmus Group, Inc. 2009. Re: Draft Remedial Investigation Review. Memo date February 20, 2009. To: Eric Sunada, Executive Director, San Gabriel Valley Oversight Group. From: Dr. Julie Blue, Senior Hydrologist, The Cadmus Group, Inc. http://www.sgvog.org/images/CadmusMemo_Sunada_V2.pdf.

CalEPA. 2000. Public Health Goals for Chemicals in Drinking Water: Carbon tetrachloride. Sacramento, CA: CalEPA Office of Environmental Health Hazard Assessment. http://oehha.ca.gov/water/phg/pdf/carbtet.pdf.

California Regional Water Quality Control Board, Central Coast Region. 2003. Staff Report for Regular Meeting of February 6, 2003. Prepared January 7, 2003. Item Number: 13. Subject: Perchlorate Sites.

California State Water Quality Control Board. 2010. Draft Underground Storage Tank Low-Threat Site Closure Policy. September 7, 2010.

Catalogue of Federal Domestic Assistance. 2011. Superfund Technical Assistance Grants for Community Groups at National Priority List Sites. https://www.cfda.gov/index?s=program&mode=form&tab=step1&id=a83b10849f1dc74453b499c5d9e0370f (accessed September 21, 2011).

CL:AIRE (Contaminated Land: Applications in Real Environments). 2011. A Framework for Assessing the Sustainability of Soil and Groundwater Remediation. Annex 1: The SuRF-UK Indicator Set for Sustainable Remediation Assessment. London: CL:AIRE.

Cohen, J. T., B. D. Beck, and R. Rudel. 1997. Life years lost at hazardous waste sites: Remediation worker fatalities vs. cancer deaths to nearby residents. Risk Analysis 17(4):419-425.

Cowan, D. M., P. Dopart, T. Ferracini, J. Sahmel, K. Merryman, S. Gaffney, and D. J. Paustenbach. 2010. A cross-sectional analysis of reported corporate environmental sustainability practices. Regulatory Toxicology and Pharmacology 58(3):524-538.

DoD (U.S. Department of Defense). 1994. Guidance on the Environmental Review Process to Reach a Finding of Suitability to Transfer for Property Where Release or Disposal Has Occurred. http://www.epa.gov/fedfac/pdf/fost_prprty_release_occurred.pdf.

DoD. 2009. Consideration of Green and Sustainable Remediation Practices in the Defense Environmental Restoration Program. Office of the Secretary of Defense (August 10, 2009).

DoD. 2010. Defense Environmental Programs Annual Report to Congress 2010. http://www.denix.osd.mil/arc/upload/08_FY09DEPARC_Restoration_DENIX.pdf.

DOE (U.S. Department of Energy). 1998. Joint DOE/EPA Interim Policy Statement on Leasing Under the "Hall Amendment." June 1998. http://www.epa.gov/fedfac/documents/hall.htm.

Doyle, C. P., J. C. Teraoka, D. Pfeffer Martin, and S. Torabi. 2012. State Water Resources Control Board Unanimously Adopts Low-Threat Case Closure Policy for Petroleum Underground Storage Tank Sites. Legal Alert. May 21, 2012.

Ellis, D. E., and P. W. Hadley. 2009. Sustainable Remediation White Paper—Integrating Sustainable Principles, Practices, and Metrics into Remediation Projects. http://www.sustainableremediation.org/library/issue-papers/SURF%20White%20Paper.pdf.

EPA (U.S. Environmental Protection Agency). 1986. Guidelines for Ground-Water Classification. Final Draft. Washington, DC: EPA Office of Ground-Water Protection. www.epa.gov/superfund/health/conmedia/gwdocs/pdfs/grndh2o.pdf.

EPA. 1988a. Comprehensive Environmental Response, Compensation and Liability Act (CERCLA) Compliance with Other Laws: Interim Final (at xvi). EPA/540/G-89/006. Washington, DC: EPA.

EPA. 1988b. Guidance for Conducting Remedial Investigations and Feasibility Studies under CERCLA, Interim Final. EPA/540/G-89/004. http://rais.ornl.gov/documents/GUIDANCE.PDF.

EPA. 1989. Risk Assessment Guidance for Superfund (RAGS). Volume I: Human Health Evaluation Manual (Part A) (Interim final). Washington, DC: EPA Office of Emergency and Remedial Response.

EPA. 1990. National Oil and Hazardous Substances Pollution Contingency Plan. 55 Fed. Reg, at 8732. Washington, DC: EPA Office of Emergency and Remedial Response.

EPA. 1991a. ARARs Q's & A's: General Policy, RCRA, CWA, SDWA, Post-ROD Information, and Contingent Waivers. Publication 9234.2-01/FS-A. Washington, DC: EPA Office of Solid Waste and Emergency Response. http://www.epa.gov/superfund/policy/remedy/pdfs/92-34201fsa-s.pdf

EPA. 1991b. Risk Assessment Guidance for Superfund: Volume I - Human Health Evaluation Manual (Part B, Development of Risk-Based Preliminary Remediation Goals) Interim. EPA/540/R-92/003. http://rais.ornl.gov/prg/HHEMB.pdf.

EPA. 1991c. A Guide to Principal Threat and Low Level Threat Wastes. Publication 9380.3-06FS. Washington, DC: EPA Office of Solid Waste and Emergency Response. http://www.epa.gov/superfund/health/conmedia/gwdocs/pdfs/threat.pdf.

EPA. 1993. Guidance for Evaluating the Technical Impracticability of Ground-Water Restoration. OSWER Dir. No. 9234.2-25. Washington, DC: EPA Office of Solid Waste and Emergency Response.

EPA. 1996a. The Role of Cost in the Superfund Remedy Selection Process. Washington, DC: EPA Office of Solid Waste and Emergency Response. http://www.epa.gov/superfund/policy/remedy/pdfs/cost_dir.pdf.

EPA. 1996b. Memorandum from Steven A. Herman, Assistant Administrator, Office of Enforcement and Compliance Assurance, and Elliot Laws, Assistant Administrator for Solid Waste and Emergency Reponses, to EPA Regions. Re: Coordination between RCRA Corrective Action and Closure and CERCLA Site Activities at 2 (September 24, 1996).

EPA. 1996c. Presumptive Response Strategy and Ex-Situ Treatment Technologies for Contaminated Ground Water at CERCLA Sites. Final Guidance at 15-18. EPA 540/R-96/023.

EPA. 1997a. Rules of Thumb for Superfund Remedy Selection at 12-13. EPA 540-R-97-013. Washington, DC: EPA Office of Solid Waste and Emergency Response. http://www.epa.gov/superfund/policy/remedy/rules/rulesthm.pdf.

EPA. 1997b. EPA Landfill Presumptive Remedy Saves Time and Cost. Directive No. 9355.0-66I, EPA 540/F-96/017. Office of Emergency and Remedial Response (5202G) Intermittent Bulletin, Volume 1 Number 1. http://www.epa.gov/superfund/policy/remedy/presump/finalpdf/landfill.pdf.

EPA. 1997c. Performance Based Measurement System. 62 Fed. Reg. 52,098 (October 6, 1997). http://www.epa.gov/fedrgstr/EPA-WASTE/1997/October/Day-06/f26443.htm.

EPA. 2001a. Comprehensive Five-Year Review Guidance. EPA 540-R-01-007; OSWER No. 9355.7-03B-P. http://www.epa.gov/superfund/accomp/5year/guidance.pdf.

EPA. 2001b. 40 CFR Parts 9, 141 and 142 [WH–FRL–6934–9] RIN 2040–AB75. National Primary Drinking Water Regulations; Arsenic and Clarifications to Compliance and New Source Contaminants Monitoring. Washington, DC: Environmental Protection Agency.

EPA. 2003. The DNAPL Remediation Challenge: Is There a Case for Source Depletion? EPA/600/R-03/143. Ada, OK: EPA NRMRL.

EPA. 2004. Handbook of Groundwater Protection and Cleanup Policies for RCRA Corrective Action for Facilities Subject to Corrective Action Under Subtitle C of the Resource Conservation and Recovery Act at xii. EPA/530/R-01/015. http://www.epa.gov/correctiveaction.

EPA. 2005. MEMORANDUM from Michael Cook, Director of Office of Superfund Remediation and Innovative Technology, TO: Superfund National Policy Managers, Regions 1–10, Re: Use of Alternate Concentration Limits (ACLs) in Superfund Cleanups. July 19, 2005. OSWER Directive 9200.4-39. http://www.epa.gov/superfund/health/conmedia/gwdocs/pdfs/aclmemo.pdf.

EPA. 2006. Guidance on Systematic Planning Using the Data Quality Objectives Process. EPA QA/G-4./240/B-06/001. http://www.epa.gov/quality/qs-docs/g4-final.pdf.

EPA. 2008. Green Remediation: Incorporating Sustainable Environmental Practices into Remediation of Contaminated Sites. EPA 542-R-08-002. http://www.clu-in.org/download/remed/green-remediation-primer.pdf.

EPA. 2009a. Memorandum from Office of Superfund Remediation and Technology Innovation and Office of Federal Facilities Restoration and Reuse Office Re: Summary of Key Existing EPA CERCLA Policies for Groundwater Restoration at 6 (June 9, 2009).

EPA. 2009b. Final Second Five-Year Review Report, Middlefield-Ellis-Whisman (MEW) Superfund Study Area, Mountain View and Moffett Field, California.

EPA. 2010a. Guidelines for Preparing Economic Analyses. Washington, DC: EPA National Center for Environmental Economics, Office of Policy.

EPA. 2010b. Overview of Primary Environmental Regulations Pertinent to BRAC Cleanup Plan Development: Appendix A. http://www.epa.gov/fedfac/documents/appenda.htm.

EPA. 2010c. EPA Activities Provide Limited Assurance of the Extent of Contamination and Risk at a North Carolina Hazardous Waste Site. Report No. 10-P-0130. May 17, 2010. Washington, DC: EPA Office of Inspector General.

EPA. 2010d. Explanation of Significant Differences to the Records of Decision for Operable Unit 7 Former Sewage Treatment Plant and Operable Unit 1 West Gate Landfill and to the Engineering Evaluation/Cost Analysis for Operable Unit 22 Area of Concern 55C Naval Air Station South Weymouth, Weymouth, Massachusetts. http://www.epa.gov/superfund/sites/-rods/fulltext/e2010010003569.pdf.

EPA. 2011a. RCRA and CERCLA cleanup programs have roughly the same approach to cleanups. CERCLA: The Hazardous Waste Cleanup Program in EPA RCRA Orientation Manual at VI-13. http://www.epa.gov/osw/inforesources/pubs/orientat/.

EPA. 2011b. RCRA Orientation Manual 2011: Resource Conservation and Recovery Act, Chapter III: Permitting of Treatment, Storage and Disposal Facilities. http://www.epa.gov/osw/inforesources/pubs/orientat/rom38.pdf. See p. III-109.

EPA. 2011c. Plan EJ 2014: Supporting Community-Based Action Programs. http://www.epa.gov/compliance/environmentaljustice/plan-ej/community-action.html.

EPA. 2011d. Groundwater Road Map Recommended Process for Restoring Contaminated Groundwater at Superfund Sites. OSWER 9283.1-34. Washington, DC: EPA OSWER.

EPA. 2011e. Contaminated Site Clean-up Information, Green Remediation. Accessed October 17, 2011. http://clu-in.org/greenremediation/regions/index.cfm.

EPA and DoD. 2005. Memorandum of Understanding Between the U.S. Environmental Protection Agency and the U.S. Department of Defense; Subject: Support for Department of Defense (DoD) Cleanup Implementation for Base Realignment and Closure (BRAC) Installations Rounds I – IV (October 5, 2005). http://www.epa.gov/fedfac/pdf/brac_mou.pdf.

EPA and USACE. 2000. A Guide to Developing and Documenting Cost Estimates during the Feasibility Study at 4-2. EPA 540-R-00-002; OSWER 9355.0-75. Washington, DC: EPA Office of Solid Waste and Emergency Response. http://www.epa.gov/superfund/policy/remedy/pdfs/finaldoc.pdf.

ESTCP (Environmental Security Technology Certification Program). 2011. Alternative Endpoints and Approaches Selected for the Remediation of Contaminated Groundwater. Washington, DC: ESTCP.

Frost, F. J., J. Chwirka, G. F. Craun, B. Thomson, and J. Stomps. 2002. Physical injury risks associated with drinking water arsenic treatment. Risk Analysis 22(2):235-243.

GAO (Government Accountability Office). 2011. Early Goals Have Been Met in EPA's Corrective Action Program, but Resource and Technical Challenges Will Constrain Future Progress Report. GAO-11-514. Washington, DC: GAO.

Greenberg, G. I., and B. D. Beck. 2011. Use of years of potential life lost (YPLL) for risk assessment at hazardous waste sites. Pp. 602-607 in Encyclopedia of Environmental Health. J. O. Nriagu (ed). Burlington, MA: Elsevier Press.

Hamilton, J. T., and W. K. Viscusi. 1999. How Costly is "Clean"? An analysis of the benefits and costs of Superfund site remediations. Journal of Policy Analysis and Management 18(1):2-27.

Holland, F. 2011. Framework for Integrating Sustainability into Remediation Projects. http://www.sustainableremediation.org/library/guidance-tools-and-other-resources.

ITRC (Interstate Technology and Regulatory Council). 2009. Evaluating LNAPL Remedial Technologies for Achieving Project Goals. Washington, DC: ITRC LNAPLs Team.

ITRC. 2011. Green and Sustainable Remediation: State of the Science and Practice. www.itrcweb.org/Documents/GSR-1.pdf.

Leigh, J. P., and A. F. Hoskin. 2000. Remediation of contaminated sediments: a comparative analysis of risks to residents vs. remedial workers. Soil and Sediment Contamination 9(3):291-309.

Mansfield, C., P. Sinha, and M. Henrion. 2009. Influence Analysis in Support of Characterizing Uncertainty in Human Health Benefits Analysis. Contract EP-D-06-00. Research Triangle Park, NC: RTI International. http://www.epa.gov/ttn/ecas/regdata/Benefits/influence_analysis_final_report_psg.pdf.

NRC (National Research Council). 1994. Alternatives for Ground Water Cleanup. Washington, DC: National Academy Press.

NRC. 1997. Innovations in Soil and Ground Water Cleanup. Washington, DC: National Academy Press.

NRC. 2003. Environmental Cleanup at Navy Facilities: Adaptive Site Management. Washington, DC: The National Academies Press.

NRC. 2005. Contaminants in the Subsurface. Washington, DC: The National Academies Press.

NRC. 2011. Sustainability and the U.S. EPA. Washington, DC: The National Academies Press.

OMB (Office of Management and Budget). 2003. Circular A-4 September 17, 2003. To the Heads of Executive Agencies and Establishments. Subject: Regulatory Analysis.

OMB. 2012. Memorandum to Federal Agencies, Re: 2012 Discount Rates for OMB Circular No. A-94. (January 3, 2012). http://www.whitehouse.gov/sites/default/files/omb/memoranda/2012/m-12-06.pdf.

Sale, T. C., and C. Newell. 2011. Guide for Selecting Remedies for Subsurface Releases of Chlorinated Solvents. ESTCP Project ER-200530.

Sweeney, D. 2010. New Jersey Site Remediation. Presentation to the NCR Committee on September 13, 2010. Washington, DC. Mr. Sweeney is Assistant Commissioner, New Jersey Department of Environmental Protection.

U.S. Air Force. 1995. Streamlined Oversight: Moving Sites Faster through Streamlined Oversight. U.S. Air Force Project MUHJ947070. Air Combat Command with the Assistance of Versar.

Zewdie, T., C. M. Smith, M. Hutcheson, and C. Rowan-West. 2010. Basis of the Massachusetts reference dose and drinking water standard for perchlorate. Environmental Health Perspectives 118:42-48.

4

Current Capabilities to Remove or Contain Contamination

INTRODUCTION

Part of the Committee's statement of task was to discuss what is technically feasible in terms of removing a certain percentage of the total contaminant mass from the subsurface (and by association, reducing concentrations of target chemicals below drinking water standards). These questions were addressed comprehensively in the 2005 National Research Council (NRC) report that focused on source removal technologies, and previous NRC reports (NRC, 1994, 1997, 1999) provided professional judgment as to the potential effectiveness of various remedial technologies. This chapter reviews more recent data and reports on the ability of currently available remedial technologies to meet remedial action objectives for groundwater restoration. It is noted at the outset that poor design, poor application, and/or improper post-application monitoring at some sites makes evaluation of these technologies challenging, and reported performance results often appear in non-peer-reviewed documents.

Since the 2005 NRC report, technologies have evolved and more pilot-scale tests and full-scale remediation system performance data are available to help determine technology effectiveness (e.g., Johnson et al., 2009; Krembs et al., 2010; Stroo and Ward, 2010; Triplett Kingston et al., 2010a,b; Siegrist et al., 2011; Stroo et al., 2012). Technical information available for relevant case studies, however, is still often inadequate, particularly post-treatment monitoring, which severely constrains our ability to reach definitive statements regarding the effectiveness of a particular technology to meet remedial action objectives (RAOs). Critical evaluations of remedial technologies have been performed in the last six years for thermal and *in situ* chemical oxidation (ISCO) applications (Triplett Kingston et al., 2009, 2010a,b; Siegrist et al., 2011). For dissolved chlorinated solvent plumes, information on remedial technologies may be found in Stroo and Ward (2010).

Based on what is known about the effectiveness of remediation technologies (as described in this chapter), the Committee concluded that regardless of the technology used, the complete removal of contaminant mass at complex sites is unlikely. Furthermore, the Committee discovered no transformational remedial technology or combination of technologies that can overcome the current challenges associated with restoring contaminated groundwater at complex sites. At these sites, some amount of residual contamination will remain in the subsurface after active remedial actions cease, requiring long-term management.

To evaluate the effectiveness of remediation, performance metrics need to be specified, along with monitoring to measure progress toward achieving the metrics. Performance metrics are discussed in several publications (e.g., see EPA, 2003; NRC, 2005; Kavanaugh and Deeb, 2011). They include metrics that are commonly used and can be reliably measured, such as (1) source mass removal and (2) change in dissolved concentrations, as well as metrics that can be measured but are not commonly used, such as (3) contaminant mass remaining, (4) change in dense nonaqueous phase liquid (DNAPL) distribution (residual versus pooled), (5) change in DNAPL composition and properties, and (6) physical, microbial, and geochemical changes. Metrics that are under development include (7) changes in contaminant mass flux distribution, (8) change in contaminant mass discharge rate downgradient from source areas, and (9) change in stable isotope ratios. Change in contaminant mass discharge in particular is receiving greater attention (see ITRC, 2010; CDM, 2009). The appropriate performance metrics for a given site are both technology and site specific.

Conceptual Model

In this report, groundwater remedial technologies are categorized based on their primary target: the contaminant source zone or the dissolved groundwater plume (see Figure 4-1). The source zone can include (1) residual DNAPL, (2) pooled DNAPL, (3) sorbed contaminants, and (4) dissolved contaminants that may have diffused into fine-grained media. All of these compartments represent long-term continuing sources of contaminants to the dissolved or aqueous plume. The dissolved plume is typically located downgradient from the source, and may be extensive (i.e., miles in length for recalcitrant chemicals). Chlorinated solvents—the primary recalcitrant organic contaminants at complex sites—can occur in four phases (organic

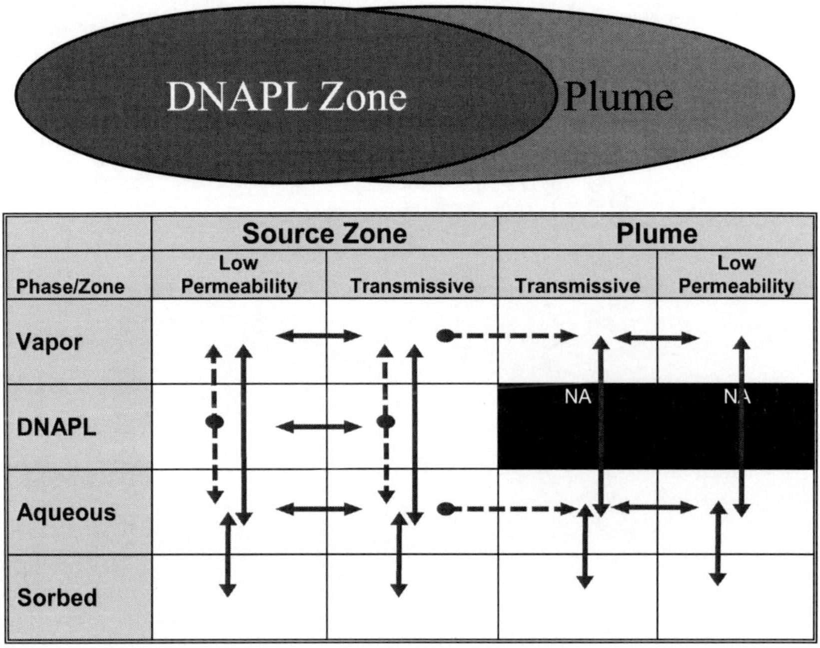

FIGURE 4-1 Conceptual model showing source zone and dissolved plume. The lower portion of the figure shows the 14-compartment model with common contaminant fluxes between compartments (solid arrows are reversible fluxes, dashed arrows are irreversible fluxes).
SOURCE: ESTCP (2011).

liquid, aqueous, solid-sorbed, and vapor) in the source zone and in three phases in the plume (there is no DNAPL phase in plumes). Each of these phases can occur in areas that can be classified as "transmissive" (mobile) or "low permeability" (immobile). This has led to a 14-compartment conceptual model depicting where contaminant mass could reside (Sale and Newell, 2011), which is discussed further in Chapter 6 of this report.

Because remedy selection and effectiveness depend, in part, on the contaminant mass distribution among phases and media (e.g., fine-grained media versus more permeable media, vadose zone versus saturated zone, DNAPL versus dissolved contaminants, etc.), a prerequisite for remediation is thorough site characterization, including the development of a conceptual site model that identifies, as much as possible, where DNAPL resides. As noted in Stroo et al. (2012), "source remediation is only as effective as the source delineation." The technology reviews found in Triplett Kingston

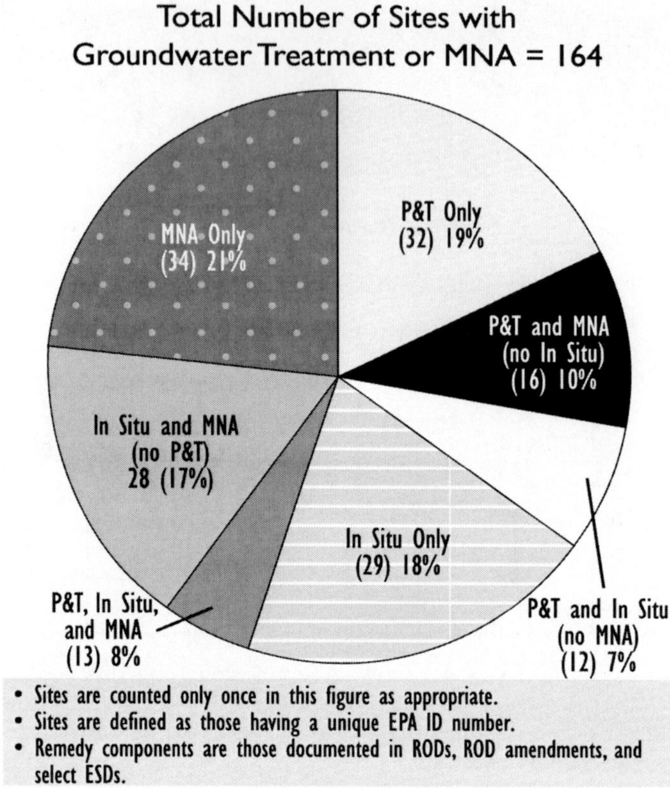

FIGURE 4-2 Sites with P&T, *in situ* treatment, or MNA as part of the groundwater remedy (FY 2005-2008).
SOURCE: EPA (2010a).

et al. (2009, 2010a,b) highlight the risks of inadequate site characterization: approximately two-thirds of the 14 thermal remediation case studies with sufficient data to evaluate technology performance ended up leaving mass in place because the treatment zone was smaller than the actual contaminant source zone. The reader is referred to Chapter 6 and particularly NRC (2005) for a more comprehensive discussion of site conceptual model development.

Dissolved plume remedies include pump and treat (P&T), bioremediation (including phytoremediation), permeable reactive barriers (PRBs), constructed wetlands (at the discharge point), monitored natural attenuation (MNA), and physical containment. As shown in Figure 4-2, MNA and P&T were used as groundwater remedies, either alone or in combination, at 82

TABLE 4-1 Generic Contaminant Removal or Containment Technologies and Common Applications

Technology	Application
Thermal	Source Zone
Chemical Oxidation	Source Zone
Surfactant Flushing	Source Zone
Cosolvent Flushing	Source Zone
Pump & Treat	Source Zone/Plume
Physical Containment	Source Zone/Plume
Bioremediation	Source Zone/Plume
Permeable Reactive Barrier	Source Zone/Plume
Monitored Natural Attenuation	Source Zone/Plume

percent of 164 Superfund facilities between 2005 and 2008. Several of the dissolved plume remedial technologies also can be applied to source zones (e.g., bioremediation, barriers, or hydraulic containment). A summary of the technologies discussed in this chapter and their most common application is provided in Table 4-1.

The goal of this chapter is to provide brief reviews of the major remedial technologies used in current remediation practice that can be applied to complex hazardous waste sites, particularly those with DNAPL source zones and/or large downgradient dissolved plumes. These reviews discuss our current knowledge regarding performance and limitations of the technologies, identify remaining gaps in knowledge, and provide case studies supporting these assessments. It is assumed that the reader is familiar with the material found in the NRC (2005) report, for which this chapter serves primarily as an update. The well-established technologies of excavation, soil vapor extraction/air sparging, and solidification/stabilization are not discussed because they have been presented in prior publications, and minimal advancements in these technologies have occurred during the past five to ten years. However, because of the potential importance of containment of source areas and plumes for long-term management, pump and treat for hydraulic containment is discussed.

THERMAL TREATMENT

In situ thermal treatment technologies, including electrical resistance heating (ERH), conductive heating, steam-based heating, radio frequency heating (RFH), and *in situ* soil mixing combined with steam and hot air injection, have continued to be developed and applied in the last five to ten years (see Table 4-2 and Baker and Bierschenk, 2001; Beyke and Fleming, 2005; Davis, 1998; de Percin, 1991; EPA, 1995a,b, 1999; Farouq Ali and

TABLE 4-2 Summary of Thermal Technology Applications by Technology Type (1988-2007)

Technology	Number of Applications	Pilot-Scale[a]	Full-Scale[a]	Number Since Year 2000
Steam-Based	46	26	19	15
Electrical Resistance Heating	87	23	56	48
Conduction	26	12	14	17
Other/Radio-Frequency	23	14	9	4
Total	182	75	98	84

[a]Some sites have an unknown application size and thus are not included in the pilot- and full-scale count.
SOURCE: Reprinted, with permission, from Triplett Kingston (2008).

Meldau, 1979; Vinegar et al., 1999). All involve raising the temperature of the subsurface to enhance the removal of contaminants by separate-phase liquid extraction, mobilization, volatilization, and *in situ* destruction. Relative to other technologies, some *in situ* thermal treatment technologies (e.g., ERH) applications result in preferential heating and contaminant removal from lower permeability media.

A review of the application of these technologies was conducted by Triplett Kingston (2008) and Triplett Kingston et al. (2009, 2010a,b, 2012). Data and documents from 182 thermal treatment applications conducted between 1988 and 2007 were reviewed, including 87 ERH, 46 steam-based heating, and 26 conductive heating applications. The applications were categorized based on the hydrogeology of the site, using the five generalized hydrogeologic scenarios developed in NRC (2005). These include relatively homogeneous and permeable unconsolidated sediments (Scenario A), largely impermeable sediments with inter-bedded layers of higher permeability material (Scenario B), largely permeable sediments with inter-bedded lenses of low-permeability material (Scenario C), competent, but fractured bedrock (Scenario D), and weathered bedrock, limestone, sandstone (Scenario E). The majority (72 percent) of thermal remediation applications reviewed were conducted in settings containing layers of high- and low-permeability media (Scenarios B and C).

ERH applications accounted for about 50 percent of all thermal applications since 2000 and outnumbered each of the other technology applications by about a factor of 3; there also appeared to be increasing use of conductive heating and decreasing use of steam-based heating (Table 4-2). These trends are reflective of underlying technical factors controlling performance, as well as design and operating challenges and vendor availability. ERH is attractive for volatile and semi-volatile chemicals in heterogeneous settings because its ability to achieve targeted energy delivery is less sensitive to subsurface heterogeneities than steam injection, and the energy delivery and contaminant recovery systems are arguably less complex to design and operate. Conductive heating has likely increased in use because it is the only thermal technology that can achieve *in situ* temperatures significantly greater than the boiling point of water and that is sometimes a desired operating condition. The study did not provide remediation costs because the cost data reviewed varied greatly and were thought to be unreliable, especially given some of the suboptimal designs.

Most relevant to this report are the post-treatment performance data from *in situ* thermal treatment sites. Interestingly, post-treatment groundwater monitoring data that could be used to evaluate technology performance were available for only 14 of the 182 sites (8 percent) reviewed by Triplett Kingston et al. (2010a,b, 2012), reflecting the overall industry-wide lack of sufficient post-treatment monitoring at many remediation sites. Most of the sites for which adequate data were available correspond to hydrogeologic setting Scenario C, with little or no performance data available for the other settings. Table 4-3 presents the estimated order-of-magnitude reductions in concentration and mass discharge for the 14 sites that had sufficient data for the analysis. Note that mass reduction data are not provided in Table 4-3 because initial mass in place was rarely known with certainty. For six of the 14 sites (43 percent), at least a 100-fold reduction in mass discharge was observed. For five of the 14 sites, detailed analysis revealed that post-treatment groundwater concentrations ranged from about 10 to 10,000 μg/L and source zone mass discharges ranged from about 0.1 to 100 kg/y.

The following factors should be considered in interpreting the widely varying performance results shown in Table 4-3:

1. As noted by Johnson et al. (2009), the criteria or rationale used to set the duration of treatment operation was usually not documented, and "in most cases it appeared that the duration was determined prior to start-up or may have been linked to a time–temperature performance criterion (i.e., operate for two months once a target temperature is reached *in situ*). There was little indication that the duration of operation was selected based on mass removal-, groundwater quality-, or soil concentration-based criteria" or performance monitoring.
2. Triplett Kingston et al. (2010a,b, 2012) discovered that treatment system footprints (areas treated) were often smaller than the source zones that had been treated. The main reason for this was that the pre-treatment extent of the source zone was larger than what it was conceptualized to be at the time that the remediation system was

TABLE 4-3 Effect of Application of *In Situ* Thermal Technology on Dissolved Groundwater Concentrations and Mass Discharge (Flux) from the Treatment Zone to the Aquifer

Site No.	Heating Technology	Generalized Scenario/Site	Dissolved Groundwater Concentration Reduction	Mass Discharge Reduction				
				<10×	10×	100×	1000×	>1000×
1	ERH	Generalized Scenario A[(SDC)]	10×			x		
2	ERH	Generalized Scenario B[a(SDC)]	<10×	x	x			
3	ERH	Generalized Scenario C	10×		x			
4	ERH	Generalized Scenario C[b(SDC)]	>10× to <100×		x			
5	ERH	Generalized Scenario C[c]	<10×	x				
6	ERH	Generalized Scenario C[c]	<10×	x		x		
7	ERH	Generalized Scenario C	<10×				x	
8	ERH	Generalized Scenario C[(SDC)]	10×		x			
9	ERH	Generalized Scenario C[(SDC)]	100×			x		
10	ERH	Generalized Scenario C	1000×		x			
11	SEE	Generalized Scenario C	100×			x		
12	SEE	Generalized Scenario C	10×	x				
13	SEE	Generalized Scenario C[c]	10000×				x	x
14	SEE	Generalized Scenario D[b]	<10×	x				

NOTE: SDC, supplemental data collection site for this project. Site 1 = Hunter Army Airfield, Site 2 = Air Force Plant 4, Site 4 = Camp LeJeune Site 89, Site 8 = Fort Lewis EGDY Area 3, Site 9 = NAS Alameda Site 5-1.

[a]Mass discharge assessment involved two calculations using first only the post-treatment field investigation data and then the post-treatment field investigation data supplemented with data from a set of monitoring wells that were directly in line with the field investigation transect.

[b]Pilot application appeared to encompass the entire source zone based on documentation reviewed.

[c]Site used two different vertical intervals to calculate mass discharge: (1) only shallow geology and (2) shallow and deep geology.

SOURCE: Triplett Kingston et al. (2010b).

designed. This points to a need to consider uncertainty in, and verification of, source zone extent when designing thermal remediation systems. It also suggests that decision makers and designers should weigh the incremental costs of additional source zone characterization data versus the costs of a larger system footprint and costs of failure of achieving remedial goals. Triplett Kingston et al. (2010a,b) found that sampling dissolved groundwater concentration transects perpendicular to groundwater flow and immediately downgradient of a source zone was a valuable approach for verifying source zone width.

In summary, the data in Table 4-3 are indicative of state-of-the-practice performance, but are likely not indicative of the technologies' true capabilities. Site No. 9 is probably most indicative of what thermal technologies can achieve in simple geologic settings because of the way it was designed and operated. At that site, dissolved chlorinated solvent concentrations were reduced from >10,000 μg/L to <100 μg/L levels, with final concentrations being <1 μg/L in many parts of the plume transect immediately downgradient of the source zone.

CHEMICAL TRANSFORMATION PROCESSES

Chemical transformation processes used for the treatment of both organic and inorganic contaminants have advanced significantly since 2005. There are three basic approaches to the use of abiotic chemical amendments to treating groundwater: (1) ISCO, (2) chemical reduction (discussed in the permeable reactive barriers section) using zero-valent iron (ZVI), bi-metallic reductants (BMRs), and other reductants (e.g., iron minerals such as magnetite), and (3) newer methods like the application of ISCO in permeable reactive barriers and the use of *in situ* generation of ozone using electrodes, which are discussed in Chapter 6. In most cases chemical transformation processes result in the formation of by-products that are either less toxic or amenable to subsequent degradation or natural attenuation. In a few cases, however, there is the potential to form undesirable and toxic by-products. Thus, multiple approaches may be required to ensure that complete detoxification can occur at the targeted site. In many cases, chemical transformation requires the injection and delivery of a reactant-containing fluid to the treatment zone, and is subject to the same limitations experienced by all flushing technologies—most notably the bypassing contaminants stored in low-permeability media.

In Situ Chemical Oxidation

ISCO relies upon the injection and activation of powerful chemical oxidants into subsurface sites that react with contaminants and oxidize them into carbon dioxide, carbon monoxide, or other substances less deleterious than the target contaminant. ISCO is relatively nonselective and capable of remediating a broad spectrum of contaminants. The technology has limitations including a finite amount of available oxidant, undesirable side reactions (e.g., oxidation of naturally occurring substances and the formation of precipitates), and the sometimes poor delivery of the oxidant into complex subsurface media. Thus, each site must be assessed for its biogeochemical complexity as well as hydrologic properties (e.g., fractured media, the presence of clay lenses, etc.) prior to implementing ISCO.

There are four oxidants routinely used in ISCO: catalyzed hydrogen peroxide (CHP or Fenton's reagent), persulfate, permanganate, and ozone (see Table 4-4 for a summary of their application, advantages, and disadvantages). Two other oxidants have received limited usage (peroxone and percarbonate). The number of ISCO applications has steadily increased for all the major oxidants over the past decade (Krembs et al., 2010, 2011).

Siegrist et al. (2011) examined all aspects of ISCO remediation including field applications, performance, and challenges at complex sites. Highlights from that report include the fact that delivery of the oxidant can be problematic, especially if more than one ingredient is required. Additionally, there is a risk that ISCO treatment will mobilize other contaminants of concern (e.g., chromate, selenate). Other limitations depend on the specific oxidant used. For example, reduction of permanganate results in the formation of manganese oxides that can alter aquifer permeability (although paradoxically this can also benefit remediation if the manganese oxides' high surface reactivity further attenuates contaminants through surface mediated oxidation—Loomer et al., 2010). Persulfate leads to generation of large amounts of sulfate, which can alter the biogeochemical environment of the aquifer through the generation of reduced sulfur (and even lead to an environment conducive to reductive dehalogenation). Lastly, the highly reactive nature and short half-life (~20 minutes in water) of ozone render it difficult to deliver in a stable form.

ISCO can be an effective treatment strategy, but like most other remediation technologies its success is dependent upon the complexity of the site and the nature of the contaminant. A 1999 ESTCP report summarizing 42 pilot- and full-scale ISCO projects deemed only 19 to be "successful." Another more recent but smaller evaluation of ISCO used at 29 chlorinated solvent sites found that mass was reduced (1) by 55 to 95 percent with a median reduction of 90 percent, and (2) by 75 to 90 percent with a median

TABLE 4-4 Chemicals Used in ISCO Applications, Including Advantages and Disadvantages

Specific ISCO	Principal Oxidant	Pathway	Advantages	Disadvantages
Catalyzed Hydrogen Peroxide Propagation (Fenton's Reaction)	Hydroxyl radical (OH•)	$Fe^{+2} + H_2O_2 \rightarrow Fe^{+3} + OH\bullet + OH^-$	Highly non-selective oxidant, no fouling, temperature increase	Nonproductive side reactions, cost, rapidly consumed, potential for excessive heating, gas generation.
Permanganate	MnO_4^- [Mn(VII)]	$MnO_4^- \rightarrow MnO_{2(s)}$ (neutral pH)	Somewhat non-selective, can be used in PRB (Ch. 6)	Aquifer fouling from MnO_2 precipitation, density issues, toxic metals mobilization
Persulfate	Sulfate radical ($SO_4\bullet$), reactive oxygen species (ROS)	Heat activated $S_2O_8^{-2} \rightarrow 2SO_4\bullet$. Base activated forms $SO_4\bullet$ and ROS	Somewhat non-selective, no fouling, fairly easy to deliver	Activation can be tricky (heat or add sodium hydroxide), quickly deactivates under acidic conditions, persistence in source dependent on delivery duration and groundwater flow
Ozone	O_3	From O_2 using corona discharge ozone generator	Non-selective toward hydrocarbons, oxygenates	Decomposes to oxygen quickly, delivered as a gas, need for on-site generator, continuous inputs

TABLE 4-5 ISCO Performance Metrics (1) for All Oxidants Organized by Contaminant Class (First Four Columns) and (2) for All Contaminants Organized by Oxidant (Last Five Columns)

Goals	Chloroethenes	BTEX	TPH	MTBE
% of closed sites	20 (n = 50)	43 (n = 7)	38 (n = 8)	63 (n = 8)
% of sites achieving MCLs	3 (n =105)	0[a] (n = 12)	25 (n =12)	60 (n = 5)
% of sites with rebound	72 (n = 54)	38 (n = 8)	43 (n = 7)	29 (n = 7)
Median % total contaminant reduction in groundwater	*nd*	*nd*	*nd*	*nd*

NOTE: Numbers in parentheses are total number of sites for each category. Not all sites had clearly stated goals for each category in Column 1. *nd* = not determined.

[a] The incidence of meeting MCLs was only entered in the affirmative after attempting to discuss the case study with the regulatory official. This percentage shown should not be interpreted to mean that ISCO has never reached MCLs at a BTEX site.

SOURCE: Adapted from Krembs et al. (2011).

reduction of 80 percent, for sites treated with Fenton's reagent and permanganate, respectively (Stroo et al., 2012).

Krembs et al. (2010, 2011) surveyed 242 sites variously contaminated with petroleum hydrocarbons and chlorinated solvents that have used some form of ISCO (five different oxidants were evaluated). The performance results, organized by contaminant class and oxidant, are summarized in Table 4-5. Of the 242 ISCO projects, only 15 percent achieved maximum contaminant levels (MCLs), while about 39 percent were able to reach alternative concentration limits. Approximately 80 percent of the applications were successful in reducing contaminant mass; typically, fuel-related contaminants were more amenable to ISCO treatment relative to chlorinated solvents and less expensive to treat. The DNAPL sites were more likely to require treatment beyond ISCO, which increases costs. Indeed, none of the sites containing free-phase DNAPLs were able to attain MCLs and only three sites with very dilute chloroethene-contaminated plumes achieved MCLs. Ultimately Krembs et al. (2011) found that site closure was attained at 24 percent of the full-scale sites, mostly through meeting alternative concentration limits. Of these closures, 89 percent required a combination of ISCO and some other technology either post- or pre-ISCO treatment, such as SVE, excavation, etc. In examining Table 4-5, it should be remembered that peroxone was only tested at two sites, while permanganate and Fenton's reagent were deployed at more and diverse types of sites,

Permanganate	CHP or Fenton's reagent	Ozone	Persulfate	Peroxone
16 (n = 32)	27 (n = 22)	50 (n =12)	0 (n = 4)	50 (n = 2)
0 (n = 55)	2 (n = 45)	31 (n =13)	0 (n = 8)	50 (n = 2)
78 (n = 32)	57 (n = 21)	27 (n =11)	50 (n = 2)	*nd*
51 (n = 27)	56 (n = 26)	96 (n = 5)	24 (n = 5)	*nd*

which likely accounts for their lower success percentages. Furthermore, the hydrogeology of the sites also played an important role in determining success. For example, 33 percent of the relatively homogeneous sites (hydraulic conductivities of > 10^{-5} cm/s) attained MCL goals, while none of more heterogeneous sites (K < 10^{-5} cm/s) attained MCLs (Krembs et al., 2011).

Select ISCO field-scale applications that removed a significant amount of contaminant mass are summarized in Table 4-6. The Kings Bay Naval Base was a targeted application using Fenton's reagent designed to achieve considerable source zone treatment so that natural attenuation could be utilized to address any remaining contaminants. Although mostly successful, the presence of iron shifted the redox conditions from sulfate- to iron-reducing, rendering natural attenuation less effective (Chapelle et al., 2005). The Nebraska Ordnance Plant is a good example of the use of permanganate to treat the chemical explosive RDX. A performance assessment demonstrated the difficulties in controlling the actual application of permanganate in the field compared to laboratory experiments. Scotch Cleaners, where a DNAPL plume was treated with permanganate, highlights the variable success rate of using permanganate, as well as the difficulties of oxidizing tetrachloroethene (PCE) that is strongly sorbed to the solid phase. LNAPL at Edwards AFB should have been amenable to ISCO, but delivery issues made remedy execution difficult. Air sparging was used to improve ISCO performance by enhancing the contact of persulfate with the target contaminants in the capillary fringe. Finally, the Pine Barrens case study

TABLE 4-6 Select Applications of ISCO Indicating Mass Reduced

Field Site	ISCO Treatment	Contaminant	Mass Reduced	References
Naval Base, Kings Bay, GA	CHP (50% H_2O_2 + ferrous sulfate)	PCE, vinyl chloride	Variable with significant PCE and VC reductions (> 90%)	Chapelle et al., 2005
Nebraska Ordnance Plant, Mead, NE	Permanganate (40% by weight)	RDX	Variable, sometimes > 70% reduction	Albano et al., 2010
The Scotch Cleaners, Topeka, KS	Permanganate (3% by weight)	PCE	Variable with ~ 90% reduction, but significant rebound of PCE in places	Heseman and Hildebrandt, 2009
Edwards Air Force Base, CA	Base activated persulfate (~8000 pounds) and air sparging	LNAPL	Decreases in total hydrocarbons, but not BTEX. BTEX dropped after air sparging	Siegal et al., 2009
Gas Station, Pine Barrens, NJ	Ozone Sparge	BTEX and MTBE	Approximately 95% reduction in both, but re-release has contaminated the site again	Dey et al., 2002

is an example of using ozone to treat a site contaminated with BTEX and MTBE where, in addition to oxidation, enhanced aerobic biodegradation contributed to the success of the remedy.

EXTRACTION TECHNOLOGIES

The application of extraction technologies for subsurface remediation typically involves the injection of an ingredient that can mobilize the contaminant in the treatment zone, which is then extracted from the subsurface along with the delivered active ingredient. When applied appropriately, extraction technologies can be extremely effective for removing large amounts of contaminant mass from the subsurface in relatively short periods of time, but they are best suited for Scenario A sites where the injected fluid can be delivered and recovered efficiently. Bypassing of contaminants in low-permeability material will occur when using extraction technologies. Additionally, active ingredient costs and the need to manage large volumes of delivered and extracted fluids render these technologies most suitable for Scenario A source zones of limited size. Two *in situ* extraction technologies, surfactant flushing and cosolvent flushing, are discussed in the following sections, with emphasis on lessons learned from field-scale applications.

Surfactant Flushing

Based on the lessons learned from laboratory and pilot-scale tests conducted in the 1980s and 1990s (e.g., see summaries in AATDF, 1997; Roote, 1998), subsequent field trials of surfactant flushing showed substantial success. A number of well-controlled, field-scale tests of surfactant flushing indicate that DNAPL recoveries in the range of 60 to 70 percent can be expected, and that mass recoveries of greater than 90 percent are achievable (e.g., Hasegawa et al., 2000; Londergan et al., 2001; Ramsburg et al., 2005). For example, after flushing a DNAPL-contaminated source zone at Hill AFB OU2 with approximately 2.5 pore volumes of surfactant solution, it was estimated that 99 percent of the DNAPL mixture was recovered from the test cell at an average cost of $793/L DNAPL (Londergan et al., 2001). Representative examples of surfactant enhanced aquifer remediation (SEAR) field demonstrations, including the surfactant formulation and estimated mass recovery, are summarized in Table 4-7. Recent field applications by commercial vendors have focused on the mobilization (displacement) of light NAPLs (e.g., gasoline) using relatively low concentration surfactant formulations (< 2% wt) to minimize active ingredient costs.

Despite the mass reductions evident in Table 4-7, the use of surfactants to treat source zones has declined markedly in the past five to ten years and is linked to problems in delivering the surfactant solution to the in-

TABLE 4-7 Representative Examples of SEAR Field Demonstrations

Field Site	Surfactant Formulation	NAPL	Amount Recovered (Estimated Mass Recovery)[a]	References
Dover Air Force Base	3.3% Aerosol® MA + 3.3% isopropanol + 0.4% $CaCl_2$	PCE	46 L (68%)	Childs et al., 2006
Bachman Road Oscoda, MI	6% Tween® 80	PCE	19 L (greater than 90%)	Abriola et al., 2005; Ramsburg et al., 2005
Alameda Point, CA	5% Dowfax® 8390 + 2% Aerosol® MA + 3% NaCl + 1% $CaCl_2$	TCA, TCE, DCA, DCE	325 kg (97%)	Hasegawa et al., 2000
Camp Lejeune Marine Corps Base	4% Alfoterra® 145-4PO sulfate + 16% propanol + 0.2 % $CaCl_2$	PCE	288 L (72%)	Delshad et al., 2000; Holzmer et al., 2000
Hill Air Force Base OU1	3% Brij® 97 + 2.5% pentanol	Jet fuel, chlorinated solvents	396 L (72%)	Jawitz et al., 1998, 2001
Hill Air Force Base OU2	7.6% Aerosol® MA + 4.5% isopropanol + 0.7% NaCl	TCE, TCE, PCE, CT	363 L (98.5%)	Londergan et al., 2001

Hill Air Force Base OU1 (Cell 6)	4.3% Dowfax®	Jet fuel, chlorinated solvents	1.5 kg (58%, 68%)[b]	Knox et al., 1999
Hill Air Force Base OU1 (Cell 5)	2.2% Aerosol® OT + 2.1% Tween® 80 + 0.4% $CaCl_2$	Jet fuel, chlorinated solvents	14.4 kg (42%, 97%)[b]	Knox et al., 1999
Coast Guard Station, Traverse City, MI	3.6% Dowfax® 8390	PCE, jet fuel	3.3 g PCE + 47 kg TH[c]	Knox et al., 1997
Thouin Sand Quarry, Quebec, Canada	9.2% butanol + 9.2 % Hostapur® SAS 60 + 13.2% toluene + 13.2% d-limonene	TCE, PCE, waste oil	532 kg (86%)	Martel et al., 1998
Canadian Forces Base Borden	2% 1:1 Rexophos® 25/97 + Alkasurf® NP10	PCE	67 L (69%)	Fountain et al., 1996

[a]Except when DNAPL was released intentionally, percent mass recovery values are based on *estimates* of the initial mass present at the site.

[b]The first value is based on pre- and post-treatment soil cores, while the second value is based on pre- and post-treatment partitioning interwell tracer tests (PITTs).

[c]Total hydrocarbons monitored but initial mass not reported.

tended target zone, and the need for subsequent extraction and treatment of the effluent waste stream (contaminant and surfactant). Injections may require Underground Injection Control permits, and extraction and treatment can be time consuming and expensive. Additionally, selection of a surfactant formulation that is both safe and effective requires laboratory and/or pilot-scale treatability tests. For example, due to their emulsifying properties, surfactants tend to disperse fine particles, which can lead to particle mobilization and pore clogging (Liu and Roy, 1995; Rao et al., 2006). Surfactants may be lost via adsorption on the solid phase and/or partitioning into NAPL, which can lead to the formation of viscous emulsions (Jain and Demond, 2002; Zimmerman et al., 1999). Low interfacial tension surfactant formations (< 1 dyne/cm) also can lead to uncontrolled mobilization of dense NAPLs (Pennell et al., 1996).

Another drawback accounting for SEAR's decrease in popularity is that the up-front costs for active ingredients and effluent treatment systems can be substantial. To minimize surfactant costs, it is advisable to minimize the amount of active ingredient required and to consider surfactants that are used in other commercial applications, such as food products, detergents, and pharmaceuticals. For example, sorbitan ethoxylates (e.g., Tween® 80), which are used in whipped toppings and other food products, typically cost less than $2/lb (Ramsburg and Pennell, 2001). In contrast, specialty surfactants can cost $20 to $40/lb.

Cosolvent Flushing

Cosolvent (alcohol) flushing is similar to SEAR in objective, mode of action, and field application, except that cosolvents increase contaminant dissolved-phase concentrations by making the aqueous phase less polar. Several field-scale demonstrations of cosolvent flushing have been conducted, most notably at Hill Air Force Base, Dover Air Force Base, and the former Sages dry cleaner site in Jacksonville, FL (Table 4-8). Two field demonstrations of cosolvent flushing were conducted at Hill Air Force Base Operable Unit 1 (OU1), which recovered approximately 80 and 85 percent of the NAPL mass from separate 3 m × 5 m test cells, which were vertically confined by interlocking sheet pile walls (Falta et al., 1999; Rao et al., 1997). At the Sages dry cleaner site, injection of a 95 percent ethanol solution was able to achieve similar PCE mass recovery (62 to 65 percent) without test cell confinement, while downgradient aqueous PCE concentrations were reduced by up to 92 percent following cosolvent flushing (Jawitz et al., 2000). A subsequent field demonstration of cosolvent flushing was conducted at the Dover Air Force Base, where a known amount of PCE was released into a test cell that was subsequently flushed with a 70 percent ethanol solution, recovered approximately 64 percent of initial PCE mass (Brooks et al., 2004).

Although cosolvent flushing showed initial promise as an effective strategy for NAPL source zone remediation, the technology faces a number challenges that have limited implementation beyond the cases shown in Table 4-8. First, the lower density of the concentrated cosolvent solution (70 to 90 percent active ingredient) relative to water requires careful design of injection and extraction systems. Gradient injection (increasing the concentration of cosolvent delivered over time) can minimize density override effects (Rao et al., 1997), while careful placement of injection wells promotes upward migration of the miscible effluent in unconfined systems (Jawitz et al., 2000). Second, due to the relatively high concentration of active ingredient required, material costs can be substantial even though ethanol can be purchased without federal and state alcohol consumption taxes. For example, ethanol typically costs approximately $3/gallon, but can increase to $5/gallon due to market fluctuations and demands for ethanol as an alternative fuel. In addition, the use of alcohols other than ethanol (e.g., iso-propanol) to enhance performance, even at relatively low concentrations (e.g., 2 to 5 percent), can greatly increase material costs. Finally, cosolvents are highly flammable liquids and require special safety procedures, both during mixing and injection as well as during treatment or disposal of the effluent waste stream.

Some of the problems associated with cosolvent flushing are illustrated by additional work that was performed at the former Sages dry cleaner site. Following a second cosolvent flush, monitoring revealed that chlorinated ethenes and ethanol had migrated from shallow groundwater to surface water in a nearby drainage ditch. This triggered the need for additional remediation consisting of an air sparge/soil vapor extraction (AS/SVE) system. Consequently, it became necessary to remove and treat the ethanol and PCE breakdown products. The SVE system was subsequently expanded and a dewatering groundwater pump-and-treat system was initiated (Levine-Fricke-Recon, 2008). Although the AS/SVE system was eventually terminated, the dewatering system continued to operate to reduce discharge to the drainage ditch with plans to expand groundwater extraction to include intermediate groundwater (Levine-Fricke-Recon, 2008). This expansion to deeper groundwater was performed because PCE DNAPL was found at depths below the original shallow treatment zone discussed in Jawitz et al. (2000).

To the Committee's knowledge there have been no new publications on surfactant or cosolvent flushing since about 2005. Almost all of the recent field-scale implementations that it is aware of have been performed by the company Surbec Environmental, LLC using low-IFT, low-concentration floods applied to LNAPLs (diesel/gasoline). While the low-IFT Surbec appli-

TABLE 4-8 Representative Field-Scale Demonstrations of Cosolvent Flushing

Field Site	Co-Solvent	DNAPL	Amount Recovered (Estimated Mass Recovery)	References
Dover Air Force Base Test Cell	70% Ethanol	PCE	53 L (64%)	Brooks et al., 2004
Sages Dry Cleaner, Jacksonville, FL	95% Ethanol	PCE	30 L (63%)	Jawitz et al., 2000
Hill Air Force Base OU1	80% t-Butanol + 15% Hexanol	Jet fuel, chlorinated solvents	99 mg/kg[a] (80%)	Falta et al., 1999
Hill Air Force Base OU1	70% Ethanol + 12% Pentanol	Jet fuel, chlorinated solvents	~300 L (85%)	Rao et al., 1997

[a] Solid phase concentration reported, based on sum of targeted compounds.

cations are gaining traction, other surfactant/cosolvent flushing applications have received limited attention since the disappointing results obtained during full-scale implementation at the former Sages dry cleaners (cosolvent flushing) and Site 88, Marine Corps Base Camp Lejeune, NC (surfactant flushing) (Battelle and Duke Engineering, 2001), both of which necessitated additional source zone treatment.

PUMP AND TREAT

Pump-and-treat (P&T) systems use extraction wells to remove groundwater containing dissolved contaminants. The extracted water is treated on-site and/or discharged to a publicly owned treatment works, then reinjected to the aquifer or reused for industrial or potable purposes. The design and operation of P&T systems are based on (1) *hydraulic containment* to prevent further plume migration, and/or (2) contaminant *mass removal* to restore the aquifer to drinking water conditions. These goals are not mutually exclusive, as many systems follow a hybrid approach that combines containment with limited mass removal.

As noted in the previous NRC study on P&T (NRC, 1994), this technology is effective at hydraulic containment, but less effective in removing mass. Of the 77 sites evaluated in that study, only six sites reported achieving MCLs, and more than half of the sites were candidates for very long term management. However, despite long-standing concerns regarding cost and performance with regards to contaminant mass removal, P&T remains one of the most widely applied groundwater remediation technologies, appearing in 20 to 30 percent of CERCLA groundwater decision documents, approximately the same proportion as *in situ* technologies (EPA, 2010b). The emphasis of this section is on containment because efficient and effective containment is a key consideration for long-term management of complex sites.

In 2000 EPA began implementing a systematic review and modification of P&T systems within Superfund, based on the Remediation System Evaluation (RSE) process developed by the U.S. Army Corps of Engineers (EPA, 2000). Over the ensuing decade, the application of this procedure to most of the approximately 75 EPA-led P&T sites has generated dozens of recommendations expected to significantly reduce O&M costs (EPA, 2008). Many of these savings are associated with more efficient operation of monitoring and above-ground treatment systems, but in some cases savings were partially offset by investments needed to increase confidence in the achievement of plume capture. In parallel with these evaluations, the EPA has produced an extensive set of guidance documents to support more cost-effective P&T design and operation (EPA, 2002, 2005a,b,c, 2007a,b,c) and support greater confidence in the achievement of successful hydraulic containment (EPA, 2008).

P&T has several desirable attributes, including its straightforward application, a relatively long track record of operation, and the ability to achieve plume hydraulic containment with a high degree of confidence. While further improvement in the design and operation of P&T containment systems is likely to be incremental, additional work is needed to (1) refine and deploy more cost-effective above-ground treatment systems and monitoring networks, (2) improve the ability to accurately predict P&T duration, and (3) develop "greener" P&T systems that more effectively utilize renewable energy sources and perform better with respect to sustainability metrics such as energy consumption and carbon footprint (e.g., Environmental Management Support Inc., 2008).

As was discussed in NRC (1994), P&T is capable of cleaning up groundwater to health-based standards in a relatively short time for dissolved, weakly sorbing contaminants in simple geology, especially if aided by biodegradation. However, at complex sites like those considered in this report, the earlier NRC report concluded that P&T is not expected to be effective in attaining health-based standards for many decades due to the slow processes of dissolution from DNAPL, desorption, and back-diffusion from fine-grained media.

PHYSICAL CONTAINMENT

Physical containment relies on a barrier that prevents transport by groundwater flow and/or contaminant diffusion. Physical containment is not a removal technique, but prevents contaminant migration from the contaminated area and passing groundwater from interacting with the source zone. Containment requires that contaminant migration is prevented both vertically and horizontally. To prevent horizontal migration, an impermeable vertical wall is installed that surrounds the contamination source. Vertical migration is prevented by ensuring that the vertical cutoff walls are embedded in a confining layer and, if desired, a cap at the ground surface also is installed. In many applications, physical containment draws on experience derived from the design of landfills, specifically the use of low-permeability clay (e.g., bentonite or soil-bentonite) and polymer geomembrane liners. Other materials are also used including sheet pile or cement/grout. Physical containment can be used for any contaminant as long as the material used to construct the wall is chemically compatible with the contaminants. Degradation of the containment barrier by either the chemical pollutants or via natural processes will lead to failure (Jefferis, 2008).

In the *Treatment Technologies for Site Cleanup: Annual Status Report (Twelfth Edition)* (EPA, 2007d), the number of cover systems and vertical walls at Superfund facilities was quantified and evaluated. For the 71 covers designed to contain source or groundwater contamination, 52 were functioning as intended. At the remaining sites, the covers had been removed, had not yet been constructed, or were too new to have reliable data. From 1982 to 2005, there were 67 installations of vertical cutoff walls at 55 facilities (or 3.6 percent of Superfund facilities). Of 16 representative containment systems evaluated (Appendix H of EPA, 2007d), 13 were functioning as intended. Of the remaining three, one had been removed, one was under construction, and one had insufficient data to evaluate its performance.

A 1998 EPA report provided design guidance and performance evaluation for subsurface engineered barriers and evaluated 36 sites (EPA, 1998). At 25 of the 36 sites, effective containment was achieved, seven sites had insufficient data, and four sites had failures (leaks). Sites reporting leakage, however, were reparable. The EPA (1998) report concludes "subsurface engineered barriers are effective containment systems for the short- and middle-term, if properly designed" and "[t]he most likely pathway for leaking of continuous subsurface barriers is in the vicinity of their keys" (i.e., their connection to the aquitard). A recent NRC review, however, concluded that available field data are insufficient to provide a robust assessment of the potential for or actual occurrence of failure in vertical barriers, particularly over times scales of 100 years or longer (NRC, 2007).

Failure assessment of physical containment systems is made difficult for two reasons. First, the vertical barrier is often installed in a manner that allows some downgradient contamination to be outside the barrier, such that downgradient detections post-barrier may be residual contamination and not representative of a failure. Second, the vertical barrier eliminates or greatly reduces groundwater flow, making groundwater velocities outside the wall so low that it could take years for a failure to reach a downgradient monitoring well only tens of feet away.

When designed, installed, and monitored properly, physical containment can be an effective technique for preventing contaminant migration. Contaminant mass is *not* reduced, but containment can be combined with treatment technologies as long as the treatment applied does not lead to physical damage of the containment barrier or alter it in a chemically adverse way. Various *in situ* techniques (ISCO, bioventing, etc.) can be performed in a contained area. Vertical barriers are often combined with low-rate pumping from inside the contained area, and hence perfect installation may not be essential. A case study of this approach is provided in Box 4-1.

BOX 4-1
Physical Containment at Former Koppers Company Wood Treating Plant, Salisbury, MD

In 1936, wood-treating operations began in Salisbury, MD; the facility closed in 1984. The site, which contained a large amount of creosote, was located adjacent to the Wicomico River. The groundwater flow is in the direction of nearby Wicomico River. The site is underlain by sand and peat containing a large, shallow DNAPL source. The DNAPL sits ~50 feet at depth on top of a confining layer of clay and slit (type III).

Sixteen years passed between the initial site environmental assessment and implementation of the remedy, the goal of which was to contain DNAPL. The remedy consisted of (1) a barrier wall encircling 41.3 acres, (2) Keens Creek reroute, (3) planting new trees (phytoremediation), (4) a shallow hydraulic gate (with an air sparging system just downgradient), (5) *in situ* biological groundwater treatment, (6) wetland mitigation, (7) product (creosote) recovery, and (8) soil cover. The main component of the remedy was the barrier; other components mainly supplemented containment. The capital cost of the remedy was about \$10–\$11 million with an annual operation and maintenance cost of approximately \$200,000.

In 2008, the barrier was found to be effective based on evaluation of (a) existing water-level data from monitoring wells inside and outside the barrier wall, and (b) tidal influences observed in monitoring wells located inside and outside the barrier wall. The successful implementation of the remedy at this site demonstrates that containment for large sources may be the most appropriate approach and that it can be combined with other technologies. It also shows that regulators are willing to accept a remedy including containment as part of the remedial plan. Containment, however, is not inexpensive, and the monitoring to ensure it is working correctly leads to recurring costs.

BIOREMEDIATION

In the context of remediation technologies, bioremediation refers to the transformation of contaminants by microorganisms, to benign by-products. Effective bioremediation requires the presence of appropriate organisms in sufficient densities for meaningful reaction rates, along with adequate supplies of electron donors or acceptors (depending on the type of reactions desired) and nutrients to support the desired biological reactions. In addition, the geochemical environment must be controlled to ensure appropriate conditions (e.g., pH, temperature, redox potential) for optimal bacterial growth. In contrast to natural attenuation, enhanced bioremediation typically involves *biostimulation* and/or *bioaugmentation*. Biostimulation can be achieved by the delivery of electron acceptors (e.g., oxygen for aerobic metabolism), electron donors (e.g., hydrogen for reductive dehalogenation), or suitable precursors that are utilized by organisms to support the biochemical transformation of the target contaminant. For example, active and passive schemes have been used to introduce oxygen into the subsurface to support the aerobic transformation processes, while hydrogen delivery methods (primarily through the injection of soluble nontoxic organic materials such as emulsified oil or molasses that serve as the microbially produced hydrogen) have been developed to enhance biological reductive dehalogenation. Bioaugmentation, which refers to the amendment of the subsurface with certain microorganisms, is used when native microbial populations are insufficient or incapable of transforming the contaminant regardless of system conditions.

Enhanced *In Situ* Bioremediation

Enhanced *in situ* bioremediation (ISB) is one of the most widely used technologies for the treatment of contaminated source zones and groundwater plumes for a variety of organic contaminants. Initially, the treatment of source zones by ISB was not considered to be feasible due to the relatively high (e.g., close to solubility level) contaminant concentrations. Laboratory studies conducted in the early 1990s indicated that biodegradation could occur in source zones contaminated with chlorinated solvents, and this bioactivity was able to enhance the rate of PCE dissolution from highly concentrated NAPL pools (Seagren et al., 1993, 1994). Subsequent studies demonstrated that the bacteria responsible for reductive dechlorination could survive in close proximity to residual TCE- and PCE-DNAPL (Cope and Hughes, 2001; Yang and McCarty, 2000, 2002). These findings served as the basis for several well-characterized pilot-scale tests (e.g., Hood et al., 2008) and full-scale remediation efforts (e.g., Wymore et al., 2006) that further demonstrated the ability of ISB to effectively treat chlorinated solvent source zones.

Under anoxic conditions (less than 0.1 to 0.5 mg/L dissolved oxygen), several bacteria (e.g., *Geobacter sp.* strain SZ) are able to chloro-respire PCE and TCE to form DCE, while only bacteria in the genus *Dehalococcoides* have been shown to obtain energy from the dechlorination of DCE and VC to ethene (Bradley and Chapelle, 2010). Due to this specificity, bioaugmentation is widely used for remediation of chlorinated solvent source zones. Several commercially available bacterial consortia are available for this purpose, including KB-1® and BDI®. In addition, an adequate supply of electron donors is necessary to support chloro-respiration, and a number of soluble (e.g., lactate) and less soluble (e.g., emulsified vegetable oil, molasses, and HRC®) electron donors are delivered to the subsurface to support reductive dechlorination. Some of the complications that can result are (1) substantially lowering of the pH, which can inhibit *Dehalococ-*

coides, (2) incomplete dechlorination due to competition from alternative electron acceptors (e.g., nitrate), (3) interspecies competition for hydrogen (methanogens), and (4) accumulation or presence of inhibitory compounds (Bradley and Chapelle, 2010).

Recent surveys of source zone remediation technologies reveal that ISB has been used at approximately 25 percent of the sites considered and is one of the most commonly applied *in situ* treatment methods along with thermal treatment and *in situ* chemical oxidation (NAVFAC, 2004; ESTCP, 2011). The primary reasons for this widespread adoption of ISB compared to other *in situ* technologies include relatively low capital costs, minimal infrastructure requirements, ability to treat a wide range of contaminants, and absence of an effluent waste stream that requires above-ground treatment and/or disposal. ISB has proven to be particularly effective for low-strength source zones (Newell et al., 2006) and was shown to have the lowest median cost ($29/yd^3) when compared to other *in situ* technologies (McDade et al., 2005; McGuire et al., 2006). Additionally, ISB has the potential to be coupled with more aggressive remediation technologies to achieve a "polishing" or "combined remedy" approach (e.g., Ramsburg et al., 2005; Sleep et al., 2006). Performance of ISB at selected sites is summarized in Table 4-9.

While much of the above discussion focused on chlorinated solvents, the range of contaminants to which the technology can be applied is quite broad, and very few common groundwater contaminants remain that cannot be treated biologically. For example, considerable success has been achieved for a range aerobic and anaerobic biological treatment methods applied to petroleum hydrocarbons (e.g., BTEX), polyaromatic hydrocarbons (PAHs), polychlorinated biphenyls (PCBs), and pesticides (see Stroo et al., 2012, for comprehensive listing of references summarizing *in situ* biological treatment methods). Nevertheless, several notable contaminants, including chloroform, and 1,4-dioxane, remain relatively recalcitrant from a bioremediation perspective. The identification and evaluation of mixed consortia and pure cultures that are capable of functioning at high contaminant concentrations and in the presence of co-contaminants is needed to further extend the applicability of ISB to source zones.

During the past five to ten years ISB has remained a widely used technology for source zone and plume treatment as costs have remained competitive compared to other treatment technologies, effectiveness has improved through more targeted applications and refinement of electron donor delivery and pH control, and successes in source zone treatment have been reported. The data shown in Table 4-9 serve to illustrate the implementation of ISB for treatment of highly contaminated chlorinated solvent source zones. Another recent review of data from 32 sites using *in situ* bioremediation for chlorinated solvent sites found that mass was

TABLE 4-9 Representative *In Situ* Bioremediation Field-Scale Demonstrations

Field Site	ISB Treatment	Contaminant(s)	Concentration Reduction	Reference(s)
Tarheel Army Missile Plan	Emulsified Oil Substrate	TCE, petroleum hydrocarbons	From 1 mg/L to < 5 μg/L[a]	ITRC, 2007
Dry Cleaner Site, Portland, OR	Hydrogen Release Compound	PCE, TCE, *cis*-DCE	From 7,000 μg/L to 50 μg/L (PCE) after 1 yr[b]	ITRC, 2007
Cape Canaveral Launch Complex 34 (LC 34)	Ethanol, KB-1®	TCE	From 155 mg/L to 0.2 mg/L[c]	Hood et al., 2008
Kelly AFB, Texas	Methanol, Acetate, KB-1®	TCE	From 1 mg/L to < 5 μg/L	Major et al., 2002
Dover National Test Site	Lactate, Ethanol, KB-1®	PCE (100 L release)	NA[d]	ESTCP, 2007
INEEL Test Area North	Lactate, Whey	TCE	From <10 mg/L to 150-550 μg/L[e]	Wymore et al., 2006

[a]Concentrations of VC remained above the MCL (0.15 μg/L) after one year.

[b]Concentrations of TCE and *cis*-DCE also exhibited substantial decreases followed by rebound and *cis*-DCE stall was apparent, all of which indicates the need for an additional electron donor treatment.

[c]Approximately 98% reduction in TCE mass in the upper unit based on pre- and post-treatment soil samples, consistent with average ethene concentrations measured.

[d]Concentrations were not provided. However, 77 kg of PCE were removed during the test, of which 15 kg were estimated to be recovered as degradation byproducts (TCE, *cis*-DCE, VC, ethene).

[e]Concentrations at injection well before and after whey injection. Concentration 25 ft downgradient from the injection well decreased to 150-400 μg/L.

reduced at these sites by 60 to 90 percent, with a median reduction of 81 percent (Stroo et al., 2012). Concentration reduction ranged from 75 to 95 percent, with a median of 91 percent. Both Stroo et al. (2012) and Table 4-9 suggest that ISB is likely to achieve substantial reductions in mass and dissolved-phase concentrations, but will have difficulty reaching drinking water standards (MCLs) in highly contaminated source zones (e.g., Cape Canaveral Launch Complex 34). As newly developed gene-based monitoring tools become more commonplace, more sophisticated and successful implementation of ISB technologies is anticipated.

PERMEABLE REACTIVE BARRIERS

Permeable reactive barriers (PRBs) are highly permeable zones comprised of material that is either abiotically reactive or that encourages the development of biological reactivity. PRBs are placed so as to capture a plume of subsurface contamination (i.e., installed to be perpendicular to the hydraulic gradient). PRBs may be a trench completely filled with reactive material or consist of a "funnel and gate" system in which impermeable barriers are used to direct flow through the reactive material. As groundwater passes through the PRB, dissolved contaminants are chemically or biologically transformed, or removed by sorption or precipitation, so that concentrations exiting the PRB are below risk-based thresholds. Because a PRB affects contaminants only as they are transported through it, its usage reflects a strategy of long-term site management, implemented because removal of source zone contamination is infeasible and/or the lack of ongoing operation and maintenance costs leads to a favorable economic evaluation. PRBs are often compared against P&T for hydraulic containment, and, depending on the lifetime of the barrier, may offer cost advantages compared to P&T (see e.g., ITRC, 2011).

PRBs used in the early 1990s relied on zero-valent iron (ZVI, i.e., iron metal or Fe^0) to reduce chlorinated solvents. The application of ZVI has gone beyond the reduction of chlorinated solvents and is now being studied as a treatment technology for a variety of subsurface contaminants including chromium (Lo et al., 2006) to ordnance chemicals such as RDX (Wanaratna et al., 2006). ZVI can also be mixed into soils to achieve contaminant reduction. In addition to ZVI, various other materials (including activated carbon, zeolites, metal oxides and other minerals, and organoclays) have been used to achieve the abiotic transformation of organic compounds and transform/sequester inorganic contaminants (Thiruvenkatachari et al., 2008). A few sorbing barriers (e.g., use of granular activated carbon or ion exchange materials) have been installed that will require replacement when the capacity is exhausted. A recent DOE installation at West Valley (which consisted of 850 feet of zeolite) was designed for a 20-year lifespan (Chamberlain et al., 2011), although future source removal activity could extend the life of the wall.

One advantage of PRBs is that the reactive material can be tailored to specific contaminants. Additionally, barriers comprised of organic material can be conducive to the development of biological consortia that are able to treat the groundwater contaminants (Davis and Patterson, 2003; ITRC, 2011). It is also possible to generate reactive zones in the subsurface by chemically altering the geologic materials themselves via *in situ* redox manipulation. For example, the injection of dithionite to reduce iron minerals present in the subsurface can be used to generate a reactive zone. Barriers of different chemistries may also be placed sequentially. Thus, treatment of a variety of organic and inorganic pollutants is possible with PRBs, as long as an appropriate reactive material is available or the necessary conditions for biological degradation can be generated in the PRB.

Critical needs in the design of PRBs (ITRC, 2005, 2011) are knowledge of the local hydrology, the nature and extent of the plume to be treated, the depth to a confining layer, and the concentration(s) of the contaminant(s) to be treated. Additionally, because groundwater chemistry (pH, alkalinity, hardness, other chemical species) may affect barrier performance and longevity, these parameters need to be known and the effects on the reactive media tested prior to installation. While PRBs remove contaminants from groundwater via various mechanisms, they are generally considered a containment technology because they prevent further migration of contaminants from the source zone.

Environmental Technologies, Inc. reports 156 PRBs have been installed to treat VOCs around the world (113 in the United States, nine in Canada, 19 in Europe, 14 in Japan, and two in Australia), but the company does not maintain a database on performance. The 2005 ITRC PRB guidance document complied data for 113 PRBs, and those data are summarized in Table 4-10.

For a properly designed barrier, it is likely that the ultimate cause of failure will be loss in media reactivity (unless it is regenerated or replaced) or failure to maintain the necessary conditions for biodegradation (ITRC, 2005). Thus, all barriers eventually will need to be excavated and replaced if the upstream contamination remains. While some ZVI barriers have been operating for over 15 years, currently no ZVI PRB has been taken out of operation, and their operational lifetime is unknown. For biowalls, performance will be dictated by the availability of substrates/nutrients, which will need to be replenished over time (ITRC, 2011).

TABLE 4-10 Summary of PRBs for Treatment of Contaminated Groundwater (1994–2004)

	Type of Site				
Barrier Media	Superfund	Industrial	U.S. Government	International	Total
Zero-valent iron	5	36	26	16	83
Non-iron reactive materials	1	6	1	1	9
Bio-barrier	0	6	5	4	15
Combination/sequenced[a]	0	4	2	0	6

[a]Five of the six in this category include zero-valent iron.
SOURCE: Based on data from ITRC (2005).

MONITORED NATURAL ATTENUATION

Monitored natural attenuation (MNA) relies on natural processes to degrade or immobilize groundwater contaminants. For source zones, natural dissolution can result in the transfer of the contaminant from the nonaqueous liquid phase to the dissolved phase, where contaminants are amenable to additional attenuation reactions. The attenuation may be biological, abiotic, or biogenic (i.e., bacteria are required to create the necessary abiotic reagent(s)). Monitoring is required to verify that transformation or immobilization is occurring (as opposed to decreases in concentration resulting from dilution or dispersion), that the contaminants are not migrating, and that attenuation continues to occur over time. The success of remediating a contaminated site using MNA is dependent upon a number of biogeochemical variables and the types of contaminants present. These include the type, amount, and distribution of terminal electron acceptors and other important chemical species (e.g., nutrients); the composition of the contaminants; the nature of the hydrogeologic media (particle composition, organic content, hydraulic conductivity, etc.); and some knowledge of the *in situ* microbial fauna.

Natural attenuation is now well established for the degradation of many petroleum based compounds such as BTEX (e.g., Wiedenmeir et al., 1999) and PAHs (e.g., Neuhauser et al., 2009). Laboratory studies have demonstrated the potential for (abiotic or biological) reduction of halogenated ethenes, under conditions in which iron sulfide or green rusts are generated (Scherer et al., 2009; Butler et al., 2009; Wilson, 2010), but these studies also show the complexity of these processes. The degradation of chlorinated solvents has been shown at a site with reduced iron minerals present (see Box 4-2). With chlorinated solvents, it must be verified that reaction products that are of similar or greater toxicity than the parent compound are not produced. For other organic compounds (halogenated aromatics, oxygenated hydrocarbons) evidence for natural attenuation is still minimal. For selected inorganic pollutants, natural attenuation via biodegradation is established (Coates and Achenbach, 2004).

Because there is no hydraulic, physical, or chemical containment of the pollutants during MNA, site hydrology and the extent of contamination must be known in detail so that it can be verified that the contamination is not spreading. As described in NRC (2000), a conceptual site model must be built that characterizes the groundwater flow, accounts for temporal and spatial variability and uncertainties in the flow, delineates the contaminant source and plume, and includes terms for contaminant loss. Site characterization needs to be carefully done and can be expensive; monitoring typically occurs over longer time spans (Kennedy et al., 2006, Vaneglas et al., 2006). As discussed in Chapter 6, new chemical and microbiological

BOX 4-2
MNA at the Twin Cities Army Ammunition Plant (TCAAP)
New Brighton/Arden Hills, Minnesota

The TCAAP site was built by the U.S. Army to manufacture ammunition for World War II. Starting in the late 1950s, Alliant Techsystems began manufacturing of ammunition at the site (MPCA, 2010). Degreasing operations are the cause of the major groundwater contamination. Up to 20 million pounds total of TCE and 1,1,1-TCA were disposed of on-site in a sandy glacial outwash (Mark Ferrey, personal communication, July 2, 2010).

The unsaturated zone under the disposal area has a depth of 150 feet to groundwater. The TCE and 1,1,1-TCA migrated rapidly to the groundwater, which is in the Prairie du Chien aquifer (a high-yielding fractured dolomite bedrock). The aquifer is highly permeable and manganese- or iron-reducing. The Jordan Sandstone, which lies below the Prairie du Chien, was also contaminated. While the solvent disposal area is known, no DNAPL sources have been identified in the overlying material or aquifer (Mark Ferrey, personal communication, July 2, 2010). The total length of the contaminated groundwater plume is approximately 5.5 miles. A pump-and-treat system is used to capture water, treat it with activated carbon to non-detect levels, and then use it in the municipal drinking water system. A soil-vapor extraction system was also installed in the suspected source zone (MPCA, 2010; EPA, 2010c; Mark Ferrey, personal communication, July 2, 2010).

Because the redox conditions in the plume were neither sulfate reducing or methanogenic, it was assumed that biodegradation would be minimal. After the installation of the pump-and-treat and soil vapor extraction systems, concentrations in the plumes were observed to fall, and concentrations at the containment

analyses (e.g., compound specific isotope analysis, genomics, proteomics) are being developed that can help verify that degradation of contaminants is actively occurring during MNA.

Ten years ago the consensus was that MNA was an option "for only a few types of contaminants" under certain circumstances. Over time, MNA has become a more prevalent component of remedial systems for contaminated groundwater. As reported in *Superfund Remedy Report Thirteenth Edition* (EPA, 2010c), the use of MNA at NPL sites steadily increased through the 1990s and is selected as a component in a remedy at approximately 30 to 40 percent of NPL facilities annually. From 2005 to 2008, 56 percent of NPL sites implementing groundwater treatment used MNA alone or as a component of the remedy. MNA is likely to be a component of the remediation plan when more aggressive treatments leave residual contamination in place.

An evaluation of 52 temporal records at 23 sites containing chlorinated

wells also decreased. Groundwater modeling predicted concentrations that were 20 to 30 times higher than those observed, suggesting that degradation of the chlorinated solvents was occurring. Investigations by scientists at the Minnesota Pollution Control Agency and EPA (Ferrey et al., 2004; Brown et al., 2007) demonstrated that abiotic reduction of chlorinated solvents (including dichloroethenes) occurred at this site due to the presence of reduced iron minerals. It was hypothesized that degradation was occurring in the groundwater, and once the plume was no longer fed by the source zones the plumes began to recede.

This was the first site to show that abiotic degradation could be an important component of monitored natural attenuation for chlorinated ethenes like TCE or DCE (which was unexpected at this site). The site altered the paradigm regarding whether and how abiotic reactions should be considered and under what prevailing redox conditions. When MNA became a component of the remediation plan (i.e., it was superimposed on the larger regulatory response), there was initial public concern from New Brighton that the Minnesota Pollution Control Agency was backing away from the goal of being protective of public health and that it would give the U.S. Army a means to argue that treatment via activated carbon at the drinking water treatment plant was no longer necessary. The agreement, however, still states that the activated carbon beds will be used until contaminants are non-detectable (at levels obviously below MCLs) in the source water. The site is currently scheduled for delisting from the NPL in 2040. Initial estimates were that cleanup would take 80 years. Models that include the observed abiotic degradation suggest that groundwater cleanup may be completed in 17 to 25 years (MPCA, 2010; EPA, 2010c; Mark Ferrey, personal communication, July 2nd, 2010).

Additional detail is available in Wilson (2010).

solvent contamination showed median decreases in contaminant concentration of 74 percent over five to 15 years (Newell et al., 2006). It should be noted, however, that these were sites where source treatment was not necessary, and it cannot be ruled out the reductions observed are due to dilution or plume migration. That said, the results are promising for MNA of chlorinated solvents at sites with low levels of contamination. A recent analysis of 35 sites with chlorinated aliphatic hydrocarbon contamination (Brauner et al., 2008) was used to develop a sustainability assessment framework for MNA, including (1) analysis of plume stability (using statistical analysis of measured concentration), and (2) estimation of remediation timeframes (mathematical modeling). The study was unable, however, to find a robust method to assess the longevity of specific degradation processes.

Given the right combination of contaminants present and site hydrology and biogeochemistry, MNA can be an effective remediation technique. However, specific chemical conditions and/or bacteria are required and the necessary reactions must occur on time scales that are *faster* than contaminant transport and must be sustainable over long periods of time (NRC, 2000). Thus, MNA might ultimately achieve restoration at many sites but probably not within a timeframe of less than 100 years. Although implementation can be straightforward if an accurate conceptual site model and appropriate site conditions exist, verification of degradation with multiples lines of evidence is necessary for MNA to survive scientific, regulatory, and public scrutiny. Indeed, NRC (2000) documented that public opinion of MNA is decidedly wary, with many communities considering it a "do nothing approach."

COMBINED REMEDIES

It is now widely recognized that even successful application of remedial technologies will not completely remove all of the contaminant mass from most DNAPL source zones. In fact, aggressive source zone treatments are likely to increase the mobility and distribution of the residual mass or stable pools, which may lead to increased aqueous phase concentrations in the short term. As a result, attention has shifted from developing the most effective stand-alone single technology to the development and testing of complementary *in situ* remediation technologies that can be combined, at the same time (in parallel) or sequentially (in series) to more efficiently treat contaminant source zones (Amos et al., 2007; Christ et al., 2005; Costanza et al., 2009; Friis et al., 2006; Ramsburg et al., 2004). In a sequential or "treatment train" approach, an aggressive *in situ* treatment technology, such as electrical resistive heating or surfactant flushing, is used to remove a large fraction of contaminant mass in a relatively short timeframe, while a second "polishing" technology such as microbial reductive dechlorination is then applied to remove or detoxify the remaining contaminant mass. Such sequential remediation strategies have the potential to take advantage of efficient mass removal achieved by aggressive treatment technologies, while addressing limitations associated with an individual technology (e.g., flow bypassing) that lead to incomplete mass removal. When designing a combined remedy, it must be kept in mind that physical-chemical treatments may alter geochemical conditions and microbial ecology, which could be either detrimental (e.g., aquifer clogging, reduced microbial diversity) or beneficial (e.g., enhanced electron donor availability) to the overall secondary remediation process (Christ et al., 2005; Stroo et al., 2003).

There are several examples of combined technologies relevant to complex sites contaminated with DNAPLs. The combination of thermal treatment and bioremediation has been recently demonstrated (Costanza et al., 2009; Friis et al., 2005). In this application, ERH was used both for source removal and to release electron donors to stimulate bioremediation. (It should be noted that in regions where the temperature exceeded 50°C for prolonged periods of time, bioaugmentation may be required to achieve meaningful levels of microbial reductive dechlorination; Friis et al., 2006). Another example involves the use of surfactants and surfactant + cosolvent formulations to enhance ISCO. The primary advantages of this approach are increased contaminant availability and stabilization of the oxidant and catalyst during delivery. For example, a patent (USPTO 7,976,421 B2) was recently issued for surfactant enhanced *in situ* chemical oxidation (S-ISCO®); however, this technology has received limited testing.

Third, pilot- and field-scale trials conducted at former dry cleaning facilities indicate that surfactants and cosolvents can enhance biological reductive dechlorination to treat residual contaminants (Ramakrishnan et al., 2005; Ramsburg et al., 2004). Post-treatment monitoring of the Bachman Road site, which was flushed with Tween® 80, provided evidence of stimulated microbial reductive dechlorination (see Box 4-3). Within the treated PCE-DNAPL source zone, residual Tween 80 provided suitable electron donors that stimulated native microbial dechlorination activity (Ramsburg et al., 2004). These findings were supported by laboratory studies that further defined the effects of Tween 80 on reductive dechlorination (Amos et al., 2007). When combining surfactant flushing and *in situ* bioremediation, selection of compatible surfactants and appropriate concentrations is critical due to potential surfactant toxicity or inhibition toward microbial populations relevant to the desired degradation pathway (e.g., McGuire and Hughes, 2003; Yeh et al., 1999). Christ et al. (2005) performed a modeling analysis of source zone treatments, demonstrating that bioremediation alone provides minimal benefits when compared to natural gradient dissolution, while aggressive treatment of the source zone with surfactants followed by bioremediation dramatically reduced source longevity by several orders of magnitude (see Box 4-3).

In the Committee's opinion, combined remedy approaches offer substantial opportunities to reduce the costs associated with the remediation of complex sites. For example, using a combined remedy can capitalize on the beneficial aspects of specific technologies, such as source zone flushing (e.g., steam, surfactant) coupled with ERH to address contamination existing in high- and low-permeability media, respectively. In some cases, it is appropriate to develop a combined remedy at the initiation of a remedial action (e.g., surfactant-enhanced *in situ* chemical oxidation or thermal treatment followed by bioremediation). When the combined remedy involves aggressive source removal activity followed by bioremediation or monitored natural attenuation for the residual/dilute contamination, it is critical to have prior knowledge of the biogeochemical environment and the potentially important reactive pathways for the target contaminants. More often, combining remedies is done sequentially at sites where the initial rem-

BOX 4-3
Combined Remedies at the Bachman Road Dry Cleaner Site

In surfactant flushing applications, some fraction of the initial contaminant mass and the introduced surfactant is likely to remain in the subsurface following source treatment. Long-term monitoring of the Bachman Road site, which was flushed with a biodegradable, food-grade, nonionic surfactant (Tween® 80), provided evidence of stimulated microbial reductive dechlorination (Figure 4-3a). Within the treated PCE-DNAPL source zone, fermentation of residual Tween 80, detected at concentrations of 50 to 2,750 mg/L 450 days after SEAR, provided suitable electron donors(s) that stimulated native microbial dechlorination activity in the oligotrophic aquifer (Ramsburg et al., 2004). These findings were complemented by laboratory-based studies that further demonstrated the ability of low levels (< 1,000 mg/L) of Tween® 80 to support reductive dechlorination of chlorinated ethenes (Amos et al., 2007).

To further evaluate the potential benefits of surfactant flushing coupled with subsequent bioremediation relative to bioremediation or natural gradient dissolution, Christ et al. (2005) performed a number of numerical simulations based on the conditions observed at the Bachman Road site. Results of the model predictions clearly demonstrate that bioremediation alone provides only minimal benefits when compared to natural gradient dissolution, while aggressive treatment of the source zone with SEAR followed by bioremediation dramatically increased the rate of mass removal and reduced source longevity by several orders of magnitude (Figure 4-3b). These findings clearly demonstrate the potential benefits of coupling aggressive source zone mass removal technologies with compatible bioremediation processes.

FIGURE 4-3 (A) Plan view diagram of the Bachman Road site source zone and downgradient PCE-contaminated groundwater plume. SOURCE: Reprinted, with permission, from Abriola et al. (2005) ©2005 by the American Chemical Society. (B) Percent DNAPL mass remaining as a function of time for three alternative remediation strategies: natural gradient dissolution alone; bioremediation alone; and SEAR (4% Tween® 80) followed by bioremediation. SOURCE: Reprinted, with permission, from Christ et al. (2005) ©2005 by Environmental Health Perspectives.

edy has reached an asymptote (in terms of performance) prior to reaching cleanup goals. In such cases, there can be substantial cost savings in transitioning to less aggressive and less expensive technologies (see Chapter 7 for further discussion of this transition). Whether planned from the beginning

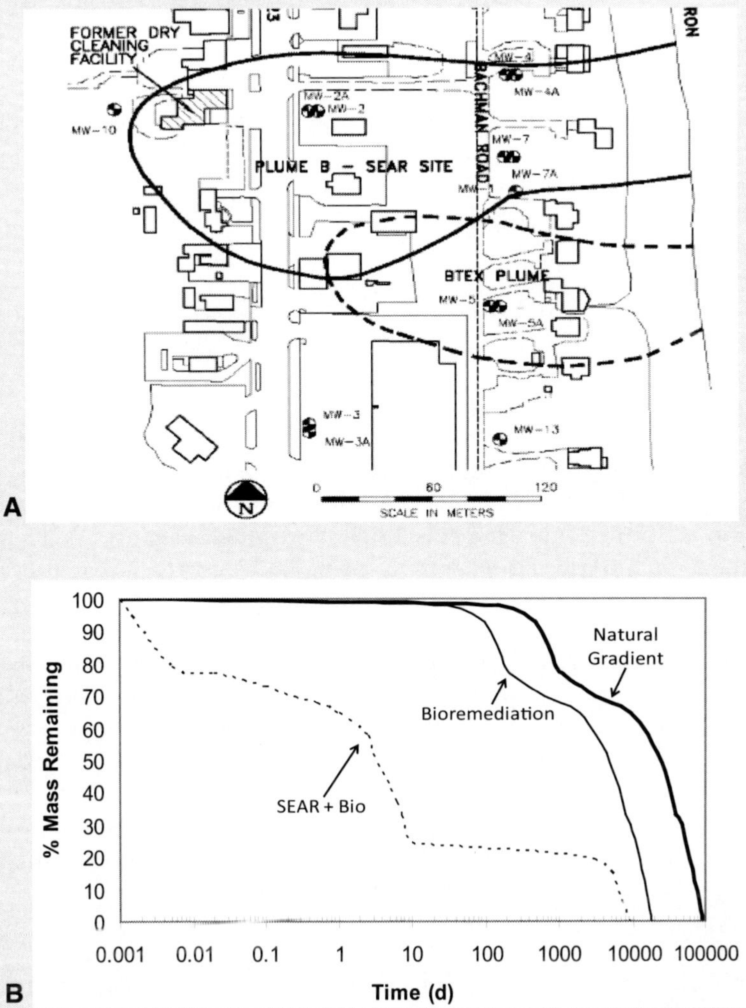

or turned to after initial remedy failure, the design and implementation of combined remedies should be undertaken on a case-by-case basis. In practice, selection of combined remedies is likely to require additional technical expertise, field sampling and analysis, and laboratory-scale treatability tests to ensure that the combination is feasible and yields added value.

CONCLUSIONS AND RECOMMENDATIONS

Over the last two decades, remedial technologies have matured and evolved, especially in the area of DNAPL source-zone remediation. Indeed, development of the technologies discussed in this chapter has advanced to the accepted-use stage (although many of the technology applications have been at simple sites or only a small portion of a complex site). Summaries of the effectiveness of source zone and plume technologies are provided in Tables 4-11 and 4-12, respectively, drawing on results discussed in the preceding sections.

Given Tables 4-11 and 4-12 and the best professional judgment of the Committee, the capabilities of the technologies described in this chapter are often not sufficient to meet the conventional objective of meeting MCLs at complex sites, such that contamination is likely to remain in place following treatment for a large number of complex sites. **That is, significant technical limitations persist that make achievement of MCLs throughout the aquifer unlikely at most complex groundwater sites for many decades.** Furthermore, future improvements in these technologies are likely to be incremental, such that long-term monitoring and stewardship at sites with groundwater contamination should be expected.

The Committee could identify only limited data upon which to base a scientifically supportable comparison of remedial technology performance for the technologies reviewed in this chapter. There have been a few well-studied demonstration projects and lab-scale research studies, but adequate performance documentation generated throughout the remedial history at sites either is not available or does not exist for the majority of completed remediation efforts. This has hindered attempts to perform empirical analyses of technology performance and how that relates to design parameters, operating conditions, monitoring and optimization plans, and site characteristics. Furthermore, poor design, poor application, and/or poor post-application monitoring at typical (i.e., non-research or demonstration) sites makes determination of the best practicably achievable performance difficult.

There is a clear need for publicly accessible databases that could be used to compare the performance of remedial technologies at complex sites (performance data could be concentration reduction, mass discharge reduction, cost, time to attain drinking water standards, etc.). The Committee envisions a database with much more comprehensive performance data than is found, for example, in the CERCLIS database of Superfund facilities. To ensure that data from different sites can be pooled to increase the statistical power of the database, a standardized technical protocol regarding data collection and analysis would be needed, although it goes beyond

TABLE 4-11 Source Zone Technology Summaries

Technology	Performance	Comments
Thermal	<10X to 100,000X concentration and flux reduction[a] 95 to 99+ percent mass reduction[b]	Can be effective in heterogeneous media; potentially high energy consumption; limited number of vendors to perform work
In Situ Chemical Oxidation	55 to 90 percent mass reduction[b]	Most applications have been at small sites in permeable media; applications to DNAPL are very challenging (potential for DNAPL mobilization and contaminant bypassing); side reactions could render ISCO less effective
Surfactant Flushing	65 to 90+ percent mass recovery[c]	Can bypass contaminants in heterogeneous media; risk of uncontrolled DNAPL mobilization and migration; low IFT formulations suitable for LNAPL recovery
Cosolvent Flushing	65 to 85 percent mass recovery[c]	Can bypass contaminants in heterogeneous media; risk of uncontrolled DNAPL mobilization and migration; requires large volumes of cosolvents thereby driving up costs
In Situ Bioremediation	60 to 90 percent mass reduction[b]	Problematic conditions include pooled DNAPL, the potential for high methane levels, and groundwater velocity <10 ft/y or >10 ft/d; potential for biofouling and metals solubilization

NOTE: The summary table includes only those technologies for which significant new performance information has become available since NRC (2005). For complete descriptions of contaminant source remediation technologies, see NRC (2005).

[a]From Table 4-3.

[b]From Stroo et al. (2012).

[c]From Tables 4-7 and 4-8. Mass recovery percentages should be interpreted with caution because initial mass in place is uncertain.

the scope of this report to provide the details of such a protocol. Federal agencies with nationwide responsibility for complex sites (EPA, DoD, DOE) should take the lead on developing such databases.

Additional independent reviews of source zone technologies are needed to summarize their performance under a wide range of site characteristics. Since NRC (2005), only thermal and ISCO technologies have undergone a thorough, independent review. Other source zone technologies should also be reviewed by an independent scientific group (e.g., SERDP/ESTCP, ITRC,

TABLE 4-12 Plume Technology Summaries

Technology	Performance	Comments
Pump & Treat	Containment reduces or eliminates downgradient mass flux; some mass removal achieved	Assessing capture can require extensive monitoring; long-term management required with associated operation-and-maintenance costs; extensive guidance available; technology is robust and flexible; treated water can be used as resource
Physical Containment	Can reduce or eliminate downgradient mass flux	Needs natural low-permeability bottom; long-term monitoring and maintenance required; water management (and possible treatment) inside the containment area likely required
Permeable Reactive Barrier	Containment reduces or eliminates downgradient mass flux; some mass removal achieved	Usually needs natural low-permeability bottom; treatment occurs in the subsurface; treatment is passive; monitoring can be focused; barrier replacement eventually required
Monitored Natural Attenuation	Significant mass reduction can be achieved, reducing mass flux downgradient	Often considered a polishing step in treatment train; can require extensive long-term monitoring to ensure requisite biogeochemical conditions persist

or ASTM). Such reviews should include a description of the state of the practice, performance metrics, and sustainability information of each type of remedial technology so that there is a trusted source of information for use in the remedial investigation/feasibility study process and optimization evaluations.

Research is needed on how to better combine existing or new remediation technologies to address complex contaminated sites. There is the potential at most complex sites to combine multiple technologies in space and time to cost-effectively remove/treat contamination in both the source zone and the downgradient dissolved plume. However, additional research is needed to examine combinations of *in situ* remediation technologies to optimize removal and cost effectiveness.

REFERENCES

AATDF (Advanced Applied Technology Demonstration Facility). 1997. Technology Practices Manual for Surfactants and Cosolvents. Houston, TX: AATDF.

Abriola, L. M., C. D. Drummond, E. J. Hahn, T. C. G. Kibbey, L. D. Lemke, K. D. Pennell, C. A. Ramsburg, and K. M. Rathfelder. 2005. A pilot-scale demonstration of surfactant-enhanced PCE solubilization at the Bachman Road site (1) Site characterization and test design. Environmental Science & Technology 39:1778-1790.

Albano, J., S. D. Comfort, V. Zlotnik, T. Halihan, M. Burbach, C. Chokejaroenrat, and W. Clayton. 2010. *In situ* chemical oxidation of RDX-contaminated groundwater with permanganate at the Nebraska Ordnance Plant. Ground Water Monitoring & Remediation 30(3):96-106.

Amos, B. K., R. C. Daprato, J. B. Hughes, K. D. Pennell, and F. E. Loffler. 2007. Effects of the nonionic surfactant Tween 80 on microbial reductive dechlorination of chlorinated ethenes. Environmental Science & Technology 41:1710-1716.

Baker, R. S., and J. M. Bierschenk. 2001. In-Situ Thermal Destruction Makes Stringent Soil and Sediment Cleanup Goals Attainable. Proceedings of the Fourth Tri-Service Environmental Technology Symposium, San Diego, CA. June 18-20, 2001.

Battelle and Duke Engineering & Services. 2001. Final Technical Report for Surfactant-Enhanced DNAPL Removal at Site 88 Marine Corps Base Camp Lejeune, North Carolina. Prepared for Naval Facilities Engineering Service Center, Post Hueneme, California.

Beyke, G., and D. Fleming. 2005. *In situ* thermal remediation of DNAPL and LNAPL using electrical resistance heating. Remediation Summer:5-22.

Bradley, P. M., and F. H. Chapelle. 2010. Biodegradation of chlorinated ethenes. Pp. 39-67 in In Situ Remediation of Chlorinated Solvent Plumes. H. F. Stroo and C. H. Ward, Eds. Springer. New York.

Brauner, J. S., D. C. Downey, and R. Miller. 2008. Using advanced analysis approaches to complete long-term evaluations of natural attenuation processes on the remediation of dissolved chlorinated solvent contamination. Final report for the Strategic Environmental Research and Development Program.

Brooks, M. C., M. D. Annable, P. S. C. Rao, K. Hatfield, J. W. Jawitz, W. R. Wise, A. L. Wood, and C. G. Enfield. 2004. Controlled release, blind test of DNAPL remediation by ethanol flushing. Journal of Contaminant Hydrology 69:281-297.

Brown, R. A., J. T. Wilson, and M. Ferrey. 2007. Monitored Natural Attenuation Forum: The case for abiotic MNA. Remediation Journal 17:127-137.

Butler, E. C., Y. Dong, X. Liang, T. Kuder, R. P. Philp, and L. R. Krumholz. 2009. Abiotic reductive dechlorination of tetrachloroethylene and trichloroethylene in anaerobic environments. Final report for the Strategic Environmental Research and Development Program.

CDM. 2009. Final Focused Feasibility Study, Well 12A Superfund Site, Tacoma, Washington, EPA Region X. http://www.epa.gov/region10/pdf/sites/cb-stc/well12a_final_ffs_apr2009.pdf.

Chamberlain, J., L. Michalczak, and S. Seneca. 2011. Design and installation of a permeable treatment wall at the West Valley Demonstration Project to mitigate expansion of Strontium-90 contaminated groundwater. Waste Management 2011 Conference, Feb. 27–March 3, 2011.

Chapelle, F. H., P. M. Bradley, and C. C. Casey. 2005. Behavior of a chlorinated ethene plume following source-area treatment with Fenton's Reagent. Ground Water Monitoring & Remediation 25(2):131-141.

Childs, J., E. Acosta, M. D. Annable, M. C. Brooks, C. G. Enfield, J. H. Harwell, M. Hasegawa, R. C. Knox, P. S. C. Rao, D. A. Sabatini, B. Shiau, E. Szekeres, and A. L. Wood. 2006. Field demonstration of surfactant-enhanced solubilization of DNAPL at Dover Air Force Base, Delaware. Journal of Contaminant Hydrology 82:1-22.

Christ, J. A., C. A. Ramsburg, L. M. Abriola, K. D. Pennell, and F. E. Löffler. 2005. Coupling aggressive mass removal with microbial reductive dechlorination for remediation of DNAPL source zones: A review and assessment. Environmental Health Perspectives 113(4):465-477. DOI: 10.1289/ehp.6932.

Coates, J. D., and L. A. Achenbach. 2004. Microbial perchlorate reduction: Rocket-fuelled metabolism. Nature Reviews Microbiology 2:569-580.

Cope, N., and J. B. Hughes. 2001. Biologically-enhanced removal of PCE from NAPL source zones. Environmental Science & Technology 35(10):2014-2020.

Costanza, J., K. E. Fletcher, F. E. Loffler, and K. D. Pennell. 2009. Fate of TCE in heated Fort Lewis soil. Environmental Science & Technology 43:909-914.

Davis, E. L. 1998. Steam Injection for Soil and Aquifer Remediation. Ground Water Issue No. EPA 540-S-97-505. Washington, DC: EPA.

Davis, G. B., and B. M. Patterson. 2003. Developments in permeable reactive barrier technology. Pp. 205-226 in Bioremediation: A Critical Review. I. M. Head, L Singleton, and M. G. Milner, eds. Horizon Scientific Press.

De Percin, P. R. 1991. Demonstration of *in situ* steam and hot-air stripping technology. Journal of Air and Waste Management 41(6):873-877.

Delshad, M., G. A. Pope, L. Yeh, and F. J. Holzmer. 2000. Design of the surfactant flood at Camp Lejeune. Second International Conference on Remediation of Chlorinated and Recalcitrant Compounds. Monterey, CA, May 22-25, 2000. Pp. 203-210.

Dey, J. C., P. Rosenwinkel, and K. Wheeler. 2002. *In situ* remediation of MTBE utilizing ozone. Remediation Journal 13(1):77-85.

Environmental Management Support, Inc. 2008. Energy Consumption and Carbon Dioxide Emissions at Superfund Cleanups. Draft report prepared for EPA Office of Superfund Remediation and Technology Innovation, Washington, D.C., EPA Contract EP W-07-037.

EPA (U.S. Environmental Protection Agency). 1995a. KAI Radio Frequency Heating Technology. EPA 540-R-94-528a. Washington, DC: EPA Government Printing Office.

EPA. 1995b. IITRI Radio Frequency Heating Technology. EPA 540-R-94-527a. Washington, DC: EPA Government Printing Office.

EPA. 1998. Evaluation of Subsurface Engineered Barriers at Waste Sites. EPA 542-R-98-005. Washington, DC: EPA Office of Solid Waste and Emergency Response.

EPA. 1999. Cost and Performance Report: Six Phase Heating (SPH) at a Former Manufacturing Facility, Skokie Illinois. Washington, DC: EPA Office of Solid Waste and Emergency Response, Technical Innovation Office.

EPA. 2000. Superfund Reform Strategy, Implementation Memorandum: Optimization of fund-lead ground water pump and treat (P&T) systems. OSWER 9283.1-13. Washington, DC: EPA Office of Solid Waste and Emergency Response.

EPA. 2002. Elements for Effective Management of Operating Pump and Treat Systems. EPA 542-R-02-009. Washington, DC: EPA Office of Solid Waste and Emergency Response.

EPA. 2003. The DNAPL Remediation Challenge: Is There a Case for Source Depletion? EPA/600/R-03/143. Ada, OK: EPA NRMRL.

EPA. 2005a. Cost-Effective Design of Pump and Treat Systems. EPA 542-R-05-008. Washington, DC: EPA Office of Solid Waste and Emergency Response.

EPA. 2005b. Effective Contracting Approaches for Operating Pump and Treat Systems. EPA 542-R-05-009. Washington, DC: EPA Office of Solid Waste and Emergency Response.

EPA. 2005c. O&M Report Template for Ground Water Remedies (with Emphasis on Pump and Treat Systems). EPA 542-R-05-010. Washington, DC: EPA Office of Solid Waste and Emergency Response.

EPA. 2007a. Options for Discharging Treated Water from Pump and Treat Systems. EPA 542-R-07-006. Washington, DC: EPA Office of Solid Waste and Emergency Response.

EPA. 2007b. Optimization Strategies for Long-Term Ground Water Remedies (with Particular Emphasis on Pump and Treat Systems). EPA 542-R-07-007. Washington, DC: EPA Office of Solid Waste and Emergency Response.

EPA. 2007c. A Cost Comparison Framework for Use in Optimizing Groundwater Pump and Treat Systems. EPA 542-R-07-005. Washington, DC: EPA Office of Solid Waste and Emergency Response.

EPA. 2007d. Treatment Technologies for Site Cleanup: Annual Status Report (Twelfth Edition). Washington, DC: EPA Office of Solid Waste and Emergency Response.

EPA. 2008. Ground Water Remedy Optimization Progress Report: 2006–2007. EPA 540/R-08/004. Washington, DC: EPA Office of Solid Waste and Emergency Response.

EPA. 2010a. Superfund Remedy Report, Thirteenth Edition. EPA-542-R-10-004. Washington, DC: EPA Office of Solid Waste and Emergency Response.

EPA. 2010b. O&M Report Template for Ground Water Remedies (with Emphasis on Pump and Treat Systems). EPA 542-R-05-010. Washington, DC: EPA Office of Solid Waste and Emergency Response.

EPA. 2010c. Summary of New Brighton/Arden Hills/TCAAP. http://www.epa.gov/region5 superfund/npl/minnesota/MN7213820908.htm.

ESTCP (Environmental Security Technology Certification Program). 1999. Technology Status Review *In Situ* Oxidation. Washington, DC: ESTCP.

ESTCP. 2007. Biodegradation of Dense Non-Aqueous Phase Liquids (DNAPLS) through Bioaugmentation of Source Areas—Dover National Test Site, Dover, Delaware. ESTCP Project ER-0008, 101 pp.

ESTCP. 2011. *In Situ* Chemical Oxidation for Groundwater Remediation—Technology Practices Manual ESTCP Project ER-200623. Washington, DC: ESTCP.

Falta, R. W., C. M. Lee, S. E. Brame, J. T. Coates, C. Wright, A. L. Wood, and C. G. Enfield. 1999. Field test of high molecular weight alcohol flushing for subsurface nonaqueous phase liquid remediation. Water Resources Research 35(7):2095-2108.

Farouq Ali, S. M., and R. F. Meldau. 1979. Current steamflood technology. Journal of Petroleum Technology 31(10):1332-1342.

Ferrey, M. L., R. T. Wilkin, R. G. Ford, and J. T. Wilson. 2004. Nonbiological removal of *cis*-dichloroethylene and 1,1-dichloroethylene in aquifer sediment containing magnetite. Environmental Science & Technology 38:1746-1752.

Fountain, J. C., R. C. Starr, T. Middleton, M. Beikirch, C. Taylor, and D. Hodge. 1996. A controlled field test of surfactant-enhanced aquifer remediation. Ground Water 34(5): 910-916.

Friis, A. K., G. Heron, H. J. Albrechtsen, K. S. Udell, and P. L. Bjerg. 2006. Anaerobic dechlorination and redox activities after full-scale electrical resistance heating (ERH) of a TCE-contaminated aquifer. Journal of Contaminant Hydrology 88:219-234.

Friis, A. K., H. J. Albrechtsen, G. Heron, and P. L. Bjerg. 2005. Redox processes and release of organic matter after thermal treatment of a TCE-contaminated aquifer. Environmental Science & Technology 39:5787-5795.

Hasegawa, M. H., B.-J. Shiau, D. A. Sabatini, R. C. Knox, J. H. Harwell, R. Lago, and L. Yeh. 2000. Surfactant enhanced subsurface remediation of DNAPLs at the Former Naval Air Station Alameda, California. Pp. 219-226 in Treating Dense Nonaqueous-Phase Liquids (DNAPLs): Remediation of Chlorinated and Recalcitrant Compounds. G. B. Wickramanayake et al. (eds.). Columbus, OH: Battelle Press.

Hesemann, J. R., and M. Hildebrandt. 2009. Successful unsaturated zone treatment of PCE with sodium permanganate. Remediation Journal 19(2):37-48.

Holzmer, F. J., G. A. Pope, and L. Yeh. 2000. Surfactant-enhanced aquifer remediation of PCE-DNAPL in low-permeability sand. Second International Conference on Remediation of Chlorinated and Recalcitrant Compounds; Monterey, CA. May, 22-25, 2000. Pp. 187-193.

Hood, E. D., D. W. Major, J. W. Quinn, W.-S. Yoon, A. Gavaskar, and E. A. Edwards. 2008. Demonstration of enhanced bioremediation in a TCE source area at Launch Complex 34, Cape Canaveral Air Force Station. Ground Water Monitoring & Remediation 28(2):98-107.

ITRC (Interstate Technology and Regulatory Council). 2005. Permeable Reactive Barriers: Lessons Learned/New Directions. PRB-4. Washington, DC: ITRC Permeable Reactive Barriers Team. http://www.itrcweb.org/documents/prb-4.pdf.

ITRC. 2007. The Vapor Intrusion Pathway: A Practical Guideline. Washington, DC: ITRC. http://www.itrcweb.org/documents/VI-1.pdf.

IRTC. 2010. Use and Measurement of Mass Flux and Mass Discharge. Washington, DC: ITRC. http://www.itrcweb.org/Documents/MASSFLUX1.pdf.

ITRC. 2011. Permeable Reactive Barrier: Technology Update. PRB-5. Washington, DC: ITRC PRB Technology Update Team.

Jain, V., and A. H. Demond. 2002. Conductivity reduction due to emulsification during surfactant enhanced-aquifer remediation. 1. Emulsion transport. Environmental Science & Technology 36(24):5426-5433.

Jawitz, J. W., M. D. Annable, P. S. C. Rao, and R. D. Rhue. 1998. Field implementation of a Winsor type I surfactant/alcohol mixture for in situ solubilization of a complex LNAPL as a single phase microemulsion. Environmental Science & Technology 32(4):523-530.

Jawitz, J. W., R. K. Sillan, M. D. Annable, P. S. C. Rao, and K. Warner. 2000. In situ alcohol flushing of a DNAPL source zone at a dry cleaner site. Environmental Science & Technology 34:3722-3729.

Jawitz, J. W., M. D. Annable, P. S. C. Rao, R. D. Rhue. 2001. Evaluation of remediation performance and cost for field-scale single-phase microemulsion (SPME) flushing. Journal of Environmental Science Health A36:1437-1450.

Jefferis, S. A. 2008. Reactive transport in cut-off walls and implications for wall durability. Geotechnical Special Publication 177(GeoCongress 2008):652-659.

Johnson, P. J., P. Dahlen, J. Triplett Kingston, E. Foote, and S. Williams. 2009. State-of-the-Practice Overview: Critical Evaluation of State-of-the-Art *In Situ* Thermal Treatment Technologies for DNAPL Source Zone Treatment. EXTCP Project ER-0312. May 2009.

Kavanaugh, M., and R. Deeb. 2011. Diagnostic Tools For Performance Evaluation of Innovative In-Situ Remediation Technologies at Chlorinated Solvent-Contaminated Sites. ESTCP Project ER-200318. Washington, DC: ESTCP.

Kennedy, L. G., J. W. Everett, and J. Gonzales. 2006. Assessment of biogeochemical natural attenuation and treatment of chlorinated solvents, Altus Air Force Base, Altus, Oklahoma. Journal of Contaminant Hydrology 83(3-4):221-236.

Knox, R. C., D. A. Sabatini, J. H. Harwell, R. E. Brown, C. C. West, F. Blaha, and C. Griffin. 1997. Surfactant remediation field demonstration using a vertical circulation well. Ground Water 35(6):948-953.

Knox, R. C., B. J. Shau, D. A. Sabatini, and J. H. Harwell. 1999. Field demonstration studies of surfactant-enhanced solubilization and mobilization at Hill Air Force Base, Utah. In Innovative Subsurface Remediation. Brusseau M. L., D. A. Sabatini, J. S. Gierke, M. D. Annable, eds. ACS Symposium Series 725, American Chemical Society, Washington, DC.

Krembs, F. J., R. L. Siegrist, M. L. Crimi, R. F. Furrer, and B. G. Petri. 2010. ISCO for groundwater remediation: Analysis of field applications and performance. Ground Water Monitoring & Remediation 30(4):42-53.

Krembs, F. J., W. S. Clayton, and M. C. Marley. 2011. Evaluation of ISCO field applications and performance. Pp. 319-353 in *In Situ* Chemical Oxidation for Groundwater Remediation. Siegrist, R. L., M. Crimi, and T. J. Simpkin, eds. New York: Springer Science + Business Media, LLC. DOI: 10.1007/978-1-4419-7826-4.

Levine-Fricke-Recon. 2008. Cosolvent Flushing Pilot Test Report, Former Sages Dry Cleaner December 4, 1998, LFR Project Number: 6006.10-08.

Liu, M. W., and D. Roy. 1995. Surfactant-induced interactions and hydraulic conductivity changes in soil. Waste Management 15(7):463-470.

Lo, I. M., C. S. Lam, and K. C. Lai. 2006. Hardness and carbonate effects on the reactivity of zero-valent iron for Cr(VI) removal. Water Research 40(3):595-605.

Londergan, J. T., H. W. Meinardus, P. E. Mariner, R. E. Jackson, C. L. Brown, V. Dwarakanath, G. A. Pope, J. S. Ginn, and S. Taffinder. 2001. DNAPL removal from a heterogeneous alluvial aquifer by surfactant-enhanced aquifer remediation. Ground Water Monitoring & Remediation 21:57-67.

Loomer, D. B., T. A. Al, V. J. Banks, B. L. Parker, and K. U. Mayer. 2010. Manganese valence in oxides formed from in situ chemical oxidation of TCE by $KMnO_4$. Environmental Science & Technology 44(15):5934-5939.

Major, D. W., M. M. McMaster, E. E. Cox, E. A. Edwards, S. M. Dworatzek, E. R. Hendrickson, M. G. Starr, J. A. Payne, and L. W. Buonamici. 2002. Field demonstration of successful bioaugmentation to achieve dechlorination of tetrachloroethene to ethene. Environmental Science & Technology 36(23):5106-5116.

Martel R., P. J. Gelinas, and L. Saumure. 1998. Aquifer washing by micellar solutions: 3. Field test at the Thouin Sand Pit (L' Assomption, Quebec, Canada). Journal of Contaminant Hydrology 30:33-48.

McDade, J. M., T. M. McGuire, and C. J. Newell. 2005. Analysis of DNAPL source-depletion costs at 36 field sites. Remediation Journal 15(2):9-18.

McGuire, T., and J. B. Hughes. 2003. Effects of surfactants on the dechlorination of chlorinated ethenes. Environmental Toxicology and Chemistry 22(11):2630-2638.

McGuire, T. M., J. M. McDade, and C. J. Newell. 2006. Performance of DNAPL source depletion technologies at 59 chlorinated solvent-impacted sites. Ground Water Monitoring & Remediation 26(1):73-84.

MPCA. 2010. Twin Cities Army Ammunition Plant (TCAAP) Superfund Site. Minnesota Pollution Control Agency. http://www.pca.state.mn.us/index.php/waste/waste-and-cleanup/cleanup-programs-and-topics/topics/remediation-sites/twin-cities-army-ammunition-plant-tcaap-superfund-site.html

NAVFAC (Naval Facilities Engineering Command). 2004. Guidance for Optimizing Remedy Evaluation, Selection, and Design.

Neuhauser, E. F., J. A. Ripp, N. A. Azzolina, E. L. Madsen, D. M. Mauro, and T. Taylor. 2009. Monitored natural attenuation of manufactured gas plant tar mono- and polycyclic aromatic hydrocarbons in ground water: A 14-year field study. Ground Water Monitoring & Remediation 29(3):66-76.

Newell, C. J., I. Cowie, T. M. McGuire, and W. McNab. 2006. Multi-year temporal changes in chlorinated solvent concentrations at 23 MNA sites. Journal of Environmental Engineering 132(6):653-663.

NRC (National Research Council). 1994. Alternatives for Groundwater Cleanup. Washington, DC: National Academy Press.

NRC. 1997. Innovations in Soil and Ground Water Cleanup. Washington, DC: National Academy Press.

NRC. 1999. Groundwater and Soil Cleanup. Washington, DC: National Academy Press.

NRC. 2000. Natural Attenuation for Groundwater Remediation. Washington, DC: National Academy Press.

NRC. 2005. Contaminants in the Subsurface: Source Zone Assessment and Remediation. Washington, DC: The National Academies Press.

NRC. 2007. Assessment of the Performance of Engineered Waste Containment Barriers. Washington, DC: The National Academies Press.

Pennell, K. D., G. A. Pope, and L. M. Abriola. 1996. Influence of viscous and buoyancy forces on the mobilization of residual tetrachloroethylene during surfactant flushing. Environmental Science & Technology 30(4):1328-1335.

Ramakrishnan, V., A. V. Ogram, and A. S. Lindner. 2005. Impacts of co-solvent flushing on microbial populations capable of degrading trichloroethylene. Environmental Health Perspectives 113(1):55-61.

Ramsburg, C. A., and K. D. Pennell. 2001. Experimental and economic assessment of two surfactant formulations for source zone remediation at a former dry cleaning facility. Ground Water Monitoring & Remediation 21:68-82.

Ramsburg, C. A., L. M. Abriola, K. D. Pennell, F. E. Loffler, M. Gamache, B. K. Amos, and E. A. Petroviskis. 2004. Stimulated microbial reductive dechlorination following surfactant treatment at the Bachman road site. Environmental Science & Technology 38:5902-5914.

Ramsburg, C. A., K. D. Pennell, L. M. Abriola, G. Daniels, C. D. Drummond, M. Gamache, H. U. Hsu, E. A. Petrovskis, C. M. Rathfelder, J. L. Ryder, and T. P. Yavaraski. 2005. Pilot-scale demonstration of surfactant-enhanced PCE solubilization at the Bachman Road site. 2. System operation and evaluation. Environmental Science & Technology 39(6):1791-1801.

Rao, P. H., M. He, X. Yang, Y.-C. Zhang, S.-Q. Sun, and J.-S. Wang. 2006. Effect of an anionic surfactant on hydraulic conductivities of sodium- and calcium-saturated soils. Pedosphere 16(5):673-680.

Rao, P. S. C., M. D. Annable, R. K. Sillan, D. Dai, K. Hatfield, W. D. Graham, A. L. Wood, and C. G. Enfield. 1997. Field-scale evaluation of in situ cosolvent flushing for enhanced aquifer remediation. Water Resources Research 33:2673-2686.

Roote, D. S. 1998. Technology Status Report *In Situ* Flushing (TS-98-01). Pittsburgh, PA: Ground-Water Remediation Technologies Analysis Center.

Sale, T. C., and C. Newell. 2011. Guide for Selecting Remedies for Subsurface Releases of Chlorinated Solvents. ESTCP Project ER-200530.

Scherer, M. M., E. O'Loughlin, G. F. Parkin, R. Valentine, H. Al-Hosney, R. Handler, C. Just, P. Larese-Casanova, T. Pasakarnis, and S. L. Smith. 2009. S.L. ER-1369 Sustainability of Long-Term Abiotic Attenuation of Chlorinated Ethenes. Final report for the Strategic Environmental Research and Development Program.

Seagren, E. A., B. E. Rittman, and A. J. Valocchi. 1993. Quantitative evaluation of flushing and biodegradation for enhancing in situ dissolution of non-aqueous phase liquids. Journal of Contaminant Hydrology 12(1-2):103-132.

Seagren, E. A., B. E. Rittmann, and A. J Valocchi. 1994. Quantitative evaluation of the enhancement of NAPL-pool dissolution by flushing and biodegradation. Environmental Science & Technology 28(5):833-839.

Siegal, J., A. A. Rees, K. W. Eggers, and R. L. Hobbs. 2009. *In situ* chemical oxidation of residual LNAPL and dissolved-phase fuel hydrocarbons and chlorinated alkenes in groundwater using activated persulfate. Remediation Journal 19(2):19-35.

Siegrist, R. L., M. Crimi, T. J. Simpkin (eds). 2011. *In Situ* Chemical Oxidation for Remediation of Contaminated Groundwater. New York: Springer Science+Business Media, LLC.

Sleep, B. E., D. J. Seepersad, K. Mo, C. M. Heidorn, L. Hrapovic, P. L. Morrill, M. L. McMaster, E. D. Hood, C. LeBron, B. Sherwood Lollar, D. W. Major, and E. A. Edwards. 2006. Biological enhancement of tetrachloroethene dissolution and associated microbial community changes. Environmental Science & Technology 40(11):3623-3633.

Stroo, H. F., M. Unger, C. H. Ward, M. C. Kavanaugh, C. Vogel, A. Leeson, J. A. Marquesee and B. P. Smith. 2003. Remediation of chlorinated solvent source zones. Environmental Science & Technology 37:224A-230A.

Stroo, H. F., and C. H. Ward. 2010. In situ remediation of chlorinated solvent plumes. New York: Springer.

Stroo, H. F., A. Leeson, J. A. Marquesee, P. C. Johnson, C. H. Ward, M. C. Kavanaugh, T. C. Sale, C. J. Newell, K. D. Pennell, C. A. Lebron, and M. Unger. 2012. Chlorinated ethene source remediation: Lessons learned. Environmental Science & Technology 46:6438-6447.

Thiruvenkatachari, R., S. Vigneswaran, and R. Naidu. 2008. Permeable reactive barrier for groundwater remediation. Journal of Industrial and Engineering Chemistry (Amsterdam, Netherlands) 14(2):145-156.

Triplett Kingston, J. L. 2008. A Critical Evaluation *of In Situ* Thermal Technologies. Ph.D. Dissertation, Department of Civil and Environmental Engineering, Arizona State University.

Triplett Kingston, J. L., P. R. Dahlen, P. C. Johnson, E. Foote, and S. Williams. 2009. State of the Practice Overview: Critical Evaluation of State-of-the-Art *In Situ* Thermal Treatment Technologies for DNAPL Source Zone Treatment. ESTCP Project ER-0314. May 2009. http://www.serdp.org/content/download/4577/67309/version/2/file/ER-0314-State-of-the-Practice+Overview.pdf.

Triplett Kingston, J. L., P. Dahlen, and P. C. Johnson. 2010a. State-of-the-practice review of *in-situ* thermal technologies. Ground Water Monitoring & Remediation 30(4):64-72.

Triplett Kingston, J. L., P. R. Dahlen, P. C. Johnson, E. Foote, and S. Williams. 2010b. Critical Evaluation of State-of-the-Art *In Situ* Thermal Treatment Technologies for DNAPL Source Zone Treatment. ESTCP Project ER-0314. January 2010. http://www.serdp.org/content/download/4576/67301/file/ER-0314-FR.pdf.

Triplett Kingston, J. L., P. R. Dahlen, and P. C. Johnson. 2012. Assessment of groundwater quality improvements and mass discharge reductions at five *in situ* electrical resistance heating remediation sites. Ground Water Monitoring & Remediation. DOI:10.1111/j.1745-6592.2011.01389.x

Vangelas, K. M., B. B. Looney, T. O. Early, T. Gilmore, F. H Chapelle, K. M. Adams, and C. H. Sink. 2006. Monitored natural attenuation of chlorinated solvents—moving beyond reductive dechlorination. Remediation Journal 16(3):5-23.

Vinegar, H. J., G. L. Stegemeier, F. G. Carl, J. D. Stevenson, and R. J. Dudley. 1999. *In Situ* Thermal Desorption of Soils Impacted with Chlorinated Solvents. Proceedings of the Annual Meetings of the Air and Waste Management Association, Paper No. 99-450.

Wanaratna, P., C. Christodoulatos, and M. Sidhoum. 2006. Kinetics of RDX degradation by zero-valent iron (ZVI). Journal of Hazardous Materials 136(1):68-74.

Wiedenmeir, T. H., H. S. Rifai, C. J. Newell, and J. T. Wilson. 1999. Natural Attenuation of Fuel Hydrocarbons and Chlorinated Solvents in the Subsurface. New York: John Wiley and Sons.

Wilson, J. T. 2010. Monitored natural attenuation of chlorinated solvent plumes. Chapter 11 in *In Situ* Remediation of Chlorinated Solvent Plumes. H. Stroo and C. H. Ward (eds.). New York: Springer.

Wymore, R. A., T. W. Macbeth, J. S. Rothermel, L. N. Peterson, L. O. Nelson, K. S. Sorenson, N. Akladiss, and I. R. Tasker. 2006. Enhanced anaerobic bioremediation in a DNAPL residual source zone: Test Area North case study. Remediation 16(4):5-22.

Yang, Y., and P. L. McCarty. 2000. Biologically enhanced dissolution of tetrachloroethene DNAPL. Environmental Science & Technology 34(14):2979-2984.

Yang, Y., and P. L. McCarty. 2002. Comparison between donor substrates for biologically enhanced tetrachloroethene DNAPL dissolution. Environmental Science & Technology 36(15):3400-3404.

Yeh, D. H., K. D. Pennell, and S. G. Pavlostathis. 1999. Effect of Tween surfactants on methanogenesis and microbial reductive dechlorination of hexachlorobenzene. Environmental Toxicology and Chemistry 18:1408-1416.

Zimmerman, J. B., T. C. G. Kibbey, M. A. Cowell, and K. F. Hayes. 1999. Partitioning of ethoxylated nonionic surfactants into nonaqueous-phase organic liquids: Influence on solubilization behavior. Environmental Science & Technology 33(1):169-176.

5

Implications of Contamination Remaining in Place

Despite the ability of some remedial technologies to remove substantial amounts of mass, at most complex sites contamination will remain in place at levels above those allowing for unlimited use and unrestricted exposure (see Chapter 4). This chapter discusses the potential technical, legal, economic, and other practical implications of this finding.

First, contamination from these sources must be contained on-site, by using either hydraulic or physical containment systems combined with institutional controls. Indeed, 65 percent of source control RODs from FY 1998–2008 included containment, and institutional controls are used at the vast majority of CERCLA source control remedial actions to enhance and ensure their effectiveness and protectiveness (EPA, 2010a). Because the failure of these systems could create new exposures, potentially responsible parties (PRPs) should weigh the robustness and potential for failure during remedy selection and implementation. Second, our understanding of the risk posed by contaminated groundwater is inherently dynamic. For example, toxicity information is regularly updated, and contaminants that were previously unregulated may become so, changing the drivers for risk assessment and cleanup decisions. In addition, pathways of exposure that were not previously under consideration can be found to be important, such as has happened with the vapor intrusion pathway over the past decade. Consideration of these new factors can change the overall protectiveness of a remedy that leaves contamination in place. Third, residual contamination necessarily reduces the amount of groundwater available for unrestricted use. Treating groundwater for drinking water purposes is very costly and, for some contaminants (e.g., 1,4-dioxane), technically challenging. Finally, leaving contamination in the subsurface may expose the landowner, property manager, or original disposer to complications that would not exist in the absence of the contamination. PRPs may be sued for natural resource damages by the resource trustee (if the underlying groundwater is no longer potable without treatment because of remaining contamination) or for personal injury and/or property damages pursuant to common law by local residents or others (if the contamination crosses property boundaries and causes injury or property damage).

At any given site, the risks and the technical, economic, and legal complications associated with residual contamination need to be compared to the time, cost, and feasibility involved in removing contamination outright. As a practical matter, the Committee did not seek to estimate the relative scope of the nontechnical impacts of leaving contamination in place, and it is probably not feasible to do so. Whether these potential consequences are likely to occur is site specific, and some implications may not materialize at some sites.

POTENTIAL FOR FAILURE OF REMEDIES AND ENGINEERED CONTROLS

The long-term management strategies for many complex sites include leaving significant amounts of contamination in place. At such sites the achievement of risk-based goals is based on a reduction of the contaminant flux (e.g., reduction in source strength) between the zone of residual contamination and the point(s) of compliance. Such flux reduction is generally accomplished by one of four approaches, possibly coupled with partial removal of source zone contamination: (1) hydraulic containment, (2) physical containment, (3) reduction of contaminant concentrations through natural processes (monitored natural attenuation), and/or (4) reduction of contaminant concentrations through an engineered reaction zone, most commonly in the form of a downgradient *permeable reactive barrier* (PRB) (see Chapter 4 for descriptions of these technologies). This section summarizes key concepts and tools for assessing the likelihood and consequences of failure for these approaches.

Each of the remedial strategies listed above is well established and is unlikely to exhibit "complete" failure in any meaningful sense. Rather, some degree of contaminant flux reduction is likely to be realized, even if the overall magnitude and/or spatial extent of the reduction is less than expected from design calculations. The consequences of such "partial failure" would depend both on the measures used to monitor performance and the corrective actions that are triggered by inadequate performance. There are few reports in the peer-reviewed literature that document both the failure of a long-term remedial strategy and the resulting response (although these issues should be addressed in the five-year review process for sites regulated under CERCLA). This lack of focused literature on the failure of remediation systems designed for long-term management may be due, in part, to the likelihood that system failure would generate incremental, rather than sharp, increases in operation costs, as discussed below for each of the four strategies.

Hydraulic Containment

Pump and treat (P&T) has increasingly been implemented as a long-term management strategy, with the primary goal of hydraulic containment to prevent further spreading of contamination. In a general sense, "failure" of hydraulic containment occurs when groundwater that originates from within the target capture zone is not completely captured by extraction wells, but instead is allowed to migrate downgradient beyond property boundaries and toward a receptor. Such failure could occur as a direct consequence of inadequate well placement and/or underspecified pumping rate(s) due to a misunderstanding of the governing hydrogeology (e.g., an incorrect or incomplete groundwater model). Even for a properly designed extraction system, containment failure could occur after startup because of temporal changes in hydrologic conditions such as recharge or regional flow conditions.

To assist with identifying potential P&T failure, the U.S. Environmental Protection Agency (EPA) has recently developed a six-step procedure for evaluating the hydraulic containment of target capture zones, with an emphasis on comparing measured water levels and concentrations against model predictions (EPA, 2008a). While establishing a formal comparison between measured and predicted capture zones still requires considerable site-specific judgment, the availability of established guidance (and an ongoing process to refine it and expand its applicability) is an important development.

As discussed in Chapter 4, EPA has applied the *Remedial System Evaluation* (RSE) process to more than 60 operating P&T systems at Superfund facilities. At many of these, field observations were unable to establish the success of hydraulic containment at the desired level of confidence. In some cases, additional monitoring was recommended to clarify the evaluation. However, for other sites, adjustment to the locations and/or operation of extraction wells was recommended. Although such midcourse corrections typically increase the cost of P&T system operation, pumping rates, monitoring programs, or even extraction wells can also be reduced if the system is overdesigned for current conditions. In general, actions to improve P&T performance are straightforward to implement and normally generate an incremental, rather than drastic change in the life-cycle cost of site management. In this regard, hydraulic containment may be regarded as an adaptive strategy that can readily be updated in response to new information about the site.

Physical Containment

Barriers are frequently used to influence groundwater flow in combined remedies that also use extraction wells and/or engineered reaction zones. From a containment standpoint, the overall remediation goal is similar to hydraulic containment: maintain control of groundwater within a target capture zone. Thus, similar monitoring and analytical approaches might be used to assess performance. Failures in physical containment may occur due to incorrect design or construction of barriers, poor seals between sections (in the case of sheet pile barriers or geomembranes), holes/defects in materials, physical or chemical damage, poor connection between a vertical barrier and underlying confining bed, and lack of control of recharge inside the contained area. These and similar expressions of "failure" are likely to occur locally at small defects in vertical walls, rather than across the full extent of the barrier system. Because the flow influence of a barrier irregularity will likely be distributed across a large area, detecting such local failure through routine groundwater monitoring is likely to be difficult.

A recent NRC review concluded, in part, that available field data are insufficient to provide a robust assessment of the potential for or actual occurrence of failure in vertical barriers (NRC, 2007), particularly over long decision horizons. However, reports from site-specific remedial system evaluations and CERCLA five-year reviews have identified instances where hydraulic monitoring indicated that physical containment systems may have "failed" (e.g., EPA, 1999; Northgate Environmental Management, 2008), although specific mechanisms are typically not identified. Even if the precise location of a barrier defect could be identified through field monitoring, effective measures for the direct repair of a flawed or cracked vertical barrier have not been developed. Instead, adjustments to other aspects of the remedial system would likely be needed. For both of the above CERCLA examples, the vertical barriers functioned as components of combined remedies that also included extraction wells, which resulted in straightforward adjustments to system operation that maintained a high degree of confidence in successful hydraulic containment.

In additional to the possibility of hydraulic failure, earthen barriers can also release contaminants by molecular diffusion. Because chemicals in most barrier materials have diffusion coefficients that are similar to those in aquifer material, and the diffusion path length is relatively short (typically one meter or less), the time for a solute to diffuse across the barrier could be relatively short, on the order of years rather than decades (e.g., Mott and

Weber, 1991; Khandelwal et al., 1998; Krol and Rowe, 2004). Although the potential for diffusion across slurry walls has been long recognized by scholars, field studies to assess this scenario have not been performed. However, even if elevated contaminant concentrations are present in the immediate vicinity of a vertical barrier, diffusive contaminant fluxes are typically several orders of magnitude less than advective fluxes, and it is plausible that molecular diffusion would constitute a significant concern at only a very small number sites (e.g., sites with both large concentrations within a containment zone and a receptor located in close proximity to a vertical barrier).

Permeable Reactive Barriers

To function successfully, a permeable reactive barrier (PRB) must provide hydraulic control of the upgradient target capture zone, such that all contaminated water flows through the PRB rather than around or below it. In addition the PRB must have sufficient reaction capacity to sustain the necessary reduction in contaminant concentrations over the appropriate design time frame. Failure to achieve either or both objectives can occur because of inadequate design (e.g., improper wall placement or reaction zone thickness) or because of changes within the PRB that occur over time (loss of permeability and/or reactivity). In addition, if a PRB was placed downgradient of a source zone but within a region that previously contained dissolved contamination, it is possible that measurable downgradient concentrations will persist due to back-diffusion, even if the PRB is functioning as designed (Sale and Newell, 2010).

The vast majority of installed PRBs are constructed of zero-valent iron, which produces redox conditions and results in pH changes that are likely to promote precipitation of groundwater minerals. This phenomenon has long been recognized as a potential problem, and numerous laboratory and modeling studies have explored the potential consequences of these processes for PRB longevity (e.g., Yabusaki et al., 2001; Kohn et al., 2005; Johnson et al., 2008; Wilkin and Puls, 2003; Sass et al., 2002; Phillips et al., 2010). However, as noted by ITRC (2005a, 2011), no PRB has "failed" due to loss of permeability or reactivity. In the most detailed published evaluation of iron-based PRB performance (Henderson and Demond, 2007), a handful of active PRB projects reported situations where improper design (insufficient depth or width) resulted in incomplete hydraulic capture. Of the 40 projects, only three exhibited post-installation performance degradation involving the loss of permeability due to precipitation and/or deceased reactivity.

As with low-permeability barrier systems, the failure of a PRB system is likely to occur locally rather than across the entire plane of interest, and it is plausible that repair, rather than replacement, could be the appropriate response action. At the time of this writing, reports where installed PRBs were repaired or replaced were not located in the literature. As with the other long-term management strategies, the operating history of PRB technology is simply too short to support a robust assessment of the potential long-term management costs. However, concerns related to back-diffusion could potentially limit the application of PRB systems to sites where substantial contamination is not initially present downgradient of the installed PRB.

Monitored Natural Attenuation

Monitored natural attenuation (MNA) is most often used in conjunction with other active or engineering remedial components and is seldom employed as a stand-alone measure (EPA, 2010a). The success of natural attenuation as a remedy depends on the site-specific ability to predict the evolution of complex biogeochemical processes over an extended period of time. Because of uncertainties in long-term predictions, natural attenuation requires confirmatory monitoring, such that MNA remedies are accompanied by a detailed program of monitoring (e.g., NRC, 2000; EPA, 2004a).

Numerous protocols exist for evaluating MNA performance including a recently proposed decision framework for evaluating MNA for inorganic or radionuclide contamination (e.g., ITRC, 2010). Although focused on inorganic contaminants, the ITRC protocol contains many elements appropriate for sites with organic contaminants. In particular, the need for a contingency plan was emphasized, which provides a cleanup approach that will be implemented if "the selected remedy fails to perform as anticipated" (EPA, 2007). For MNA remedies, a suitable contingency plan might include optimization of source or plume treatments, implementation of an *enhanced attenuation* (EA) technology, pursuit of a technical impracticability waiver, or the use of institutional controls.

MNA systems could fail for many reasons, including temporal changes in site-specific hydrologic or geochemical conditions, the depletion of natural sources of nutrients or electron acceptors/donors, and lower-than-anticipated transformation rates. Further, the regulators may believe that there is insufficient evidence that MNA is occurring in the intended fashion. For example, it may be difficult to verify that the presence of daughter products is due to parent compound degradation and not co-contamination.

It is difficult to generalize regarding the potential cost of MNA failure, which will depend on site-specific conditions, the nature of the contingency actions, and the degree of conservatism built into the monitoring program. A properly designed monitoring program should provide "early detection" that allows for the implementation of a contingency plan prior to the point when a migrating plume would present elevated risks to receptors.

However, if contaminant migration and/or plume expansion occurs prior to the detection of failure, additional costs may be incurred. In certain circumstances, the combined cost of failed MNA and implementation of an additional remedy may exceed the cost that would have accrued had the remedy originally been put in place instead of MNA. To avoid such occurrences, the monitoring program should be directed at providing confirmation of the assumptions used to extrapolate the performance of MNA, in an adaptive management mode.

* * *

Common to all the remedies discussed above are unplanned and catastrophic events that may lead to failure of the proposed containment/treatment techniques, potentially for long periods of time. For example, natural disasters (e.g., earthquakes, floods, or other events) could cause changes in local hydrology, damage the remediation/containment system, or cause a loss of power to an active containment process. Flooding or other events could spread contamination to new areas and/or create new exposure pathways (e.g., vapor intrusion). Because contaminant migration from source zones or the plume is often slow, none of these events is likely to lead to catastrophic failure of the remedial system, but such events could lead to contaminant releases from the target capture zone if the failure is not identified and remedied. In summary, at sites where contamination remains in place, an evaluation of potential events that could lead to a failure of the long-term management approach should be performed and contingency plans developed.

IMPLICATIONS OF THE LONG-TERM NEED FOR INSTITUTIONAL CONTROLS

At every site where contaminants will be left in place (for any significant length of time), institutional controls are necessary to prevent the exposure of local residents to chemicals in groundwater and soil. At groundwater sites, institutional controls play three roles. First, they can restrict the use of contaminated groundwater. Second, they can protect the occupants of overlying buildings (or proposed buildings) from exposure to chemicals from contaminated groundwater through vapor intrusion (e.g., by requiring systems and barriers to prevent vapor from entering buildings). And third, they can prevent activities that might compromise remedies, such as penetration of landfill caps where the landfill is a source of groundwater contamination or pumping that is likely to spread contamination. If properly implemented and enforced, institutional controls allow a groundwater remedy to be protective in cases where residual contamination remains above unrestricted use level. From 1986 to 1996, 3 to 20 percent of groundwater remedies at Superfund facilities had institutional controls. However, by 2008, 93 percent of the groundwater remedies selected that year included institutional controls (EPA, 2010a) and current guidance is likely to require such controls at every groundwater contamination site.

Types of Institutional Controls

Institutional controls (ICs) are administrative and/or legal controls that minimize the potential for human exposure to contamination and/or protect the integrity of a remedy, generally by attempting to modify human behavior. For example, *proprietary controls* represent a private agreement between the current property owner and, in this situation, EPA, a state, or a federal agency that has transferred or plans to transfer property that has use restrictions. The control is generally authorized by state law. An easement or restrictive covenant prohibiting the extraction of groundwater for drinking water on property containing the contaminated groundwater plume is an example of this type of instrument.

There are also *direct governmental controls* on the use of property, such as zoning laws, building codes, or state, tribal, or local groundwater use regulations. Federal agencies such as the Army may possess the authority to enforce institutional controls on their property, e.g., in Base Master Plans, facility construction review processes, facility digging permit systems, and/or the facility well permitting systems.

The third category of institutional controls are *components of enforcement instruments* or permits issued by federal or state regulators to private or federal PRPs (e.g., administrative orders, permits, Federal Facility Agreements, and judicial consent decrees). These legally enforceable instruments may limit site activities or require the performance of specific activities like the monitoring of IC effectiveness.

Finally, there are *informational devices* such as recording site cleanup documents in property records and providing advisories to local communities, tourists, recreational users, or other interested persons that residual contamination remains on-site. Although informational devices are not enforceable, they may be required by an enforceable consent decree or other enforceable instrument.

Each type of institutional control has advantages and disadvantages, which revolve around, for example, how the control enables or restricts future economic development, whether the control is enforceable, and at which level of government it is enforced (e.g., zoning is traditionally a function of local government and generally, EPA and federal agencies have little or no direct role in local zoning). Different institutional controls differ with respect to who pays to maintain and enforce the control. At CERCLA-funded cleanups, EPA does not pay for monitoring or enforcing institutional controls because the statute requires states to ensure the payment of all future routine operations and monitoring following CERCLA-financed remedial actions. However, at sites where private companies or other federal agencies perform the cleanup, they, not the states, pay for monitoring or enforcing institutional controls (see discussion below). The degree to which the public is involved in establishing, monitoring, and ensuring that institutional controls are enforced differs by type (EPA, 2010b), as does the length of time over which the institutional control must be maintained.

Past Experience with Institutional Controls

Not surprisingly, past experience suggests that institutional controls have been effective at some sites and have failed other sites (ELI, 1999, and see Box 5-1 for three prominent failures). Institutional controls "rely heavily on humans to implement, oversee, and administer them" and it is human "to ignore tasks that no one else seems to care about or where the purpose is not readily apparent and indeed is often buried underground" (ELI, 1999). A specific problem is the fact that zoning requirements can be modified by political bodies (ELI, 1999; Spina, 2008). Furthermore, environmental regulatory agencies may not be able to enforce restrictions on subsequent property owners (Spina, 2008; Probst, 2006), although increasingly states have adopted statutes that allow enforcement of land use restrictions on subsequent owners. Finally, where EPA does not regularly consult with local authorities about institutional controls, remedies may be selected, including a specific institutional control, without determining whether it can be implemented by the local government (ELI, 1999; Probst, 2006).

The New Emphasis and Direction on Institutional Controls

EPA has substantially improved its process of developing, implementing, and enforcing institutional controls. Each Superfund facility is supposed to have an Institutional Control Implementation and Assurance Plan (ICIAP) "prior to, or at the same time as, the remedial design phase under CERCLA and finalize it with the completion of the response action" (EPA, 2010b), and coordination between states, tribes, and local land use planning jurisdictions is required. Institutional controls at "construction-completion" sites have begun to be recorded within the Superfund Enterprise Management System to help ensure the long-term effectiveness of the controls (EPA, 2011a).

EPA has clarified that institutional control documents and instruments should clearly articulate the substantive restrictions that are needed at a

BOX 5-1
Examples of the Failure of Institutional Controls

At Love Canal (one the first hazardous waste sites of general public concern), the City of Niagara Falls built a school on a landfill in 1954, even though there was a 1953 deed from a chemical company to the city (i.e., both a proprietary and informational IC) disclosing that chemical production waste was buried on the property and disclaiming responsibility for any injuries that might result (Technical Review Committee, 1988). In the mid-1970s after residential housing was built around the landfill, heavy rain caused the groundwater to mobilize and release the chemicals onto residential properties and into local storm sewers, resulting in the first Presidential Declaration of a man-made national disaster (Technical Review Committee, 1988).

At the Cannons Engineering Corporation Superfund facility in Bridgewater, MA, the ROD required that a Declaration of Restrictions (i.e., a proprietary IC) be recorded with the deeds to the affected properties, along with zoning ordinances (a direct governmental control) and public education programs (an informational IC) (ELI, 1999). In 1998, a company, without prior approval of the environmental agencies, excavated soil below the water table, dewatered the excavation, and discharged the water on the property while erecting a telecommunications relay tower, in violation of the Declaration of Restrictions (EPA, 2010c). EPA issued a written notice of violation of the deed restriction to the property owner, lessee, and the Town of Bridgewater. In response, the leases and subleases have been modified (EPA, 2010c). In addition, the Town of Bridgewater has incorporated the deed restriction and the requirement to notify EPA prior to work at the tower into its site plan approval process. The deed restriction currently remains in place and there have been no additional violations. The education program apparently was never carried out because of lack of public interest (ELI, 1999).

At the Sharon Steel Superfund facility in Midvale, Utah, the ICs included (1) regulations governing excavations on private property within a residential area where some contaminated soils were left in place (a governmental control) and (2) education programs (ELI, 1999). The education programs were not successful, in large part due to lack of cooperation between the city, state, and EPA (ELI, 1999). As a result, one property owner who did not know about the ordinance began unpermitted construction of a new sewer line, another property owner removed his patio exposing unremediated soils for a day and half until the City learned of the activity, and another property owner and the state Department of Transportation failed to coordinate with state environmental regulators concerning the excavation of a city right of way (ELI, 1999).

property to ensure that the land use assumptions that were made as part of the remedy decision continue to remain accurate (EPA, 2011b). Where residential properties are located over a contaminated groundwater plume and the properties are not the source of contamination, well drilling restrictions may be put in place to limit the use of groundwater rather than negotiating covenants or easements with a large number of parties (EPA, 2010b). EPA (2010b, 2011b) requires that each institutional control instrument be reviewed annually to consider such things as their long-term effectiveness and enforceability, and whether the property owner/lessee is aware of and complying with the institutional controls when they change land uses, perform new construction, or transfer the property.

Costs

EPA recognizes that institutional controls, maintenance, and enforcement costs "may extend beyond the 30-year period traditionally used in many response cost calculations," and that these continuing costs should be acknowledged when developing response cost estimates because they "can be important in evaluating long-term effectiveness" (EPA, 2010b). Indeed, the IC development process should begin with estimating the cost for monitoring and reporting activities over the full life cycle of the control.

At Superfund-financed sites (i.e., those without viable PRPs), EPA does not pay for monitoring or enforcing institutional controls because CERCLA Section 104(c)(3) requires *states* to ensure the payment of all future routine operations and monitoring following remedial actions. At sites where there are viable PRPs or federal RPs, EPA has long negotiated settlement agreements or consent orders with such parties, and where necessary obtained a court order, to require a PRP to perform work necessary to achieve and maintain performance standards or the effectiveness of the remedy (e.g., five-year review, additional remedy work, and/or new information or unknown condition reopener consent decree) (EPA, 2006). Recent EPA guidance explicitly directs EPA staff to have the settling parties in such settlement agreements or consent orders gather and submit data and analyses about institutional controls in conjunction with requests for monitoring data (EPA, 2011b). Additionally, EPA now recommends the use of direct payments from PRPs, settling party trust funds, surety bonds, letters of credit, insurance, and settlement proceeds to fund site-specific accounts for institutional controls (EPA, 2010b). Federal agencies, including DoD, generally pay for long-term monitoring and perform oversight of institutional controls at their sites (DoD, 2001).

* * *

EPA has improved its institutional control program so that it encourages cooperation among federal, state, and local governments; incorporates independent oversight of the entities that implement institutional controls; includes redundancy; mandates monitoring; and increases the amount of information available to the public about controls that are in place and the public's involvement in oversight. Because this guidance is new, there is limited information about the effectiveness of current institutional control efforts and their costs, and there is no documentation of injury being caused by the failure of institutional controls. Conceptually one can predict that the cost of institutional controls will increase substantially under the new guidance, particularly as settling and responsible parties pay local or state governments to oversee and enforce institutional controls.

The primary risk from a failure to establish or enforce an adequate set of institutional controls are that the public (residents, workers, and visitors) may be exposed unknowingly to contaminated groundwater or vapor intrusion at levels above those allowing unlimited use and unrestricted exposure. For example, individuals may drill wells for potable use in contaminated aquifers. Developers may perform construction in contaminated soil or water that results in worker exposure. The failure of institutional controls may lead to property damage or personal injury lawsuits. Finally, each failure undermines the credibility of the regulators and local officials with the public, which is likely to make long-term management of the site more difficult and expensive—costs that could be avoided if contaminants are not left in place. It is for this reason that DoD guidance explicitly requires that "where a use restriction will be imposed through the environmental restoration process" the DoD must "ensure that the evaluation of response alternatives includes an analysis of an alternative with a use restriction, as well as an analysis . . . of a response *not* requiring a use restriction" (DERP, 2001).

EMERGENCE OF UNREGULATED AND UNANTICIPATED CONTAMINANTS

Leaving contamination in place can become problematic when the contamination includes compounds that were previously unregulated or unanticipated or when the toxicity information for known contaminants changes. Conventional contaminants of concern have been known for some time, are widespread at many sites, are generally detected at elevated concentrations, and can be readily detected and measured using standard laboratory instruments. These include many organics (e.g., volatile organic compounds [VOCs], semi-volatile organic compounds [SVOCs]), inorganics (e.g., metals, explosives, propellants), and radionuclides. Contaminants of emerging concern are chemical substances that have been identified recently or that have been known for some time, but for which it has previously been thought that action was not necessary. It could be that they are now detected because of advances in analytical techniques (e.g., detectable to parts per trillion, nanograms per liter [ng/L]) or that a newly discovered adverse impact on humans or the environment has been identified. An example is perchlorate, where after 1997 the lowering of the detection limit led to the recognition that this rocket fuel oxidizer had been broadly released to the environment at sites of aerospace manufacture and testing, both initially in California and thereafter throughout the nation. Subsequent studies documented the presence of perchlorate associated with blasting and fertilizer use, as well as naturally high background concentrations (ITRC, 2005b). Contaminants of emerging concern in groundwater, discussed individually in Box 5-2, may present challenges not only from a risk assessment perspective but also from the perspective of remedial technologies. Existing remedial technologies may be inadequate for these compounds, necessitating either adaptation of existing methods or development of new methods.

Under the Safe Drinking Water Act (SDWA), EPA develops a list of contaminants of potential human health concern from drinking water exposure called the Contaminant Candidate List (CCL). The most recent list (EPA, 2009b) contains 104 chemicals or chemical groups and 12 microbiological contaminants including pesticides, antibiotics, and other pharmaceuticals or their degradation products, disinfection by-products, industrial chemicals, and waterborne pathogens that are not currently regulated under national primary drinking water standards. Not all contaminants on the list are likely regulatory targets for drinking water limits; i.e., MCLs are expected to be developed for only a small subset of the contaminants on the CCL. Identification of candidates for which an MCL may be appropriate is based on a number of factors including prevalence of the contaminant in drinking water supplies (including both groundwater and surface water sources), the magnitude of risk presented by the contaminant, and the potential for meaningful risk reduction through additional regulation (NRC, 2001). It should also be emphasized that EPA's decision not to develop an MCL for a chemical does not indicate that a particular chemical will not be of toxicological interest at any particular site. Nonetheless, the CCL can provide a roadmap for site managers with respect to potential site reopeners.

New toxicity information can change the way a contaminant is regulated. In particular, if the cancer potency or non-cancer toxicity value for a chemical changes (even if the MCL does not), sites regulated under CERCLA will have to be evaluated during the five-year review process to make sure the remedy is still protective. One of the most important developments for groundwater remediation is the increase in the cancer potency, and the issuance of a non-cancer reference dose, for tricholorethene or TCE (EPA, 2011h). It is not clear whether these changes will alter groundwater remediation goals for TCE or the MCL of 5 µg/L, but it is certain that EPA will evaluate what, if any, modifications may be appropriate. The decision on changing the MCL will depend on many factors, including the chosen

BOX 5-2
Contaminants of Emerging Concern in Groundwater

By definition, contaminants of emerging concern have the potential to present a risk not previously known. The examples provided below highlight compounds that could potentially be found at subsurface hazardous waste sites and may soon come under more intense scrutiny.

1,4-Dioxane. 1,4-Dioxane, a probable human carcinogen, is a stabilizer added to some chlorinated solvents, and thus is a co-contaminant at sites contaminated with these compounds (primarily 1,1,1-trichloroethane). 1,4-Dioxane is miscible with water and has low volatility, and thus is poorly retained by the solid media in groundwater systems. This makes air stripping and sorption to activated carbon used with pump-and-treat systems ineffective as treatment techniques, although advanced oxidation techniques are effective. Recent work has shown that while 1,4-dioxane is degraded via co-metabolism under specialized conditions, sustained biodegradation using enrichment cultures or aquifer materials under a variety of redox conditions were not successful (Vainberg et al., 2006; Steffan, 2007). Phytoremediation is able to remove 1,4-dioxane, but this would be limited to shallow water tables or wetland systems (Dietz and Schnoor, 2001; Aitchison et al., 2000). Oxidation of 1,4-dioxane using permanganate (Waldemer and Tratnyek, 2006) and in advanced oxidation processes (Bowman et al., 2003; Suh and Mohseni, 2004) effectively destroys 1,4-dioxane, but there is a clearly a need for the development of additional *in situ* treatment techniques.

A comprehensive review of the environmental issues surrounding this compound is available (Mohr, 2010). Recent increases in the frequency of detection of this chemical in municipal wells in California suggest that 1,4-dioxane is a chemical that may lead to many reopeners at hazardous waste sites where 1,1,1-TCA was released.

Naphthalene. Naphthalene, a polycyclic aromatic hydrocarbon consisting of two benzene rings, has been found in groundwater, particularly associated with coal tar production and distillation from former manufactured gas plan (MGP) operations. Although naphthalene is rarely found in drinking water supplies, at some MGP sites concentrations of naphthalene in the thousands of µg/L have been found (ATSDR, 2005), creating the potential for vapor intrusion given the high volatility of naphthalene. There is currently no MCL for naphthalene.

Although an EPA assessment in 2004 concluded that naphthalene was likely to be carcinogenic in humans via the inhalation pathway (EPA, 2004b), that assessment is no longer considered up to date and there are ongoing analyses within the agency and by outside scientists (e.g., Rhomberg et al., 2011). Nonetheless, the California EPA has designated naphthalene a carcinogen via the inhalation pathway and concluded that, at least at one site, cancer risks from naphthalene were comparable to those from benzene (Christopher et al., 2005). The recent toxicology findings for naphthalene, combined with increasing interest in the vapor intrusion pathway, provide an example of how emerging contaminants of concern may arise.

N-nitrosodimethylamine (NDMA). NDMA is a potential carcinogen and known to cause liver damage in humans and animals (ATSDR, 1989). NDMA pollution of groundwater may arise from two sources: manufacture of rocket propellants and recharge of disinfected wastewater. While numerous studies have explored advanced oxidation processes to treat NDMA in drinking water (raw or finished) and wastewater, only limited studies have been performed to assess means to treat NDMA in contaminated groundwater. Zero-valent iron is able to reduce NDMA, albeit slowly, and the reaction is accelerated with a nickel-catalyst (Gui et al., 2000; Odziemkowski et al., 2000). There is also evidence of biodegradation at a water recharge site (Zhou et al., 2009). A recent SERDP report found that under oxic conditions, NDMA is mineralized via a co-metabolic processes, and that under anaerobic conditions (either natural or generated via *in situ* redox manipulation), abiotic reduction of NDMA occurred that was mediated by ferrous iron adsorbed to the aquifer material (Szecsody et al., 2009). Depending on the scope of the NDMA contamination problem, additional work is necessary to identify potential treatment techniques for NDMA-contaminated sites.

Perchlorate. Perchlorate (ClO_4^-) contamination of groundwater is a result of its use as an oxidizer in solid propellants, explosives, and pyrotechnics. Compared to the other contaminants in this section, there has been substantially more research on the treatment of perchlorate contaminated groundwater. Both laboratory studies and a field demonstration (Naval Surface Warfare Center at Indian Head, MD) have shown that *in situ* biostimulation using low-cost carbon amendments (e.g., lactate, acetate, molasses, canola oil) is an effective means to treat perchlorate (GeoSyntec Consultants, 2002; Cramer et al., 2004). The bacteria responsible have been characterized in terms of their biochemistry and genetics (Coates and Achenbach, 2004, 2006). Biostimulation by plant-produced electron donors has also been demonstrated (Shrout et al., 2006), and a permeable barrier comprised of wood particles designed to allow bacteria to create reducing conditions also lead to removal of perchlorate (Robertson et al., 2007).

The potential of zero-valent iron to reduce perchlorate has been demonstrated in laboratory studies (Huang and Sorial, 2007), but the presence of other anions slows or inhibits perchlorate removal. Results from systems containing both bacteria and zero-valent iron have suggested that the combination leads to improved removal (Son et al., 2006; Yu et al., 2006), while another study suggests that iron metal inhibits bacterial removal (Shrout et al., 2005).

EPA has decided to regulate perchlorate pursuant to the Safe Drinking Water Act due to its effect on thyroid function (EPA, 2011c).

Perfluoroalkyl substances (PFAS). PFAS are a family of organic compounds that consist of a perfluorocarbon chain (typically 4 to 14 in length) and an anionic head group (Simcik, 2005; Lau et al., 2007; Ranye and Forest, 2009). They have been synthesized and used in a broad range of industrial and commercial applications for over 50 years, but their environmental risk was not highlighted until a decade ago (Simcik, 2005; Fujii et al., 2007; Richardson and Ternes, 2005). Up until 2002, the total historical production of PFAS reached approximately 100,000 tons, of which the majority will ultimately gain their entry into the environment (Pistochi and Loos, 2009). PFAS are also components of aqueous film-forming foams (AFFFs) that are used in fire fighting. Training with AFFFs has been conducted for more than 30 years, resulting in releases in fire fighting training areas, including military bases. A recent survey of the Department of Defense Knowledge Based Corporate Reporting System found 594 fire and crash training sites operated or managed by military organizations. It was found that 353 of these sites were

continued

BOX 5-2 Continued

operated by the Air Force, the Navy operated 132, and the Army had 94 (Rak and Vogel, 2009). The rest were under the purview of the Formerly Used Defense Sites. Any military site where AFFFs have been used to fight fires or to train in firefighting may have a dissolved plume containing PFAS.

In January 2009, the EPA announced Provisional Health Advisory values for perfluorooctanoic acid (PFOA) and perfluorooctane sulfonate (PFOS) of 0.4 μg/L in drinking water to protect humans from a variety of noncarcinogenic health effects (EPA, 2009a). To date, only pump-and-treat systems using activated carbon as a sorbent are effective at removing these compounds. The stability of the C-F bond makes chemical and biological treatment difficult. There is clearly a need to develop new techniques to treat groundwater contaminated with this class of chemicals.

Pharmaceuticals and Personal Care Products (PPCP). This class of compounds includes prescription and non-prescription medications, antibiotics, hormones, and others. Recently 35 different PPCP compounds were detected in groundwater samples (Barnes et al., 2008). Possible pathways to groundwater include landfill leachate (Buszka et al., 2009); indirect or direct aquifer recharge with wastewater effluent (Kreuzinger et al., 2004; Rabiet, et al., 2006); irrigation with wastewater effluent; leaking sewer pipes (Nakada et al., 2008); application of biosolids to soils; leakage from septic tanks (Standley et al., 2008; Carrara et al., 2008); or leakage from animal waste lagoons at confined animal feeding operations—suggesting that the detection of these compounds at hazardous waste sites is unlikely.

There is evidence of compound attenuation in aquifer recharge systems (Drewes et al., 2003; Snyder et al., 2004; Greskowiak et al., 2006; Soliman et al., 2007) and also of compound persistence (Drewes et al., 2003; Kreuzinger et al., 2004). Further studies are needed to determine if the environmental levels of pharmaceuticals and personal care products present a significant human or environmental risk. Any evaluation of the risks presented by such releases to the environment should consider the advantages and disadvantages of (1) limiting their entrance into groundwater systems through improved treatment prior to dis-

acceptable risk level, whether the cancer or non-cancer risk is considered, the detection limit, costs, and technical feasibility. The present MCL of 5 μg/L is now associated with a 10^{-5} risk over a 70-year exposure, which falls within EPA's acceptable risk range of 10^{-6} to 10^{-4}. The non-cancer reference dose has led to a regional screening level for TCE in drinking water of 2.6 μg/L[1] (although it should be noted that the screening level is not a cleanup goal, and site-specific factors could result in a non-cancer TCE cleanup goal that exceeds 2.6 μg/L).

[1] http://www.epa.gov/reg3hwmd/risk/human/rb-concentration_table/.

charge; (2) verifying that natural attenuation processes in surface soils degrade them to acceptable levels; or (3) verifying removal during drinking water treatment.

Nanomaterials. Nanomaterials are between 1 and 100 nm in size and are used widely in personal care products, clothing, and electronics. However, the issue of importance to this report is the use, at more than 36 sites, of nanomaterials to remediate environmental contaminants such as organohalides, trinitrotoluene, and phenols (EPA, 2011d,e; Obare and Meyer, 2005). For example, nanomaterials can serve as semiconductors in the photocatalysis of chlorinated phenols, and sorbents developed from nanomaterials can be used to remove arsenic and cadmium from water (Vaseashta et al., 2006; Husain, 2008). However, little is known about the human or environmental health effects of nanomaterials, although they are increasingly used in consumer applications.

EPA regulates new chemicals prior to widespread commercial use by requiring a pre-manufacturing review pursuant to the Toxic Substance Control Act. Regulation is based on a finding that the chemical may present an "unreasonable risk." Historically EPA did "not consider a nanoscale version of a chemical to be a different chemical substance from the macroscale version," i.e., it was not a new chemical subject to pre-manufacturing review. This interpretation, however, has changed, and EPA issued significant new use notices requiring reporting and testing of siloxane-modified silica nanoparticles, carbon nanotubes, and other nanomoterials (EPA 2008b,c, 2011e,f,g,). These actions include requiring testing to assess inhalation risks, ecotoxicity, and aquatic toxicity. EPA also is regulating nanomaterials in pesticides through the Federal Insecticide, Fungicide, and Rodenticide Act.

Siloxanes. Siloxanes are organosilicon chemicals (containing an R_2SiO unit, where R can be hydrogen or a hydrocarbon group). Chemicals with multiple R_2SiO units are called polysiloxanes. Siloxanes have high vapor pressure, low water solubility, and high Henry's law constant (McBean, 2008), and are commonly found in personal care products such as shampoo and antiperspirant. Thus, they are unlikely to be found at hazardous waste sites. The toxicity of siloxanes is variable.

The impact of the new TCE toxicity information on the vapor intrusion pathway is important because of the potential longer exposure periods, although again the nature of this impact is still uncertain. The new cancer potency value yields a 10^{-4} risk at 20 μg/m^3, a 10^{-5} risk at 2 μg/m^3, and a 10^{-6} risk at 0.2 μg/m^3 for a lifetime-year exposure.[2] The new non-cancer reference concentration (RfC) is 2 μg/m^3. Thus, indoor air concentrations

[2]For a 30-year exposure, as is typically used in Superfund, the corresponding concentrations are 41, 4.1, and 0.41 μg/m^3. http://www.epa.gov/reg3hwmd/risk/human/rb-concentration_table/Generic_Tables/pdf/master_sl_table_run_MAY2012.pdf.

of TCE arising from vapor intrusion may fall within the acceptable cancer risk range, yet exceed the non-cancer criterion.

Adding further complexity to the entire TCE situation is the question of the appropriate averaging time for non-cancer risk from TCE. Typically the exposure concentration would be an average over the period of exposure. However, the endpoints on which the RfC is based are developmental. For some developmental toxicants, EPA may find it appropriate to average over a shorter duration, because of concern that a single exposure during a critical developmental period may produce an adverse effect (EPA, 1991a). While there may be scientific debate about the appropriate averaging time, at least one EPA region is considering an averaging time for TCE over a shorter exposure duration (e.g., Lee, 2012) which, if implemented, could have significant ramifications for determining compliance. At a given site, alternative averaging periods should be determined based on specific exposure elements and chemical-specific properties.

Going forward, any review of the protectiveness of a TCE-related remedy at a Superfund facility during a five-year review is likely to consider both cancer and non-cancer risks using these new TCE toxicity criteria. EPA may conclude that some remedies meeting the current MCL are no longer protective (e.g., if the long-term concentration is above the non-cancer screening level of 2.6 μg/L). At sites where the cleanup level has not yet been set, EPA may choose a goal more stringent than the current MCL of 5 μg/L because the non-cancer level (2.6 μg/L) is lower. Because of its prevalence in contaminated groundwater, the potential reduction in the MCL for TCE has been estimated by some to have a major impact on the costs of remedial actions at DoD sites, on the order of $15 billion (Yonkers, 2012). However, there is much debate about the likely costs, and the Committee has insufficient information to reach a conclusion concerning the magnitude of the impact.

Hexavalent chromium Cr (VI) has been detected in groundwater at sites associated with waste disposal. For example, ATSDR (2008) describes several examples of sites in the United States with elevated groundwater concentrations of Cr (VI), with some concentrations exceeding 10,000 μg/L; the sources of the contamination were variable and included disposal of chromate used as a corrosion inhibitor and wastes from aircraft plants. Although inhaled Cr (VI) is clearly carcinogenic, studies of ingested Cr (VI) in humans have been conflicting, and overall, do not confirm Cr (VI) as a human carcinogen (e.g., Beaumont et al., 2008; Kerger et al., 2009). However, recent studies of high concentrations of Cr (VI) in drinking water in rodents provide evidence of carcinogenicity in the oral cavity and gastrointestinal tract (NTP, 2008). The significance of such findings for humans exposed to Cr (VI) in drinking water, albeit at much lower concentrations than in the rodent studies, is being considered. California and New Jersey have already set advisory levels, and EPA is assessing whether to set an MCL for Cr (VI). EPA has decided to require monitoring for Cr (VI) in drinking water, even in the absence of an MCL (LaRoss, 2012). [Interestingly, recent mode-of-action studies suggest that the dose-response relationship may be nonlinear (Thompson et al., 2012), which could have important ramifications for setting an MCL.]

New quantitative dose-response analyses have elevated the importance of inorganic arsenic. In 2001 the drinking water limit for inorganic arsenic was lowered from 50 µg/L to 10 µg/L based on evidence that inorganic arsenic could cause internal cancers in addition to skin cancer. The MCL analysis employed a dose-response assessment in which the potency was effectively equivalent to 3.7 mg/kg/d based on bladder cancer, *versus* the prior limit of 1.5 mg/kg/d based on skin cancer. The reduction in the arsenic MCL from 50 µg/L to 10 µg/L has already resulted in a remedy reopener at the Arsenic Trioxide Site in North Dakota, as a result of the determination that the drinking water limit of 50 ppb was no longer protective of public health.[3] Sampling determined that 84 percent of the well water users in the 26 townships were drinking water containing arsenic concentration at or over the arsenic MCL of 10 ppb (EPA, 2010d). A number of new actions were taken, including supplying some residents with bottled water, connecting other residents to a rural water supply system, and enhancing the water treatment system. EPA is continuing to evaluate the carcinogenicity of inorganic arsenic, which could have further implications for the MCL and determinations of the protectiveness of remedies.

It should be noted that, at times, new understandings of chemical toxicity can indicate lower risk values than had previously been established. This is the case with tetrachloroethene (PCE), where the recent cancer toxicity criteria are well below (i.e., indicating less potency) the prior values.[4] Specifically, the oral cancer slope factor is more than 200-fold lower and the inhalation cancer risk value is more than 20-fold lower than prior values. Assuming a 30-year exposure duration, the air concentration for 10^{-6} risk would be 9.4 µg/m^3, and the corresponding tap water concentration (considering ingestion, inhalation, and dermal exposure) would be 9.7 µg/L (above the MCL of 5 µg/L). While it is highly unlikely that the MCL would be increased, these findings could lead to reduced focus on vapor intrusion of PCE.

Thus, contaminants of emerging concern may be new, previously unrecognized contaminants, or previously recognized contaminants where new toxicity understanding or new exposure limits have developed, leading to concerns at lower concentrations. Contaminants of emerging concern

[3]See http://www.epa.gov/region8/superfund/nd/arsenic/index.html.

[4]See http://www.epa.gov/iris/subst/0106.htm.

present challenges for a number of reasons, including the lack of generally accepted toxicological criteria that can be used for risk assessment purposes. The absence of such criteria complicates the determination of health-protective limits and can result in different limits in different regions. Such contaminants also add uncertainty to risk management decisions, leaving open the possibility of future site reopeners. The Department of Defense's program for contaminants of emerging concern, called "Scan-Watch-Action" (Murnyak et al., 2011) employs a multi-tier process to identify contaminants and weigh them in terms of their likelihood of having an impact on multiple DoD functional areas, such as acquisition, training and readiness, and remediation. For the subset of contaminants of emerging concern that could have an impact on these key functional areas, risk management options are developed.

NEW PATHWAYS/RECEPTORS

In addition to identification of new constituents of potential toxicological concern as well as development of new toxicity information for "old" chemicals, new exposure pathways may be identified or recognized as more significant than had previously been considered. Even when the existence of a pathway is recognized, there may be insufficient information as to the parameters that influence that pathway and hence the lack of clearly articulated guidance for how to incorporate the pathway in a risk assessment. These deficiencies can result in failure to adequately incorporate the pathway into site-specific risk analyses or result in significant uncertainty in calculated risk results. New pathways raise institutional and public health concerns that need to be considered in remedial decision making.

Vapor Intrusion

Although not "new" in that it has been increasingly recognized over the last ten tears, the vapor intrusion pathway is of particular interest with respect to subsurface contamination. The sources of these vapors may be free phase NAPLs entrapped in the vadose zone (from a spill and/or exposed during water table fluctuations) or dissolved constituents that partition from the groundwater plume to soil gas. Vapors emitted from these contaminant sources disperse into air-filled void spaces within the soil and migrate below surface structures, leading to the intrusion of contaminant vapors into indoor air. The exposure can be substantial even at low indoor air contaminant concentrations because the average American spends more than 21 hours per day indoors (Olson and Corosi, 2002). Note that there is the potential for volatile organic compounds from coal- and oil-derived DNAPL sites to present a risk via the vapor intrusion pathway (see Singh et al., 2011, and Doroski et al., 2011). However because such investigations are ongoing with no definitive results available, the following discussion focuses on chlorinated solvents.

It is unknown how many chlorinated solvent sites in the United States have vapor intrusion levels considered by federal or state policies to be unsafe. Thousands of dry cleaners in the United States, many of which have long been out of business, may have generated plumes of PCE in the subsurface, but it is unknown what percentage of such sites is contaminated because no federal or state regulatory program requires their investigation. Several documented cases of vapor intrusion involving chlorinated solvents and their breakdown products, such as TCE and DCE, have received national attention. For example, the Redfield and Colorado Department of Transportation sites in Colorado were two of the first major sites to bring the vapor intrusion issue to the regulators' and public attention in the late 1990s. At the Redfield site, the indoor air in about half of the 800 single-family homes overlying a dissolved groundwater plume of VOCs exceeded 0.49 $\mu g/m^3$ (Folkes et al., 2009). Hill Air Force Base, Utah, was one of the first DoD sites to address the vapor intrusion pathway (with details provided by Erik Dettenmaier, Hill AFB, personal communication). There are approximately 3,000 residences overlying dilute chlorinated solvent groundwater plumes outside the base, and about half agreed to participate in an indoor air monitoring program. Roughly 4,500 indoor air samples have been collected, with TCE being detected in 235 homes. 123 homes have indoor air concentrations in excess of 2.3 $\mu g/m^3$, which correlates with a cancer risk of approximately 5×10^{-6} over 30 years (EPA, 2012a). Of the homes with concentrations above 2.3 $\mu g/m^3$, about 25 percent are suspected to be affected by indoor sources of TCE (such as gun cleaner, craft glue, and cleaning products). Sub-slab depressurization systems have been installed at homes where vapor intrusion is known or suspected to be occurring. As a result of this site, the DoD has gained considerable experience over the past decade with designing indoor air monitoring programs, managing community relations, and interpreting and communicating data on vapor intrusion. Box 5-3 describes how EPA and the states have altered cleanup programs to consider the vapor intrusion pathway.

Vapor intrusion investigations are generally required anywhere chlorinated solvents are detected in shallow aquifers or the vadose zone. Multiple lines of evidence, such as sampling of indoor air, outdoor air, and soil gas, are needed to determine the extent of indoor air contamination and its source. Some jurisdictions, such as New York State, require action based solely upon high levels of contamination in the soil gas under a building, recognizing the potential for intrusion should a crack or hole develop in the building's slab; others base responses upon current indoor air levels.

Although the recognition of vapor intrusion by EPA is at least ten years

BOX 5-3
Consideration of Vapor Intrusion in Federal and State Regulatory Programs

Contaminated sites were rarely evaluated for vapor intrusion until 2002, when the *Denver Post* published a landmark series of articles on the subject (Obmascik, 2002) documenting the Redfield site, the Colorado Department of Transportation site, and other sites in Colorado. At that time, EPA's RCRA program had issued a draft vapor intrusion pathway assessment guidance (EPA, 2001a) as part of its Environmental Indicator program, but awareness of vapor intrusion was only just growing. In the wake of the *Post* series, EPA's Office of Solid Waste and Emergency Response issued *Draft Guidance for Evaluation the Vapor Intrusion to Indoor Air Pathway from Groundwater and Soils* (EPA, 2002a), and several states followed suit. The EPA Region 9 Superfund division systematically reviewed their portfolio of sites for potential vapor intrusion. The New York Department of Environmental Conservation conducted a desktop review of 421 inactive sites with pre-2003 remedial decisions where disposal of chlorinated hydrocarbons occurred (NYS DEC, 2005), reopening numerous "closed sites" for investigation of vapor intrusion. However, in other jurisdictions (such as Texas, Michigan, and EPA Regions 4 and 6), regulators have been reluctant to reopen sites or broaden investigations to assess the vapor intrusion pathway. Finally, certain other jurisdictions have dealt with the new exposure pathway by issuing action levels (target concentration levels) that resulted in no further action.

There are a substantial number of sites where interim remediation and ongoing investigation have been under way, but where the vapor intrusion pathway was not considered. For example, the Motorola 52nd Street semiconductor factory in Phoenix, Arizona, was placed on the NPL in 1984. There is continuing investigation of the high levels of TCE in the deep bedrock under the neighborhood immediately downgradient from the plant, but Arizona's Department of Environmental Quality—lead agency for that operable unit—has not evaluated potential vapor intrusion under the moderate income, predominantly Latino neighborhood, despite repeated requests from community leaders. Only in 2010 did EPA Region 9 announce a vapor intrusion investigation and form a Community Information Group to oversee it (EPA Region 9, 2010).

Just recently EPA has initiated a process whereby vapor intrusion may become a component of the NPL listing process (EPA, 2011i).

old, the magnitude of vapor intrusion as a national issue remains poorly understood. This is partly because of significant uncertainty in understanding the vapor intrusion pathway. That is, contaminant concentrations in indoor air are variously determined by concentrations in the subsurface, the subsurface geology, environmental conditions (such as wind speed and temperature), and building characteristics (e.g., presence or absence of a basement, foundation type and condition, heating/cooling). In addition, background sources such as ambient (outdoor) air and consumer products may contribute to the indoor air levels of contaminants of concern. Chapter 6 discusses both the current limits in our understanding of vapor intrusion as well as advances in real-time sensor technologies that will make it easier to detect indoor air contamination from subsurface sources and determine the best mitigation strategies.

EPA has compiled an empirical data set for hazardous waste sites with paired groundwater, soil gas, and indoor air data on chlorinated solvents (EPA, 2012b). The raw data demonstrate that there are sites where there is reasonable potential for exceeding indoor air screening levels and, depending upon regulators' final risk management decisions, the non-cancer risk management level in particular. Furthermore, detailed field studies under way in residences overlying dilute plumes suggest that the contaminant intrusion rate may be highly variable with time and the most significant contributions to longer-term exposure might happen through infrequent and short-duration periods of high vapor intrusion activity (Johnson et al., 2012). Thus, there are two choices for dealing with a possible vapor intrusion pathway at a given site: (1) invest in sampling and analyses to confirm whether or not the potential exposure is of concern, or (2) install a vapor mitigation system. The former is problematic because the number of samples likely needed to be confident in pathway assessment evaluation may not be practicable, and costs could easily exceed $10,000 per building when considering sampling spread out over multiple seasons and years (DoD, 2009). The latter also is challenging because installation costs are likely to be $2,000 or greater per residential mitigation system with annual electrical costs of $100 or more (EPA, 2008d; ITRC, 2007), and there are uncertainties surrounding continued operation and performance validation of mitigation systems. These issues are magnified when considering larger plumes with tens to hundreds of overlying buildings.

The current methods for disruption of the vapor intrusion pathway are based on substructure depressurization systems originally developed for residential radon mitigation (EPA, 2012c). In fact, one of the benefits of vapor mitigation is that it also prevents radon exposure (which, in certain regions of the United States, can pose a greater cancer risk than the levels of solvent vapors that require mitigation). Some local governments, such as Mountain View, California, require the installation of vapor mitigation systems in developments on or near solvent contamination plumes (Woodhouse, 2009). New Jersey requires Radon Resistant Construction for all homes and schools in high-radon areas.[5] In general, though, state and local governments rarely require that VOC mitigation systems be installed in existing buildings in the absence of an environmental investigation.

[5] See http://www.state.nj.us/dca/divisions/codes/codreg/pdf_regs/njac_5_23_10.pdf.

Mitigation techniques usually consist of perforated plastic pipes installed in the subsurface that are linked to venting systems propelled by electric blower fans. These must be operated, maintained, and monitored to ensure that they are working properly over the long term. Installing vapor mitigation systems is easiest and most inexpensive during initial building construction. To retrofit existing buildings, it is usually necessary to drill holes through floors and run vent pipes through rooms, while for new construction, horizontal pipes and plastic or rubber vapor barriers can be laid beneath foundations, with venting systems hidden, for a fraction of the cost of a retrofit. Some regulatory agencies suggest installing systems without fans, and then testing the indoor air after construction but before occupancy, at least where the threat of vapor intrusion is uncertain. If sampling shows elevated levels of VOCs indoors, fans are then installed and operated (e.g., see CalEPA, 2011). There are also commercially available passive barriers that mitigate vapor intrusion, including concrete sealants for industrial structures.

At many vapor intrusion sites involving residential property owners, some homeowners refuse to allow access for sampling or they reject mitigation (EPA, 2012c). There are a number of reasons for this, including owners' fear that vapor intrusion will stigmatize their property and drive down its value, or a distrust of the agencies and companies that are conducting the response. Regulators do not insist that non-cooperating homeowners take part in investigations in the belief that they have a right to refuse protection. This presents challenges when property is sold or leased because buyers or occupants may not have the opportunity to have mitigation or even sampling conducted. They may be unaware that a vapor intrusion response has taken place in the area. EPA (2012c) addresses steps to be taken if the property owner refuses access, ranging from providing information on the consequences to requiring access "in the interest of protecting the occupants, for evaluating the need for response, choosing a response action, taking a response action, or otherwise enforcing CERCLA."

It is difficult to enforce proper maintenance of the mitigation system over the life of the contamination and building. Building occupants may not understand why or how the systems work or otherwise fail to maintain them over the long run; some building occupants simply unplug them because of the energy cost or noise. Furthermore, there is often no protection for new occupants of buildings previously owned or occupied by people who chose not to cooperate with the vapor intrusion response. Therefore, regulatory agencies with active vapor intrusion programs consider mitigation an interim response (NJ DEP, 2005).

LITIGATION RISKS

Leaving residual contamination in place above unrestricted groundwater and land use levels has several important legal ramifications, as discussed in detail below. The Committee recognizes that litigation is contentious and, by definition, there are at least two diametrically opposed versions of the facts and views of the law. The purpose of this section is not to argue that litigation will occur in every case, but rather that the potential is likely to exist as long as chemicals remain in the groundwater above drinking water levels. The reader should not consider the examples and illustrations given to be a comprehensive review.

Environmental Liability Disclosure Requirements

Leaving contaminants in place may trigger a variety of public notification and/or disclosure requirements, some of which are illustrated below. The Securities and Exchange Commission (SEC) requires companies to disclose, among other things, (1) administrative or judicial environmental protection proceedings arising under any federal, state, or local law, including hazardous waste cleanup laws; (2) when it is probable that an asset is impaired; and (3) trends in cleanup liabilities that are reasonably likely to affect its liquidity or its capital expenditures (SEC, 1982, 2011). Similarly, the Financial Accounting Standards Board has published requirements on reporting about environmental remediation liabilities in financial statements (Coate et al., 2012). These requirements exist to satisfy U.S. securities laws and general accounting practices, which are based on the principle that investors should have sufficient information about the financial status of companies to make an informed decision on whether to invest in company A versus company B (Rogers, 2007). The determination of exactly what environmental cleanup liabilities are required to be reported is based on the specific facts, e.g., the size of the company and the uncertainty associated with the cost and liability. The DoD regularly reports its environmental liabilities to Congress in order to provide reliable financial and budgetary information, among other purposes (Army, 2011).

A variety of evaluations have suggested that not all companies were reporting cleanup liabilities (NRC, 1997; SEC, 2003; Cox and Douthett, 2009; EPA, 2008e). SEC (2003) claimed that "many companies did not provide adequate disclosure relating to" environmental liabilities and "urged companies with material contingent liabilities to carefully review their disclosures and ensure that they include all required information." There have been few SEC actions enforcing these requirements in the last two decades (EPA, 2008e).

In summary, companies are required to report long-term hazardous substance cleanup liabilities under certain circumstances. Some believe that such reporting may provide an incentive for private and governmental PRPs to minimize their long-term liabilities by reducing the extent of residual contamination at their sites (NRC, 1997; EPA, 2008e).

Potential Personal Injury and Property Damage Liability

If chemicals remain in soil or groundwater in concentrations above unrestricted use levels, there is a potential that the active remedy, engineering controls, and institutional controls may fail and result in lawsuits alleging personal injury and/or property damage. In fact, even if the remedy does not fail and the government regulators determine the remedy is protective, private citizens may still file a lawsuit seeking alleged damages. [The reader is referred to Federal Judicial Center and NRC (2011) for a discussion of causes of action and burdens of proof.]

Each personal injury or property damage case depends upon the unique facts of the site and state law. Most contaminated groundwater sites do not result in litigation, although in the Committee's opinion there appears to be a trend toward increasing litigation. There is no central repository of toxic tort cases and their outcomes, as most cases are likely to settle out of court rather than be litigated to a verdict, and many settlements are confidential or not readily accessible. Box 5-4 briefly illustrates a few of the major groundwater cases.

There are many defenses to such personal injury and property damage suits, not the least of which is a lack of causation. As a result, there are many cases that are either not brought or where the defendants prevail. Nonetheless, there is a possibility of personal injury or property damage litigation if residual contamination remains (or even if no contamination remains since personal injury lawsuits can be filed for past exposure). The cost of such litigation is highly uncertain, but the transaction costs (even if the defendant prevails) can be substantial. The uncertainty arises from variable transaction costs, difficulties measuring the decrease in property value and the extent of personal injury, and limited information about when groundwater contamination (and hence the damages) began or when it is considered "over." Thus, it is not clear how responsible parties should factor the potential risk of litigation into a particular decision concerning remedy selection at a site.

Potential Liability for Natural Resource Damage Claims

At large or complex sites where chemicals are left above unrestricted use levels, there is the potential for natural resource damage (NRD) claims. CERCLA provides the United States, states, or, in certain circumstances, Indian Tribes (when they are acting as "trustees" of the resources in question) a statutory cause of action for damages, which refers to injury to, destruction of, or loss of natural resources [CERCLA § 107(f)(1)]. The natural resources covered are "land, fish, wildlife, biota, air, water, ground water, drinking water supplies, and other such resources belonging to, managed by, held in trust by, appertaining to, or otherwise controlled by the United States . . . any State, local government, or any foreign government, any Indian tribe . . ." [CERCLA § 101(16)]. Damage caused to such natural resources may result in a federal or state lawsuit to restore the resource or receive compensation for such damage. The legal basis of NRD lawsuits varies from state to state and has evolved over time (Israel, 2009).

NRDs include (a) the reasonable costs of assessing the extent of the damage (e.g., the field studies and reports); (b) "direct and indirect injury, destruction, or loss;" and (c) interim loss of use from the time of the release until the time of restoration (taking into consideration "replacement value, use value, and ability of the ecosystem or resource to recover") [CERCLA § 107(a)(4)(C)]. One of the losses recoverable in such lawsuits is the amount of money required to compensate the public for the loss of use of the resource. This includes both commercial uses of the resource (e.g., supplying drinking water, supplying fish, or a wetland providing flood control) as well as nonuse damages (i.e., a value from the mere existence of a resource even in the absence of use) [43 CFR § 11.83(c)(1)].[6] Indeed, the nonuse damages are explicitly provided for in Department of Interior and National Oceanic and Atmospheric Administration NRD assessment rules, which have been upheld from challenge (*Ohio v. United States Department of the Interior*, 880 F.2d 432, D.C. Cir. 1987). Most direct use values of groundwater can be approximated based on water prices (NRC, 1994); however, there is controversy over whether one can reliably estimate nonuse values.

Many of the larger NRD lawsuits involve contaminated sediment (Commencement Bay in Washington) or oil discharges to surface waters (such as the Exxon-Valdez and Gulf Oil spills) (Hutton and Edwards, 2011) and therefore are not relevant to contaminated groundwater sites. NRD settlements for mining sites, large oil spills, and contaminated sediment tend to be higher than other NRD settlements. Table 5-1 shows several large groundwater NRD settlements or actions to illustrate the types of historic claims that have been made. In summarizing 32 groundwater sites, Ando

---

[6] The clearest example is New Jersey NRD guidance, which states that the damage calculation may be based on the current water rate derived from the NJ Board of Public Utilities (that is, how much it costs a home owner to buy water) times the volume of water (areal extent and thickness) times the duration that the water cannot be used (i.e., the time that it would take for the remedy to reduce the concentration to below the NJ Groundwater Quality standards set for the site).

BOX 5-4
Select Examples of Litigation Involving Contaminated Groundwater Sites

Love Canal. Hooker Chemical Company disposed of approximately 21,000 tons of a wide variety of chlorinated chemicals (including 2,3,7,8-tetrachlorodibenzo-p-dioxin) in an excavation in an undeveloped area of Niagara Falls (called Love Canal) from the 1940s to 1953 (EPA, 1991b). The City required Hooker to transfer the property to the School Board in order to construct a school, which soon acted to attract additional residents to the area. In the early 1970s, unusually heavy rainfall resulted in the migration of chemicals from the original disposal. President Carter declared the area a national disaster area in 1976. In 1979, a group of local citizens sued Hooker Chemical, the city, the county, and the Board of Education seeking $11 billion in damages (University of Buffalo, 2011). 1,336 Love Canal residents settled their personal injury suit for approximately $20 million in one settlement, and in another settlement 899 residents settled for $6.75 million (Center for Justice & Democracy, 2007).

Woburn, MA. A class action lawsuit was brought by 19 residents of Woburn, Massachusetts against two industrial companies for contaminating groundwater with chlorinated compounds (including TCE) alleging that exposure to groundwater caused injury to the local residents because the city used the groundwater as a source of drinking water. The case was settled, without an admission of liability, for $8 million (Center for Justice & Democracy, 2007).

Hinkey, California (the Erin Brockavitch case). Hexavalent chromium that was used by a utility in cooling towers for corrosion control contaminated the groundwater in Hinkley, California, prompting approximately 650 area residents to sue the utility. The utility ultimately agreed, without admitting liability, to pay $333 million to the residents, to clean up the groundwater, and to cease use of hexavalent chromium (Center for Justice & Democracy, 2007). Interestingly, in 2011 the California Desert Sierra Cancer Surveillance Program found "cancer occurrence in the Hinkley Census Tract that is slightly, but not significantly, below the number of new cases expected for an average risk population having the same demographic characteristics as the Hinkley Census Tract population" (Morgan, 2011).

et al. (2004) found the average groundwater NRD settlement to be $15 million as of 2004 (or on average, a half a million dollars per site).

New Jersey is unique in the large volume of NRD cases that have been filed—approximately 123 lawsuits involving over 1,500 sites (NJAG, 2011). The settlements resulted in payments of approximately $72.9 million ($45,617 per site) and thousands of acres of open space preserved as wildlife habitat and groundwater recharge areas (NJ DEP, 2007, 2008), although some of the settlements involved larger payments and the cost of

MTBE Groundwater Class Action. Public drinking water providers filed product liability lawsuits against a number of oil companies concerning contamination of groundwater by methyl tert-butyl ether (MTBE), which leaked from underground storage tanks. MTBE is an oxygenate added to gasoline to meet the requirements of the Clean Air Act. In 2006, a district court held that injury to the water suppliers did not require an exceedence of state and federal drinking water standards (BNA, 2006). Most of these cases have been settled without admitting liability. For example, in May 2008, a dozen defendants agreed to pay $422 million to 153 water suppliers and to pay 70 percent of the future treatment costs that arise over the next 30 years (BNA, 2010). On August 4, 2010, 48 defendants agreed to a $42 million settlement with 27 water districts and towns on New York's Long Island and in Florida (BNA, 2010). One case was tried and resulted in a $105 million jury verdict in favor of New York City, although the case is being appealed.

Sunburst, Montana. In a 4-3 vote, the Montana Supreme Court upheld a jury award of $15 million (more than seven times the pre-contamination value of the property) in compensatory damages for restoration of the groundwater contaminated by benzene releases from an oil refinery. The defendant had proposed, and the Montana Department of Environmental Quality (MDEQ) approved, a monitored natural attenuation remedy with a 20- to 100-year time horizon to attain the benzene drinking water standard (Tanner, 2009, summarizing the facts in Sunburst Sch. Dist. 2 v. Texaco, Inc., 165 P.3d 1079, 1090). MDEQ based its approval of MNA on the fact there was no exposure pathway and, therefore, no risk. However, the trial court had found (and the Montana Supreme Court assumed) that the defendant had omitted data and provided other data "to create the misleading impression that the benzene level in groundwater had been declining."

Tuscon Airport TCE Plume. Approximately 1,618 area residents living within or near a TCE groundwater plume emanating from the Tucson Airport brought a lawsuit against the City of Tucson and the Tucson Airport Authority claiming that the TCE used to clean airplanes in the area from the mid-1940s through the early 1950s caused the plume and resulted in bodily injury. After years of litigation, insurance companies for the City and Airport authority agreed, without admitting liability, to pay $35 million to resolve the personal injury lawsuits (Claims Journal, 2004).

the land preservation exceed the payments in some cases. The State of Massachusetts has recovered a total of $46.4 million in natural resources damages at ten sites, i.e., an average of $464,000 per site (MAOEEA, 2012).

In summary, the potential for NRD claims to be brought is site specific, and an NRD payment is not obtained in every case. The remedy and at least a portion of the NRD claim are interrelated, such that if more contamination is removed during the remedy, it is likely to reduce the damages calculated.

TABLE 5-1 Select Natural Resource Damage Settlements

Site	Impacts	NRD Action and Amount
Kennecott Utah Copper mining site in West Jordan Valley, UT	Surface and groundwater southwest of Salt Lake Valley contaminated by metals from mining	1. Kennecott paid \$37 million (\$9 million plus letter of credit for up to \$28 million). 2. Extracted and treated high metals concentration portion of groundwater plume (i.e., performed more remediation than might be required by a CERCLA remedial action). 3. Conveyed 617 acres of land to The Nature Conservancy to resolve its liability for NRD (331 acres and an additional ~286 acres of land as a donation) and paid TNC \$175,000 for the purpose of preserving and managing such land. 4. The final restoration plan is to construct two drinking water treatment plants using reverse-osmosis technology, provide a minimum of 8,235 acre feet per year of drinking water, contain the spread of contamination in the deep aquifer and remediate it within the "affected area," and integrate the CERCLA remedial response with the NRD Consent Decree provisions.
Ensign-Bickford Trojan Facility, Utah	Explosives manufacturing allegedly created a groundwater plume extending approximately three miles from the plant	Ensign-Bickford paid Utah \$2.58 million to replace the water that would have been available had the groundwater not been contaminated.
Six landfills near Bloomington, Indiana	Groundwater contamination	CBS Corp. paid the Interior Department \$1.88 million to restore natural resources that have been injured by ongoing releases of PCBs from the six sites.
South Valley Superfund facility, Albuquerque, NM	Groundwater contamination	New Mexico sought nearly \$1.2 billion in damages from General Electric, but the 10th circuit court of appeals dismissed the state's claim in part because they had difficulty envisioning any significant loss-of-use damage.

TABLE 5-1 Continued

Site	Impacts	NRD Action and Amount
Industri-Plex Superfund facility in Woburn, MA	Landfill/waste disposal site contaminated the groundwater	Bayer Crop Science Inc. paid $4.25 million to federal and state natural resource trustees to resolve claims for damages.
Mass. Military Reservation	Munitions testing contaminated a portion of the Cape Cod sole source aquifer	Textron Systems Corp., a contractor to the Army, agreed to a $1 million NRD settlement (http://www.mass.gov/eea/land-use-habitats/antural-resource-damages/mass-nrd-cases/textron-nrd-settlement.html).

CONSEQUENCES FOR WATER UTILITIES

Over 50 percent of potable water in the United States is derived from underground sources (USGS, 2003). In general, groundwater offers a relatively clean water resource that is low in organic content (i.e., natural organic matter) and often suitable for drinking purposes with minimal treatment. Exceptions to this are sources that are affected by high levels of natural organic matter (e.g., some Florida aquifers), naturally occurring elevated levels of dissolved inorganics (e.g., some Arizona aquifers), and leached nitrogen from agricultural and animal raising activities (e.g., many shallow aquifers across the country). In addition, releases of hazardous substances to the subsurface have resulted in deleterious impacts to groundwater resources used as a source of potable water, as discussed in Chapter 2. The magnitude of the impacts to water utilities from these sources is not well defined. However, there are many examples of significant impacts to sole source aquifers from hazardous waste sites where treatment is required if the groundwater is to be used as a potable supply. In a number of cases, wellhead treatment is an integral part of the remedial action approved under CERCLA (e.g., San Gabriel Valley, San Fernando Valley, and Bethpage Navy Plant). Box 5-5 discusses two of these cases, along with two others from Southern California, that illustrate the range of situations in which drinking water utilities find themselves with respect to a contaminated groundwater source.

The U.S. Bureau of the Census projects that the U.S. population will grow from 310 million in 2010 to 439 million in 2050 (Vincent and Velkoff, 2010). This growth may force many potable water providers to use impaired water sources, including aquifers contaminated with anthropogenic hazardous chemicals. The added treatment costs are often fully or partially offset by the parties responsible for the contamination, when they can be identified.

BOX 5-5
Comparison of Southern California Groundwater Contamination Cases

Four large-scale groundwater contamination cases in Southern California illustrate the diversity of issues that are faced by water utilities drawing from contaminated aquifers. This box compares the San Fernando Valley, the San Gabriel Valley, the Orange County, and the Santa Monica cases in terms of water supply and demand; the extent of contamination; the choice of remedy and cost; and the rationale and factors affecting the selected remedy. Litigation is ongoing in some of these cases, which have not reached resolution. The summaries are drawn from four presentations given at the second meeting of the Committee in May 2010, and do not reflect more recent happenings.

The common features of the four basins include their geographic location (all are within a 50-mile radius); the production of water from unconsolidated sand, silt, and clay strata; and the servicing of millions of people from groundwater resources. Also, VOCs have been the principal contaminants of concern among these cases, with multiple PRPs named in each case. The contamination discovery dates in all four cases were in the late 1970s and early 1980s, except for the Santa Monica case which was in the mid 1990s. The entities responsible for delivering water to the public have been faced with the urgency of the task on the one hand and with the slow progress in carrying out the investigation and the design, construction, and operation of the groundwater remediation systems on the other hand, especially when the pace of the required activities depends on funding and regulatory requirements.

In the San Gabriel Valley case, it was alleged that numerous companies were PRPs and the site was placed on the NPL. While EPA was working with the California Regional Water Quality Control Board to carry out the CERCLA process, the water supply problem needed to be solved. Consequently, wellhead treatment was instituted for treating VOCs in the groundwater. When other contaminants such as perchlorate and nitrates were detected, many systems were shut down and the San Gabriel Watermaster took charge of the activities to avoid interruption in water delivery to the public. The Watermaster was faced with challenges including how to obtain funding from the PRPs to pay for the project (at an approximate cost of $200 million) and receiving approval from the California Department of Public Health under Policy Memo 97-005, which is required after the treatment system is built. Progress has been made at the San Gabriel Valley Operable Units including treatment facilities being built and operated and ongoing cost recovery.

In the Santa Monica case, 45 companies were tentatively identified as PRPs

The systems used by water utilities to treat anthropogenic chemicals in groundwater are required to meet all applicable drinking water standards mandated by the SDWA as well as any additional requirements from the primacy agency (i.e., the state or local authority). As discussed earlier in the chapter, drinking water standards continue to evolve as more chemicals

including several oil companies that were named as the principal contributors of the MTBE and tert-butyl alcohol (TBA) detected in the water supply. After extensive investigations, the City attempted to fund the project by bringing the PRPs into the process voluntarily and working through engineering committees. When this process failed, the City sued and settled with the major PRPs for $231 million to pay for water replacement, monitoring, and the design, construction, and operation of the groundwater treatment system. The GAC treatment system became operational in December 2010 with the objective of treating MTBE to less than 3 μg/L and TBA to less than 1 μg/L.

In Orange County, when VOC contamination was detected in the shallow aquifers, the Orange County Water District (OCWD) became involved in the investigation and development of remedial approaches because of its ultimate responsibility in delivering water to the public. The OCWD's approach relied on containing the plumes to prevent migration laterally and to the deeper drinking water aquifers by extracting 5,000 to 6,000 AF; treating the water by GAC, advanced oxidation, and ion exchange; and discharging in upgradient areas—all at a capital cost of $40 million and annual O&M cost of $6 million. The OCWD passes this cost to its customers but will attempt to recover the cost from PRPs through litigation. The OCWD has not gone through the Policy Memo 97-005 process because it is trying to capture the contaminants before they enter the production wells.

Finally, in the San Fernando Valley Basin where the Los Angeles Department of Public Works is responsible for delivering water to 4 million people, numerous potential sources of groundwater contamination and the emergence of new chemicals of concern such as hexavalent chromium and 1,4-dioxane have been identified. Although only approximately 11 percent of the drinking water is supplied from groundwater, the majority of this amount is produced from the San Fernando Valley Ground Water Basin. The investigations (according to the RI/FS) have shown that uncontaminated zones for drinking water extraction are scarce and under threat of impact in the next six to ten years. Thus, wellhead treatment is planned at a capital cost of approximately $704 million and an annual O&M cost of $53 million (which although high, is less than the cost of imported water). The 97-005 Policy process will need to be followed once the treatment system is built.

In the Southern California cases, the urgency of water delivery to the public combined with funding and regulatory requirements has driven the projects toward the implementation of wellhead treatment. The Orange County case is an exception because contaminant plumes have not reportedly penetrated into the deep aquifers. In the other three cases, however, there is no choice other than wellhead treatment to recover the use of the groundwater, as most of the producing aquifers are already contaminated.

are detected due to the improvement of analytical methods, as occurrence surveys indicate the presence of these compounds in potable sources (e.g., USGS studies), and as toxicological research uncovers acute or chronic human health effects at lower concentrations. Although occurring at a relatively slow pace (only a few new chemicals are regulated under the SDWA every five years), this continuous evolution of drinking water standards poses a challenge for wellhead treatment system configuration. For example, at sites in the San Fernando Valley, treatment systems were initially designed to remove TCE and PCE, using either air stripping or granular activated carbon (GAC). However, subsequent testing revealed the presence of less volatile and less adsorbable chemicals at the Glendale NPL site (e.g., 1,2,3-trichoropropane), which required modifications to the existing groundwater treatment systems and caused an increase in operating costs.

In addition to future changes in the drinking water standards for individual chemicals, EPA is also considering alternative regulatory strategies that may affect both remedial objectives as well as well-head treatment standards. For example, in early 2010 EPA administrator Lisa Jackson announced a new drinking water strategy to expand public health protection more cost effectively by addressing contaminants with similar modes of action in groups (EPA, 2010e). The first group proposed for regulation under this strategy focuses on carcinogenic VOCs, including eight that have been previously regulated by EPA and up to eight more VOCs that were on the CCL. The most recent information from EPA indicates that the drinking water standards for the previously regulated compounds may be made more strict within the new regulatory strategy to recognize improved analytical methods (e.g., the TCE standard as discussed earlier) (EPA, 2011j).

State Policies for Wellhead Treatment for Vulnerable or Impacted Water Supply Aquifers

In the context of groundwater contamination from a wide range of anthropogenic sources, various states have developed policies to establish appropriate methods to protect aquifers so they can continue to be used as water supplies. For example, in 1997 California developed an effective policy for use of groundwater resources considered to be "extremely impaired," defined as having levels of chemical contaminants several times higher than established drinking water standards (California Policy Memo 97-005, Policy Guidance for Direct Domestic Use of Extremely Impaired Sources). The general philosophy of this guidance is that the best quality sources of water reasonably available to a water utility should be used for drinking, and wherever possible lower quality water source waters should be used for non-consumptive uses that pose lower health risks such as irrigation, recreation, or industrial uses. If a permit for the use of a severely impaired source is to be approved by the California Department of Public Health, the source of the contamination must be controlled to prevent contaminant concentrations from rising, and dependence upon treatment should be minimized. Water treatment facilities should be designed and operated to meet the MCLG (when lower than the MCL) where this can be achieved in a cost-effective manner.

Shortly after this policy was created, California adopted a program to assess and protect sources of drinking water (called Policy Memo 97-005, CDHS, 1997). In the drinking water source assessment, an area around a drinking water source is delineated through which contaminants could move and reach the drinking water supply. In addition, the drinking water source assessment includes an inventory within the delineated area of activities that could lead to the release of contaminants. In California, less than a dozen permits have been approved since the initiation of the 97-005 policy, and the approval process can be lengthy (Collins, 2011). It should be noted that California recently (2012) declared that the 97-005 policy is no longer considered obligatory, and is considered as guidance only, since the policy was not subjected to legislative review and approval. Nonetheless, the risk analysis components of the 97-005 process provide the state with some assurance that the well-head treatment system will consistently remove all chemicals of concern to appropriate drinking water standards or below, and that the approved treatment system is capable of dealing with chemicals likely to be present in the aquifer, in addition to the chemicals designated for removal in the permit.

Impacts of Groundwater Contamination on Aqueduct Storage and Recovery or Conjunctive Use Programs

In the more arid areas of the country, groundwater aquifers are increasingly being considered as potential sites for storage of excess water that may become available during wet periods or transportable from wetter areas for subsequent use during periods of shortage (so-called water banking or conjunctive use). Traditionally, water agencies have built percolation ponds to divert spring stream runoff for groundwater recharge. More recently municipal wastewater that has been treated sufficiently can be used to recharge aquifers, although the increased mineral concentrations of wastewater can limit the future use of the groundwater. Although expensive and energy intensive, aquifer storage and recovery (ASR) is an increasingly attractive strategy for many areas. In 2005 there were over 70 active ASR projects in 15 states (NRC, 2008). Various forms of ASR are being practiced in the Kern, San Gabriel, and San Fernando basins of California; Texas and Arizona have similar programs.

Conjunctive use can expose the stored water to potential contamination from anthropogenic chemicals that are present in the unsaturated zone above the aquifer, posing additional treatment challenges and disputes on liability for cleanup when the water is withdrawn from the aquifer by the utility. Extraction wells must have wellhead treatment if the groundwater is contaminated, and such treatment strategies must be considered at the time a basin is planned for such use. Even if the water extracted is used for non-potable purposes, current state and federal laws support the objective of cleanup to drinking water standards.

It is highly likely that a growing number of groundwater basins throughout the United States with detectable amounts of contaminants will be considered for ASR. It is in the public interest to develop an approach to managing these aquifers that can establish the human health risk across the whole range of potential uses of the groundwater.

Wellhead Treatment

The objective of wellhead treatment is to remove natural and anthropogenic contaminants to levels consistent with standards established for the intended use. If the intended use is human consumption, the water must be rendered safe as defined by the SDWA (i.e., meet all federal MCLs or other state drinking water standards). The permitting process for the treatment system by the local authorities is expected to ensure that the treatment system is capable of complying with all the applicable regulations under SDWA. In this type of application of wellhead treatment, the water also needs to be acceptable to the consuming public and thereby may require the removal of chemicals to levels below MCLs. The burden is on the treatment system owner to obtain such regulatory approval for the design, construction and operation of a groundwater treatment system.

The Current State of Technologies for Groundwater Treatment

Above-ground treatment of contaminated groundwater is often carried out at the well head or at a location where several wells are pumped into a central treatment facility. The latter option, often utilized to achieve economies of scale, is more typical for treatment systems where the intended end use is human consumption. Many different treatment technologies are available depending on the target contaminant. Removal of organic pollutants is often accomplished through volatilization, adsorption, or oxidation whereas the removal of inorganic contaminants is accomplished through ion exchange, adsorption, or precipitation followed by solids separation. The following is a list of commonly applied treatment technologies used for cleanup of groundwater impacted with organic chemicals and toxic metals:

- Packed Tower Aeration
- Diffused Bubble Aeration
- Tray Aeration
- Granular Activated Carbon Adsorption

- Ion Exchange
- Advanced Oxidation using Ozone
- Advanced Oxidation using Ultraviolet Light

Table 5-2 lists commonly applied groundwater treatment technologies along with some limitations to their use. Among the listed technologies, packed tower aeration and granular activated carbon are most often used for *ex situ* groundwater treatment for the most common chlorinated solvents, including TCE and PCE. Anion exchange, advanced oxidation, and biological processes are becoming more common as the treatment objectives for groundwater are being expanded to include multiple contaminants as well as due to the reduction in drinking water standards for treated water.

TABLE 5-2 List of Commonly Used Groundwater Treatment Technologies

Technology	Relative Cost	Technology Applicability	Technology Limitations
Packed Tower Aeration	Low	Moderately to highly volatile compounds characterized with high Henry's Law Constant (e.g., TCE and PCE)	• Tower height usually limited to 20 or 30 ft • Air-to-water ratio usually limited to 150:1 • Scaling control (acid addition) needed for highly alkaline and hard waters
Multi Stage Bubble Aeration	Low	Highly volatile compounds only	• High degree of removal is more difficult to achieve
Tray Aeration	Low	Highly volatile compounds only	• High degree of removal is more difficult to achieve
Granular Activated Carbon Adsorption (GAC)	High	Moderately to strongly adsorbable compounds characterized with favorable Freunlich Constant (e.g., PCE)	• Natural organic matter reduces removal efficiency • Cost is high • High degree of removal may not be achieved at low influent concentrations

continued

TABLE 5-2 Continued

Technology	Relative Cost	Technology Applicability	Technology Limitations
Advanced Oxidation with Ozone (O_3-AOP)	High	Smaller organic molecules (e.g., 1,4-dioxane)	• Presence of bromide in the water results in bromate formation • Excess peroxide must be quenched
Advanced Oxidation with Ultra Violet Light (UV-AOP)	High	Smaller organic molecules (e.g., 1,4-dioxane)	• Higher dosages required compared to UV disinfection • Excess peroxide must be quenched
Biological Treatment	Low to Moderate	Anaerobic processes can be used to remove perchlorate	• Complex treatment process • Requires substrate (e.g., ethanol) addition • Requires polishing treatment
Strong Base Anion Exchange	Moderate to High	Charged anions (e.g., chromate, arsenate, or perchlorate)	• Generates concentrated brine stream that may be hazardous • Interference from other anions (e.g., sulfate)
Weak Base Anion Exchange	Moderate to High	Charged anions (e.g., chromate, arsenate, or perchlorate)	• May require pH adjustment • Generates concentrated brine stream that may be hazardous • Exhausted resins in single pass applications may be hazardous
Reduction Coagulation Filtration	High	Chromate	• Significant residual stream
Electrodialysis/ Electrodialysis Reversal	High	Moderate degree of removal for a large variety of contaminants	• Significant concentrate stream (up to 25%) • May require pH adjustment

Treatment facilities targeted for the production of potable water from a contaminated groundwater source are generally designed with technologies capable of removing contaminants well below drinking water standards and are thus designed conservatively. For example, some water utilities target non-detect levels of TCE in the treated water even though the applicable federal standard in 2012 is 5 μg/L; current analytical methods are routinely capable of measuring concentrations below 1.0 μg/L. In some instances, water utilities will attempt to the remove contaminants that are not currently regulated, but where state standards or compliance levels have been established or are likely in the future. For example, at a groundwater treatment facility located near the Tucson Airport where TCE-contaminated groundwater has been treated for several years using packed tower aeration, operators are installing a UV-AOP treatment process for removal of 1,4-dioxane, which has been recently detected at low levels. Similarly, the groundwater treatment facilities located in the San Fernando Valley in California (Box 5-5) may be upgraded to include treatment for perchlorate and hexavalent chromium.

Challenges for Wellhead Treatment

Wellhead treatment for removal of chemicals typically present at hazardous waste sites is not a new challenge for water utilities. Treatment systems have been in place for more than 30 years in locations where VOCs impact groundwater sources (e.g., New Jersey, California, Arizona). Despite a long history of using specific treatment technologies (such as packed tower aeration or GAC) for removing organic chemicals from groundwater, the presence of multiple organic chemicals in groundwater resources poses a number of treatment challenges. First, the design of water treatment systems must be based on estimates of the number of chemicals of concern present and their likely maximum concentrations. Accurate estimates of these design values are sometimes difficult to come by, given limited site characterization data or limited knowledge of the nature and extent of the contamination present in the aquifer and within the capture zone of the water supply well. Groundwater fate and transport modeling is inherently limited in terms of accurate predictions, but is currently the best tool available to estimate these design parameters.

Each of the individual treatment systems now in use are subject to operational upsets. For example, an air stripping tower may be effective at removing TCE but should a less volatile organic chemical occur, such as 1,2-DCA, the removal efficiency of the system may not be sufficient to meet the treatment objectives. For GAC, the appearance of compounds less strongly adsorbed may result in more rapid breakthrough of the chemical and thus increase the usage rate for GAC, leading to more frequent replacement and increased costs.

More recently, some organic chemicals have been found in groundwater that are not efficiently removed by either air stripping or GAC. For example, 1,4-dioxane has become a very prevalent contaminant due to its widespread use as a solvent stabilizer (Mohr, 2010). For those utilities who have already installed GAC, this chemical will not be removed effectively, and an additional treatment process is needed. The most widely adopted technology for removing 1,4-dioxane is advanced oxidation, using either the UV/hydrogen peroxide or the ozone/hydrogen peroxide technology. AOP is a more recent addition to the suite of wellhead treatment options for organics, and limited operational experience is available. An additional issue with the use of AOP technologies is the formation of oxidation byproducts that can be problematic. For example, the ozone/hydrogen peroxide technology will convert bromine ion to bromate, which has a 10 μg/L MCL. Other byproducts could also be problematic. Finally, the excess hydrogen peroxide must be destroyed prior to transfer to the water distribution system.

The reliability of any wellhead treatment system to meet treatment goals (usually below MCLs) on a consistent basis is uncertain. Water treatment plants are usually monitored at intervals that reflect the relative constant influent water quality. Often, for VOCs and other regulated contaminants, a running average based on quarterly sampling frequency is used to assess performance. Given the potential variability in influent water quality characteristics, greater frequency of sampling may be appropriate until the reliability of the treatment systems can be confirmed.

* * *

The decision of whether to perform wellhead treatment versus conducting more conventional groundwater remediation goes beyond the ability of technologies to remove contaminants at the point of use. Indeed, there are significant economic considerations, as wellhead treatment can be roughly as expensive as source or plume remediation, depending on the contaminants involved and the site hydrogeology. Second, it is not yet clear that wellhead treatment is a reliable, long-term strategy for accomplishing exposure cutoff (and it certainly cannot prevent exposure via vapor intrusion). These tensions are illustrated at the Northrop-Grumman/Navy facility in Bethpage, Long Island, New York (see Appendix B). In this case, wellhead treatment of a groundwater plume used as a primary water supply was deemed necessary because full containment of the groundwater plume was thought to be technically infeasible and not cost effective. Nonetheless, affected water districts have lobbied for plume remediation because the plume is spreading and there is the chance that MCLs for the contaminants of concern will be lowered in the near future. This will make wellhead treatment, currently estimated by one water utility to be $128 million, significantly more expensive than plume remediation (estimated to be about $91 million, by the same utility) (Massapequa Water District, 2011). State regulatory authorities have come up with different cost estimates and continue to recommend wellhead treatment over plume remediation because some wells are already contaminated and will have to be treated in perpetuity. It is too soon to tell how the dispute over the proper remediation strategy at the Navy/Northrop-Grumman plume will be resolved.

ECONOMIC IMPACTS

Sites with subsurface contamination by hazardous materials present economic challenges to responsible parties and people and businesses affected by the contamination. When contamination remains in place above unrestricted use levels, there are direct, out-of-pocket costs of controlling contamination. There are also opportunity costs such as forgone revenue from not being able to use the contaminated site and nearby areas for other endeavors.

Figure 5-1 provides a stylized example of costs associated with a contaminated groundwater site. "*Services under baseline*" represents water quality in an aquifer absent of contamination. Suppose a release of a hazardous substance occurs at time T_0. In this example, it is assumed that 25 percent of the groundwater remains accessible for use with contaminant levels below MCLs, i.e., pumping can continue in some parts of the aquifer. Once contamination occurs, the *Recovery Path* represents improvements in groundwater quality due to natural attenuation of the hazardous substance.

A treatment program will enhance recovery and is indicated by the *Recovery Path with Remedial Action* in Figure 5-1. For simplicity, the recovery path begins at time T_0, although this would only be the case where a contaminant release was instantaneously known and remediation efforts were initiated immediately to control damages. Losses in groundwater services after remediation are shown by the shaded area and can only be fully revealed over time. Potentially responsible parties can compensate affected parties for residual losses (shaded area) or they can provide equivalent services known as compensatory restoration (dotted area). Compensatory restoration is typically designed so that the benefits of the compensatory restoration just offset the loss (the dotted area just equals the shaded area). Note, the hatched area constitutes losses that occurred prior to the implementation of CERCLA; some trustees submit claims for these losses and others just submit claims for losses subsequent to the implementation of CERCLA.

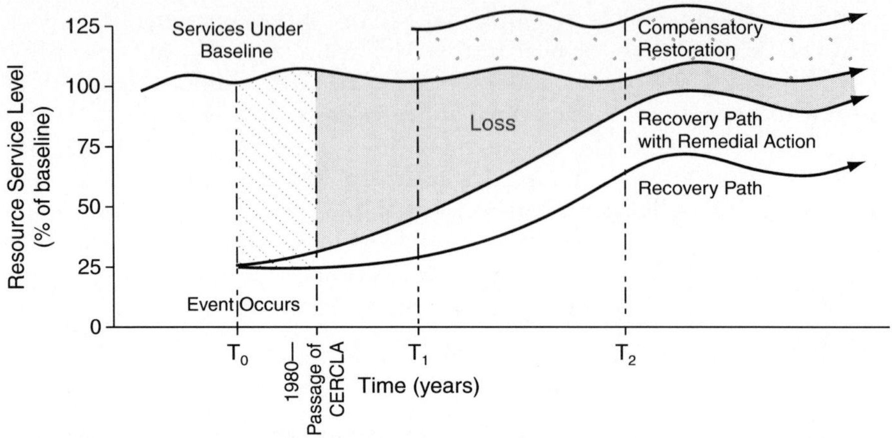

FIGURE 5-1 Stylized diagram of economic losses.

A key issue at sites that have recalcitrant and/or poorly accessible contaminants that prevent site closure is that costs to the potentially responsible party and affected parties can continue indefinitely, indicated by the arrows on the right side of Figure 5-1. Even with a remedial action plan, groundwater quality may not be returned to baseline conditions for a very long time. This residual and recalcitrant contamination increases the cost uncertainties faced by all parties associated with such a site.

Potentially Responsible Party Costs

The costs faced by a potentially responsible party are the costs of remedial actions and litigation expenses (discussed in a previous section). Remedial action costs are influenced by available cleanup technologies, the extent of contamination, and the health effect pathways of the contaminants. The cost of conducting remediation is closely tied to the cost of energy and to the cost of landfill space. As the cost of energy rises, more energy-intensive remediation methods like excavation and thermal treatment will become less competitive with methods that use or enhance natural processes.

The time it takes from the initiation of the first cleanup action until construction completion has varied since the initiation of Superfund in 1980, but was typically about nine years for NPL facilities in the late 1990s (EPA, 2002b), while the average economic cycle in the United States from 1945 to 2001 has been about 67 months or about 5.6 years (NBER, 2010). Thus, a typical cleanup at a Superfund facility can bridge two or three cycles of the national economy, during which annual cleanup and monitoring costs can increase or decrease. One factor that could induce such variation is energy costs. For example, energy costs to operate a pump-and-treat system could have started out high, then dropped, then increased substantially toward the end of the treatment project (EPA, 2001b).

New information about the site can arise after a ROD has been approved, which can affect cleanup costs. This new information can take a number of forms, such as a reduction in the MCL in the face of new toxicity values, the identification of an additional contaminant due to enhanced detection technology, or a new arm of the contamination plume could be discovered. These occurrences could increase cleanup and management costs beyond what the PRP expected at the time a ROD is approved.

Costs are only one of the nine criteria of the NCP, and the cost criterion does not require any consideration of incremental or marginal costs for additional cleanup or risk reduction. Thus, at complex sites where UU/UE levels cannot be attained, costs may increase over time with little or no reduction in risks. It is proposed in Chapter 7 that when a site reaches the point where the marginal cost of additional cleanup actions increases but there is no corresponding reduction in contaminant concentrations and hence risk, then the cleanup expenditures could be better spent elsewhere in the economy (such as investing in developing more effective remediation technologies or to improve conditions and reduce risks at other sites).

Measurement of the costs discussed above is hampered by limitations in available data. Barth and McNichols (1994) note that "actual cleanup costs generally are not known; at most sites cleanup is incomplete and there is no central tracking of cleanup costs" (p. 177). Hamilton and Viscusi (1999) reported cleanup costs for a sample of 19 NPL facilities and found the average cost per site with soil and or groundwater contamination was $25.7 million (in 1993 dollars). These cleanup costs may be lower-bound costs because Hamilton and Viscusi (1999) did not include legal and management expenses incurred by the responsible parties and regulators in securing acceptance of the ROD. For comparison, in 2009 GAO reported that, based on EPA data, the average expenditure for Superfund facilities through 2007 was $9.9 million per ROD (note that there can be multiple RODs at one Superfund facility) (GAO, 2009). These are total costs and not the key cost information—marginal costs—which indicate how costs vary with the amount of remediation accomplished.

In addition to the paucity of peer-reviewed literature on the cost of cleaning up NPL sites, firm-specific and site-specific cleanup costs are difficult to obtain. Barth and McNichols (1994) reported that 44 percent of Superfund facilities have more than one potentially responsible party and 52 percent of potentially responsible parties have more than one site where they have responsibility. These conditions make it difficult to obtain site-specific cost data for individual firms from public records and it is also difficult to aggregate cost data across firms for specific sites.

Despite the limited number of empirical cost studies, the available literature does present a few insights. First, Hamilton and Viscusi (1999) found that 20 percent of the sites account for 50 percent of the cleanup costs. Gupta et al. (1996) suggest that the higher cost sites may be associated with greater risks to human health and the environment. Overall, these results suggest cost uncertainties may be more important for complex sites that are likely to incur large cleanup costs or with large risks to human health and the environment. The number of studies, however, is too few to draw any firm conclusions.

Affected Parties

Costs to affected parties, be they businesses or individuals, can arise during cleanup activities and from residual contamination that is left after completion of active cleanup activities. These costs include claims for health effects, property damage, and lost business income. Furthermore, because complete protection from the contaminants might not occur instantaneously and compensation paid by responsible parties may not occur for years, affected parties can be faced with interim, out-of-pocket costs of providing self-protection from the contamination.

Costs for dealing with health effects can, if the concentrations of a contaminant are of sufficient magnitude, potentially be incurred where firms or households use groundwater that is contaminated by a site and/or if the subsurface contamination results in vapor intrusion into nearby buildings where people work or live. Immediate costs may include protective actions such as buying bottled water or installing ventilation systems in buildings, while long-term costs may arise from health care expenses as a consequence of past exposure to contaminants.

Property costs occur because nearby property values may be diminished by the contamination at a site. Note that if health risks are even a perceived concern, this might serve to diminish property values. Conversely, if the potentially responsible parties' protection efforts are effective, or effective self-protection was installed in a property, this may remove some of the property value diminution (Boyle et al., 2010; Kiel, 1995). However, property values still may be diminished due to a price stigma from the nearby contamination site (McCluskey and Rausser, 2003; Messer et al., 2006). This price diminution due to the stigma can be in addition to any associated health costs.

As with the costs to potentially responsible parties, the costs of residual contamination to affected parties described above are challenging to address. The goal of any damage claim is to measure these costs as accurately as possible and to seek compensation as quickly as possible. However, accurate measurement is a topic of considerable academic debate.

Site Development Impacts

The link between future economic development on a site and residual groundwater and soil contamination was dramatically illustrated when the local banks refused to lend money to Love Canal home owners and commercial entities (Niagara Gazette, 1979). Similarly, in 1991 widespread contamination of groundwater by chlorinated solvents was discovered in Wichita, Kansas, encompassing approximately 5½ square miles and 8,000 parcels of land and affecting more than 550 businesses and hundreds of residential properties (ICMA, 2006). This contamination undercut the City's tax base, and banks were reluctant to finance transactions within the site (Weinfield, 2007). At El Toro Marine Corps Air Station in Irvine, CA, the Navy was able to commence transferring the closing base to the public only after carving out the more contaminated areas and precluding them from sale.

Some of these risks have been lessened by federal legislation. For example, CERCLA was explicitly amended to exempt banks that hold mortgages on property as secured lenders from CERCLA cleanup liability, if they do not exercise decision-making control over a property's environmental compliance (EPA, 2009c). In addition, the Small Business Liability Relief and Brownfields Revitalization Act exempts from liability bona fide prospective purchasers who purchase property with knowledge of existing contamination if they (1) are not responsible parties; (2) acquire the property after January 11, 2002; (3) establish that all disposal of hazardous substances occurred before they acquired the facility; (4) make all appropriate inquiries into previous ownership and prior uses of the property; and (5) are not affiliated with a responsible party (EPA, 2009c).

It should be noted that environmental due diligence prior to property acquisition is now standard practice in everyday transactions and when mergers and acquisitions are contemplated (ASTM, 2006). Typically, the acquiring company and lenders seek certainty concerning the magnitude of future liabilities from the selling company. As a result, mergers and acquisitions may trigger action to obtain regulatory certainty concerning cleanup efforts, the adoption of indemnification allocating future liability among the purchaser and seller, renegotiation of the sales price, or set asides of funds for cleanup or other activities. Furthermore, in 2011 six federal financial institution regulatory agencies proposed credit risk retention rules for commercial real estate property where environmental hazards, such as groundwater contamination, may potentially jeopardize the value of commercial real estate as well as the borrower's ability to repay a loan (OCC et al., 2011). Thus, even though there is no longer the extreme avoidance of lending for contaminated properties that existed when hazardous waste issues first arose, the continuing presence of chemicals above unrestricted use levels is considered in transactions and may impact future economic development.

Fortunately, the liabilities discussed above that arise from having residual contamination in place can be offset by benefits that may result from redevelopment of contaminated land (EPA, 2011k). Regulators tend to use residential land use assumptions at sites because the goal of groundwater cleanup is often to return groundwater to beneficial uses (EPA, 1990, 2009d). Reuse of contaminated properties (so-called Brownfields redevelopment) or for some other use that does not require residential land use is becoming more common and is supported by federal policies (EPA, 2011L; ASTM, 2010). At Brownfields, regulatory policies and financial grants are often complemented by state and local incentives for property use or even employment.[7] Indeed, some states promote Brownfields redevelopment even in the absence of significant environmental problems.

CONCLUSIONS AND RECOMMENDATIONS

This chapter has focused on the implications of leaving contamination in place at hazardous waste sites at levels above those allowing for unlimited use and unrestricted exposure. All of the risks discussed in this chapter should be considered carefully before choosing a remedy that leaves contamination in place.

Research is needed to build a more robust understanding of the potential for failure in hydraulic and physical containment systems. Two types of studies are recommended: (1) a "data mining" study in which site-specific evaluations such as RSEs and five-year reviews are analyzed to provide a general characterization of the frequency, mechanisms, responses, and costs associated with the failure of hydraulic and physical containment, permeable reactive barriers, and MNA, and (2) consistent with previous NRC recommendations (2007), establish field experimental stations at operating waste sites to assess the long-term performance of physical containment systems.

Implementing institutional controls at complex sites is likely to be difficult because it requires long-term monitoring of site conditions and potential exposures and maintenance of land use restrictions that may be in tension with the development preferences of local authorities. Although EPA has developed a number of measures to improve the reliability, enforceability, and funding of institutional controls, their long-term efficacy has yet to be definitely determined. The public who live and work near a

[7] See http://epa.gov/superfund/programs/recycle.

given site have a direct and enduring interest in ensuring that institutional controls are followed and are more likely than others to know whether institutional controls are working.

Regulators and federal responsible parties should incorporate a more significant role for local citizens in the long-term oversight (i.e., monitoring and enforcement) of institutional controls. At a minimum, those who live, work, study, play, or otherwise occupy residences, commercial buildings, or industrial facilities near hazardous waste sites should be notified of institutional controls, including the specific reasons for them. **A national, searchable, geo-referenced institutional control database covering as many regulatory programs as practical as well as all federal sites would ensure that the public is notified of institutional controls.** Methods can also be developed to encourage citizens to report to regulators any developments that might undermine the protectiveness of such institutional controls and thus improve oversight (see Chapter 7).

New toxicological understanding and revisions to dose-response relationships will continue to be developed for existing chemicals, such as TCE and PCE, and for new chemicals of concern, such as perchlorate and perfluorinated compounds. The implications of such evolving understanding are multi-fold, including identification of new or revised MCLs (either more or less restrictive than existing ones), potentially leading to a determination that the existing remedy at some hazardous waste sites is no longer protective of human health and the environment.

Modification of EPA's existing CERCLA five-year review guidance would allow for more expeditious assessment of the protectiveness of the remedy based on any changes in EPA toxicity factors, drinking water standards, or other risk-based standards. Although EPA's existing five-year review guidance provides that cleanup levels will be evaluated when a new ARAR for drinking water has been developed and/or new or revised toxicological information is available, the existing process is slow. Furthermore, as EPA accelerates its national toxicity assessment process, there will be a large number of chemicals at hazardous waste sites whose protectiveness will need to be reassessed. EPA should ensure that it has adequate resources dedicated to this process.

Careful consideration of the vapor intrusion pathway is needed at all sites where VOCs are present in the soil or groundwater aquifer. Although it has been recognized for more than a decade that vapor intrusion of volatile chemicals is a potential exposure pathway of concern, a full understanding of the risks over time and appropriate methods for characterizing them are still evolving.

Mitigation strategies such as subslab depressurization can prevent vapor intrusion exposure. In most cases, the cost of building mitigation into new construction is significantly less than the cost of repetitive sampling. Furthermore, such systems reduce exposure to naturally occurring radon. As a precautionary measure, vapor mitigation could be built into all new construction on or near known VOC groundwater plumes; this could be imposed proactively as part of local or state building codes or other requirements or imposed as institutional controls at regulated sites. In either situation, vapor mitigation systems require monitoring over the long-term to ensure that they are operating properly.

As populations increase and industrial demands for high-quality water also increase, the demands placed on groundwater supplies will increase. Contaminated aquifers have been and may well be used more extensively in the future to augment supplies of uncontaminated water. Wellhead treatment may be an optimal remedy for low concentrations of contaminants in potential water supplies. Current wellhead treatment technologies are materials intensive and are not energy efficient. **Improved and efficient water treatment technologies should be developed both for more cost-effectively destroying VOCs and recalcitrant organic compounds as well as for removing toxic metals. An emphasis should be placed on technologies that treat a broad spectrum of chemicals.**

REFERENCES

Aitchison, E. W., S. L. Kelley, J. J. Pedro-Alvarez, and J. L. Schnoor. 2000. Phytoremediation of 1,4-dioxane by hybrid poplar trees. Water Environment Research 72(3):313-321.

Ando, A. W., M. Khanna, A. Wildermuth, and S. Vig. 2004. Natural Resource Damage Assessment: Methods and Cases. Waste Management and Resource Center, Illinois Department of Natural Resources. http://www.istc.illinois.edu/info/library_docs/RR/RR-108.pdf.

Army. 2011. Environmental Liabilities. http://aec.army.mil/usaec/cleanup/e100.html.

ASTM (American Society for Testing and Materials). 2006. ASTM Standards on Environmental Site Assessments For Commercial Real Estate, 5th Edition, which includes ASTM E 1527–00, Standard Practice for Environmental Site Assessments: Phase I Environmental Site Assessment Process, ASTM E1903-11 Standard Practice for Environmental Site Assessments: Phase II Environmental Site Assessment Process, and ASTM E 1528 Transaction Screen Process.

ASTM. 2010. ASTM Standard Classification of Environmental Condition of Property Area Types for Defense Base Closure and Realignment Facilities.

ATSDR (Agency for Toxic Substances and Disease Registry). 1989. Public Health Statement for *n*-Nitrosodimethylamine. CAS 62-75-0. Altanta, GA: ATSDR.

ATSDR. 2005. Toxicological Profile for Naphthalene, 1-Methylnaphthalene, and 2-Methylnaphthalene. http://www.atsdr.cdc.gov/toxprofiles/tp67.pdf.

ATSDR. 2008. Draft Toxicological Profile for Chromium. http://www.atsdr.cdc.gov/toxprofiles/tp7.pdf.

Barnes, K. K., D. W. Kolpin, E. T. Furlong, S. D. Zaugg, M. T. Meyer, and L. B. Barber. 2008. A national reconnaissance of pharmaceuticals and other organic wastewater contaminants in the United States - (I) Groundwater. Science of the Total Environment 402(2-3):192-200.

Barth, M. E., and M. F. McNichols. 1994. Estimation and market valuation of environmental liabilities relating to superfund sites. Journal of Accounting Research 32(Sup.):177-209.

Beaumont, J. J., R. M. Sedman, S. D. Reynolds, C. D. Sherman, L. H. Li, R. A. Howd, M. S. Sandy, L. Zeise, and G. V. Alexeeff. 2008. Cancer mortality in a Chinese population exposed to hexavalent chromium in drinking water. Epidemiology 19(1):12-23.

BNA. 2006. Daily Environment Report. News 11/15/2006. Drinking Water: Judge Trying Federal MTBE Litigation Rejects Regulatory Standards as Baseline for Injury. http://news.bna.com/deln/display/batch_print_display.adp?searchid=15581702.

BNA. 2010. Daily Environment Report. News 08/06/2010. Drinking Water: Oil Companies Agree to Settle Claims Over MTBE Contamination for $42 Million. http://news.bna.com/deln/display/batch_print_display.adp?searchid=15581702.

Bowman, R. H., P. Miller, M. Purchase, and R. Schoellerman. 2003. Ozone-peroxide advanced oxidation water treatment system for treatment of chlorinated solvents and 1,4-dioxane. Contaminated Soils 8:1-12.

Boyle, K. J., N. Kuminoff, C. Zhang, M. Devanney, and K. P. Bell. 2010. Does a property-specific environmental health risk create a 'neighborhood' housing-price stigma? Arsenic in private well water. Water Resources Research 46:W03507.

Buszka, P. M., D. J. Yeskis, D. W. Kolpin, E. T. Furlong, S. D. Zaugg, and M. T. Meyer. 2009. Waste-indicator and pharmaceutical compounds in landfill-leachate-affected ground water near Elkhart, Indiana, 2000-2002. Bulletin of Environmental Contamination and Toxicology 82(6):653-659.

CalEPA. 2011. California's Vapor Intrusion Mitigation Advisory. Sacramento, CA: CalEPA Department of Toxic Substances Control.

Carrara, C., C. J. Ptacek, W. D. Robertson, D. W. Blowes, M. C. Moncur, E. Sverko, and S. Backus. 2008. Fate of pharmaceutical and trace organic compounds in three septic system plumes, Ontario, Canada. Environmental Science & Technology 42(8):2805-2811.

CDHS (California Department of Health Services). 1997. Policy Memo 97-005 Policy Guidance for Direct Domestic Use of Extremely Impaired Sources. www.cdph.ca.gov/certlic/drinkingwater/Documents/DWdocuments/memo97-005.pdf.

Center for Justice & Democracy. 2007. Fact Sheet: Environmental Tort Lawsuits: Holding Polluters Accountable. January 23, 2007. http://centerjd.org/content/fact-sheet-environmental-tort-lawsuits-holding-polluters-accountable.

Christopher, J. P., B. K. Davis, J. M. Polisini, and M. J. Wade. 2005. Designation of Naphthalene as a Carcinogen: Risk Assessment for Inhalation Exposure Pathways at Hazardous Waste Sites. http://www.dtsc.ca.gov/AssessingRisk/upload/Naphthalene_Handout.pdf.

Claims Journal. 2004. Tucson-Area Water Contamination Victims Triumph in Ariz. Appeals Court Ruling Against Insurer, October 1, 2004. http://www.claimsjournal.com/news/west/2004/10/01/46444.htm.

Coate, C. J., K. Frey, and N. Sakuvich. 2012. Environmental Accounting and Reporting 101. September 5, 2012. http://www.newaccountantusa.com/newsfeat/ip/ip_environmental.html.

Coates, J. D., and L. A. Achenbach. 2004. Microbial perchlorate reduction: Rocket-fuelled metabolism. Nature Reviews Microbiology 2:569-580.

Coates, J. D., and L. A. Achenbach. 2006. *In situ* bioreduction and removal of ammonium perchlorate. Final report for the Strategic Environmental Research and Development Program.

Collins, H. 2011. Water Quality Regulations and Permitting Challenges for Groundwater Treatment Systems: California Policy 97-005. Presented at the AWWA Water Quality Technology Conference, November 2011.

Cox, C., and E. Douthett. 2009. Further evidence on the factors and valuation associated with the level of environmental liability disclosures. July 1, 2009. http://www.freepatentsonline.com/article/Academy-Accounting-Financial-Studies-Journal/209238515.html.

Cramer, R. J., C. Yates, P. Hatzinger, and J. Diebold. 2004. Field demonstration of in situ perchlorate bioremediation at building 1419. Final report for the Strategic Environmental Research and Development Program. CU-1163-TR-01.

DERP. 2001. Management Guidance for the Defense Environmental Restoration Program. September 21, 2001. http://aec.army.mil/usaec/cleanup/derpguidance0411.pdf.

Dietz, A. C., and J. L. Schnoor. 2001. Advances in phytoremediation. Environmental Health Perspectives Supplements 109(1):163-168.

DoD (Department of Defense). 2001. Policy on Land Use Controls Associated with Environmental Restoration Activities. Memorandum for Assistant Secretaries of the Army, Navy, Air Force, and Director, Defense Logistics Agency. January 17, 2001.

DoD. 2009. DoD Vapor Intrusion Handbook. Washington, DC: DoD Tri-Service Environmental Risk Assessment Workgroup. http://www.clu-in.org/download/char/dodvihdbk200901.pdf.

Doroski, M., K. Anders, and G. Heitzman. 2011. Environmental Energy Alliance of New York State / New York State MGP Soil Vapor Intrusion Database. 27th Annual Conference on Soil, Sediments, Water, and Energy, Amherst, MA, October 17-21.

Drewes, J. E., T. Heberer, T. Rouch, and K. Reddersen. 2003. Fate of pharmaceuticals during groundwater recharge. Ground Water Monitoring & Remediation 23(3):64-72.

ELI. 1999. Environmental Law Institute, Protecting Public Health of Superfund Sites: Can Institutional Controls Meet the Challenge? http://www.elistore.org/reports_popup.asp?did=543&file=/Data/products/d10.01.pdf.

EPA (U.S. Environmental Protection Agency). 1990. NCP Preamble. Federal Register 55:8732.

EPA. 1991a. Guidelines for Developmental Toxicity Risk Assessment. Federal Register 56(234):63798-63826. http://www.epa.gov/raf/publications/pdfs/DEVTOX.PDF.

EPA. 1991b. EPA Superfund Record of Decision: Love Canal OU 9. http://www.epa.gov/superfund/sites/rods/fulltext/r0291165.pdf.

EPA. 1999. Second five-year review: Delaware Sand and Gravel Site, New Castle Delaware. Philadelphia, PA: EPA Region 3.

EPA. 2001a. RCRA Draft Supplemental Guidance for Evaluating the Vapor Intrusion to Indoor Air Pathway. EPA/600/SR-93-140. http://www.epa.gov/wastes/hazard/correctiveaction/eis/vapor/vapor.pdf.

EPA. 2001b. Cost Analyses for Selected Groundwater Cleanup Projects: Pump and Treat Systems and Permeable Reactive Barriers. EPA 542-R-00-013. http://www.epa.gov/tio/download/remed/542r00013.pdf.

EPA. 2002a. Draft Guidance for Evaluation the Vapor Intrusion to Indoor Air Pathway from Groundwater and Soils. Washington, DC: EPA OSWER.

EPA. 2002b. The Superfund process, from site listing to cleanup construction, on average has taken roughly 8 to 10 years. Statement of Marianne Lamont Horinko, Assistant Administrator, Office of Solid Waste And Emergency Response, U.S. Environmental Protection Agency, before the Subcommittee on Superfund, Toxics, Risk And Waste Management. United States Senate (July 31, 2002). http://epw.senate.gov/107th/Horinko_073102.htm.

EPA. 2004a. Performance monitoring of MNA remedies for VOCs in groundwater. EPA/600/R-04/027. Washington, DC: EPA Office of Research and Development.

EPA. 2004b. IRIS Toxicological Review of Naphthalene. http://cfpub.epa.gov/ncea/iris_drafts/recordisplay.cfm?deid=84403.

EPA. 2006. Enforcement First to Ensure Effective Institutional Controls at Superfund Sites at 3-4, OSWER 9208.2.

EPA. 2007. Monitored Natural Attenuation of Inorganic Contaminants in Groundwater, Vol. 1: Technical Basis for Assessment. EPA/600/R-07/139. Ada, OK: National Risk Management Research Laboratory, Groundwater and Ecosystems Restoration Division.

EPA. 2008a. A systematic approach for evaluation of capture zones at pump and treat systems. EPA 600/R-08/003. http://www.epa.gov/nrmrl/pubs/600R08003/600R08003.pdf.

EPA. 2008b. Significant New Use Rules on Certain Chemical Substances. 73 Fed. Reg. 65743-65,752 (November 5, 2008).

EPA. 2008c. Toxic Substances Control Act Inventory Status of Carbon Nanotubes Manufacturers. 73 Fed. Reg. 64,946-64,947 (October 31, 2008).

EPA. 2008d. EPA Engineering Issue: Indoor Air Vapor Intrusion Mitigation Approaches. www.epa.gov/nrmrl/pubs/600r08115/600r08115.html.

EPA. 2008e. Preliminary Summary of Financial Accounting Standards for Environmental Liabilities: Intangible Assets and Climate Change Risk. DRAFT Report April 28, 2008. http://www.epa.gov/osem/financial/e-disclosure.pdf.

EPA. 2009a. Perfluorooctanoic Acid (PFOA) and Perfluorooctane Sulfonate (PFOS). Provisional Health Advisory information. http://www.epa.gov/waterscience/criteria/drinking/pha-PFOA_PFOS.pdf.

EPA. 2009b. Contaminant Candidate List. Federal Register 74(194):51850.

EPA. 2009c. EPA, CERCLA, Brownfields, and Lender Liability Fact Sheet. http://www.epa.gov/brownfields/aai/lenders_factsheet.pdf.

EPA. 2009d. Memorandum from James E. Woolford, Director, Office of Superfund Remediation and Technology Innovation and John E. Reeder, Director a Federal Facilities Restoration and Reuse Office to Superfund National Policy Managers, Regions 1-10. Subject: Summary of Key Existing EPA CERCLA Policies for Groundwater Restoration at 3 (June 26, 2009). http://www.epa.gov/superfund/health/conmedia/gwdocs/pdfs/9283_1-33.pdf, citing the 1990 NCP.

EPA. 2010a. Superfund Remedy Report, Thirteenth Edition, at 10 Figure 6. EPA-542-R-10-004. Washington, DC: EPA OSWER.

EPA. 2010b. Institutional Controls: A Guide to Planning, Implementing, Maintaining, and Enforcing Institutional Controls at Contaminated Sites. Washington, DC: EPA OSWER. http://www.epa.gov/superfund/policy/ic/pdfs/PIME-IC-Guidance-Interim.pdf.

EPA. 2010c. Fourth Five-Year Review Report for Cannons Engineering Bridgewater Superfund Site Town of Bridgewater Plymouth County, Massachusetts, September 2010. http://www.epa.gov/region1/superfund/sites/cannon/470176.pdf.

EPA. 2010d. February 2010 Update: Arsenic Trioxide Superfund Site, Richland and Sargent Counties, North Dakota. http://www.epa.gov/region8/superfund/nd/arsenic/AT_5YrUpdate2010.pdf.

EPA. 2010e. A New Approach to Protecting Drinking Water and Public Health. EPA 815F10001. Washington, DC: EPA Office of Water. http://water.epa.gov/lawsregs/rulesregs/sdwa/dwstrategy/upload/Drinking_Water_Strategyfs.pdf.

EPA. 2011a. Integrated Cleanup Initiative at 18. http://www.epa.gov/oswer/docs/ici_implementation_plan.pdf.

EPA. 2011b. Recommended Evaluation of Institutional Controls: Supplement to the Comprehensive Five Year Review Guidance. Washington, DC: EPA OSWER. www.epa.gov/superfund/cleanup/postconstruction/641333.pdf.

EPA. 2011c. Regulatory Determination on Perchlorate. February 2, 2011. https://www.federalregister.gov/articles/2011/02/11/2011-2603/drinking-water-regulatory-determination-on-perchlorate2011j.

EPA. 2011d. Nanotechnology: Applications for Environmental Remediation CLU-IN Technology Focus Area Fact Sheet. Washington, DC: EPA OSWER.

EPA. 2011e. Selected Sites Using or Testing Nanoparticles for Remediation. www.clu-in.org/download/remed/nano-site-list.pdf.

EPA. 2011f. Proposed Significant New Use Rules on Certain Chemical Substances. 76 Federal Register 81,447. http://www.gpo.gov/fdsys/pkg/FR-2011-12-28/pdf/2011-33261.pdf.

EPA. 2011g. EPA Needs to Manage Nanomaterial Risks More Effectively. Report No. 12-P-0162 December 29, 2011. Washington, DC: EPA Office of the Inspector General.

EPA. 2011h. Integrated Risk Information System (IRIS), Trichloroethlyene (September 28, 2011). http://www.epa.gov/IRIS/subst/0199.htm, which is based on EPA's *Toxicological Review of Trichloroethylene* and related Excel spreadsheet.

EPA. 2011i. Potential Addition of Vapor Intrusion Component to the Hazard Ranking System. Federal Register 76(20):5370–5374.

EPA. 2011j. Basic Questions and Answers for the Drinking Water Strategy Contaminant Groups Effort. http://water.epa.gov/lawsregs/rulesregs/sdwa/dwstrategy/upload/FactSheet_DrinkingWaterStrategy_VOCs.pdf.

EPA. 2011k. EPA Handbook on the Benefits, Costs and Impacts of Land Cleanup and Reuse. http://yosemite.epa.gov/ee/epa/eerm.nsf/vwAN/EE-0569-02.pdf/$file/EE-0569-02.pdf.

EPA. 2011L. Revitalizing Contaminated Sites: Addressing Liability Concerns (2011 Edition). http://www.epa.gov/compliance/resources/publications/cleanup/brownfields/handbook/index.html.

EPA. 2012a. Regional Screening Level (RSL) Summary Table April 2012. http://www.epa.gov/reg3hwmd/risk/human/rb-concentration_table/Generic_Tables/pdf/master_sl_table_run_MAY2012.pdf.

EPA. 2012b. EPA's Vapor Intrusion Database: Evaluation and Characterization of Attenuation Factors for Chlorinated Volatile Organic Compounds and Residential Buildings. EPA 530-R-10-002. Washington, DC: Office of Solid Waste and Emergency Response.

EPA. 2012c. Superfund Vapor Intrusion FAQs. http://www.epa.gov/superfund/sites/npl/Vapor_Intrusion_FAQs_Feb2012.pdf.

EPA Region 9. 2010. Formation of a Community Informational Group (CIG) for the Motorola 52nd St. Superfund Site. May 2010 Important Announcements Fact Sheet.

Federal Judicial Center and NRC. 2011. Third Edition of the Reference Manual on Scientific Evidence. Washington, DC: The National Academies Press.

Folkes, D., W. Wertz, J. Kurtz, and T. Kuehster. 2009. Observed spatial and temporal distributions of CVOCs at Colorado and New York vapor intrusion sites. Ground Water Monitoring & Remediation 29(1):70-80.

Fujii, S., C. Polprasert, S. Tanaka, N. P. H. Lien, and Y. Qiu. 2007. New POPs in the water environment: Distribution, bioaccumulation and treatment of perfluorinated compounds—a review paper. Journal of Water Supply Research and Technology-Aqua 56(5):313-326.

GAO (Government Accountability Office). 2009. Superfund Litigation Has Decreased and EPA Needs Better Information on Site Cleanup and Cost Issues to Estimate Future Program Funding Requirements, Table 11. GAO-09-656. Washington, DC: GAO. http://www.gao.gov/assets/300/292299.pdf.

GeoSyntec Consultants. 2002. *In situ* bioremediation of perchlorate impacted groundwater. Final report for the Strategic Environmental Research and Development Program. CU-1164-FR-01.

Greskowiak, J., H. Prommer, G. Massmann, and G. Nuetzmann. 2006. Modeling seasonal redox dynamics and the corresponding fate of the pharmaceutical residue phenazone during artificial recharge of groundwater. Environmental Science & Technology 40(21):6615-6621.

Gui, L., R. W. Gillham, and M. S. Odziemkowski. 2000. Reduction of N-nitrosodimethylamine with granular iron and nickel-enhanced iron. 1. Pathways and kinetics. Environmental Science & Technology 34:3489-3494.

Gupta, S., G. Van Houtven, and M. Cropper. 1996. Paying for performance: An economic analysis of EPA's cleanup decisions at superfund sites. Rand Journal of Economics 27:563-582.

Hamilton, J. T., and W. K. Viscusi. 1999. How Costly is "Clean"? An analysis of the benefits and costs of superfund site remediations. Journal of Policy Analysis and Management 18(1):2-27.

Henderson, A. D., and A. H. Demond. 2007. Long-term performance of zero-valent iron permeable reactive barriers: A critical review. Environmental Engineering Science 24(4):401-423.

Huang, H., and G. A. Sorial. 2007. Perchlorate remediation in aquatic systems by zero valent iron. Environmental Engineering Science 24(7):917-926.

Husain, S. 2008. Literature overview: Emerging organic contaminants in water and their remediation. Remediation 18:91-105.

Hutton, D., and A. Edwards. 2011. Recent developments in Natural Resource Damages claims. Trends 42(6) July/August 2011. http://www.winston.com/siteFiles/Publications/HuttonEdwardsarticleTrends42_6July_August2011.pdf.

ICMA. 2006. International City/County Management Association ICMA Program Excellence Award Nomination, GILBERT-MOSLEY Setting a Precedent in Environmental Cleanup. http://icma.org/Documents/Document/Document/5796.

Israel, B. D. 2009. State-by-State Guide to NRD Programs in All 50 States. http://www.mass.gov/Eoeea/docs/eea/nrd/nrd_state_by_state.pdf.

ITRC (Interstate Technology & Regulatory Council). 2005a. Permeable Reactive Barriers: Lessons Learned/New Directions. PRB-4. Washington, DC: ITRC Permeable Reactive Barriers Team.

ITRC. 2005b. Perchlorate: Overview of Issues, Status, and Remedial Options. http://www.itrcweb.org/Documents/PERC-1.pdf.

ITRC. 2007. Vapor Intrusion Pathway: A Practical Guideline. Washington, DC: ITRC Vapor Intrusion Team.

ITRC. 2010. A Decision Framework for Applying Monitored Natural Attenuation Processes to Metals and Radionuclides in Groundwater. APMR-1. Washington, DC: ITRC Attenuation Processes for Metals and Radionuclides Team.

ITRC. 2011. Permeable Reactive Barrier: Technology Update. PRB-5. Washington, DC: ITRC PRB: Technology Update Team.

Johnson, P. C., H. Luo, C. Holton, P. Dahlen, and Y. Guo. 2012. Vapor Intrusion above a Dilute CHC Plume: Lessons-Learned from Two Years of Monitoring. EPA-AEHS Workshop, Recent Advances to VI Application & Implementation 20 March 2012, San Diego. https://iavi.rti.org/WorkshopsAndConferences.cfm.

Johnson, R. L., R. B. Thoms, R. O'Brien Johnson, and T. Krug. 2008. Field evidence for flow reduction through a zero-valent iron permeable reactive barrier. Ground Water Monitoring & Remediation 28(3):47-55.

Kerger, B. D., W. J. Butler, D. J. Paustenbach, J. Zhang, and S. Li. 2009. Cancer mortality in Chinese populations surrounding an alloy plant with chromium smelting operations. Journal of Toxicology and Environmental Health A 72(5):329-344.

Khandelwal, A., A. J. Rabideau, and P. Shen. 1998. Analysis of diffusion and sorption of volatile organic contaminants in soil/bentonite barrier materials. Environmental Science & Technology 32(9):1333-1339.

Kiel, K. A. 1995. Measuring the impact of the discovery and cleaning of identified hazardous waste sites on house values. Land Economics 71(4):428-435.

Kohn T., K. J. T. Livi, A. L. Roberts, and P. J. Vikesland. 2005. Longevity of granular iron in groundwater treatment processes: Corrosion product development. Environmental Science & Technology 39(8):2867-2879.

Kreuzinger, N., M. Clara, B. Strenn, and B. Vogel. 2004. Investigation on the behavior of selected pharmaceuticals in the groundwater after infiltration of treated wastewater. Water Science and Technology 50(2):221-228.

Krol, M. M., and R. K. Rowe. 2004. Diffusion of TCE through soil-bentonite slurry walls. Soil and Sediment Contamination 13:81-101.

LaRoss, D. 2012. EPA Requires Cr6 Monitoring in New List of Drinking Water Contaminants. Inside EPA, May 1, 2012.

Lau, C., K. Anitole, C. Hodes, D. Lai, A. Pfahles-Hutchens, and J. Seed. 2007. Perfluoroalkyl acids: A review of monitoring and toxicological findings. Toxicological Sciences 99(2):366-394.

Lee, A. 2012. EPA's Comments on January 3, 2012 revision to Air Monitoring and Soil Sampling Plan for Google-Fairchild Plumbing Excavation Activities, 313 and 323 Fairchild Drive, Mountain View, CA; Middlefield-Ellis-Whisman (MEW) Superfund Study Area.

MAOEEA (Massachusetts Office of Energy and Environmental Affairs). 2012. Massachusetts NRD Case Settlements. http://www.mass.gov/eea/land-use-habitats/antural-resource-damages/mass-nrd-cases/nrd-settlements.html [accessed December 17, 2011].

Massapequa Water District. 2011. Massapequa Water District Case in Opposition to NYSDEC Navy ROD OU-2. http://www.massapequawater.com.

McBean, E. A. 2008. Siloxanes in biogas from landfills and wastewater digesters. Canadian Journal of Civil Engineering 35(4):431-436.

McCluskey, J. J., and G. C. Rausser. 2003. Stigmatized asset value: Is it temporary or long-term? Review of Economics and Statistics 85(2):276-285.

Messer, K. D., W. D. Schulze, K. F. Hackett, T. A. Cameron, and G. H. McClelland. 2006. Can stigma explain large property value losses? The psychology and economics of superfund. Environmental and Resource Economics 33:299-324.

Mohr, T. 2010. Environmental Investigation and Remediation: 1,4-Dioxane and Other Solvent Stabilizers. London: CRC Press.

Morgan, J. W. 2011. Update on Hinkley Cancer Incidence. California Cancer Registry. The Criterion 4(1). http://www.ccrcal.org/DSQC_Pubs/Criterion/2011_03_Criterion.htm.

Mott, H. V., and W. J. Weber. 1991. Factors influencing organic contaminant diffusivities in soil-bentonite cutoff barriers. Environmental Science & Technology 25(10):1708–1715.

Murnyak, G., J. Vandenberg, P. J. Yaroschak, L. Williams, K. Prabhakaran, and J. Hinz. 2011. Emerging contaminants: Presentations at the 2009 Toxicology and Risk Assessment Conference. Toxicology and Applied Pharmacology 254(2):167-169.

Nakada, N., K. Kiri, H. Shinohara, A. Harada, K. Kuroda, S. Takizawa, and H. Takada. 2008. Evaluation of pharmaceuticals and personal care products as water-soluble molecular markers of sewage. Environmental Science & Technology 42(17):6347-6353.

NBER (National Bureau of Economic Research). 2010. U.S. Business Cycle Expansions and Contractions. http://www.nber.org/cycles.html, accessed April 22, 2010.

Niagara Gazette. 1979. Glynn, Don, Banks halt FHA, VA mortgage loans. Niagara Gazette (11/16/1979).

NJAG. 2011. Natural Resource Damage (NRD) Lawsuits. The State of New Jersey Department of Law & Public Safety, Office of the Attorney General. http://www.nj.gov/oag/newsreleases07/NRD-lawsuits-07/ [accessed November 8, 2011].

NJ DEP (New Jersey Department of Environmental Protection). 2005. Vapor Intrusion Guidance. Pp. 135.

NJ DEP. 2007. State Files Lawsuits Seeking Compensation for Damages Polluters Caused to Natural Resources. State of New Jersey Department of Environmental Protection. http://www.state.nj.us/dep/newsrel/2007/07_0037.htm [accessed November 8, 2011].

NJ DEP. 2008. State Secures $4 Million in Natural-Resource Damage Settlements, Preserves more than 700 Acres. State of New Jersey Department of Environmental Protection. http://www.state.nj.us/dep/newsrel/2008/08_0038.htm [accessed November 8, 2011].

Northgate Environmental Management Inc. 2008. Fairchild Buildings Slurry Wall System Efficiency Study Report, Middlefield-Ellis-Whisman Study Area Mountain View, California, Oakland CA.

NRC (National Research Council). 1994. Alternatives for Ground Water Cleanup at 245. Washington, DC: National Academy Press.

NRC. 1997. Innovations in Soil and Ground Water Cleanup. Washington, DC: National Academy Press.

NRC. 2000. Natural Attenuation for Groundwater Remediation. Washington, DC: The National Academy Press.

NRC. 2001. Classifying Drinking Water Contaminants for Regulatory Consideration. Washington, DC: National Academy Press.

NRC. 2007. Assessment of the Performance of Engineered Waste Containment Barriers. Washington, DC: The National Academies Press.

NRC. 2008. Prospects for Managed Underground Storage of Recoverable Water. Washington, DC: The National Academies Press.

NTP (National Toxicology Program). 2008. Technical Report on the Toxicology and Carcinogenesis Studies of Sodium Dichromate Dihydrate (CAS No. 7789-12-0) in F344/N Rats and B6C3F1 Mice (Drinking Water Studies). NIH Publication 08-5887; NTP TR-546. http://ntp.niehs.nih.gov/files/546_WEB_FINAL.pdf.

NYS DEC. 2005. List of Inactive Hazardous Waste Sites with Pre-2003 Remedial Decisions Where Disposal of Chlorinated Hydrocarbons Occurred. Albany, NY: New York State Department of Environmental Conservation. http://www.dec.ny.gov/regulations/51715.html and http://www.dec.ny.gov/environmentdec/52789.html.

Obare, S. O., and G. J. Meyer. 2004. Nanostructured materials for environmental remediation of organic contaminants in water. Journal of Environmental Science and Health, Part A: Toxic/Hazardous Substances and Environmental Engineering 39(10):2549-2582.

Obmascik, M. 2002. Home Deadly Home: Toxins in Air. Denver Post, January 6, 2002.

OCC (Office of the Comptroller of the Currency, Treasury), Board of Governors of the Federal Reserve System, Federal Deposit Insurance Corporation, U.S. Securities and Exchange Commission, Federal Housing Finance Agency, and Department of Housing and Urban Development. 2011. Proposed Credit Risk Retention Rule. http://www.sec.gov/rules/proposed/2011/34-64148.pdf.

Odziemkowski, M. S., L. Gui, and R. H. Gillham. 2000. Reduction of N-nitrosodimethylamine with granular iron and nickel-enhanced iron. 2. Mechanistic Studies. Environmental Science & Technology 34:3495-3500.

Olson, D. A., and R. L. Corsi. 2002. Fate and transport of contaminants in indoor air. Soil and Sediment Contamination 11(4):583-601.

Phillips, D. H., T. V. Nooten, M. I. Russell, K. Dickson, S. Plant, J. M. E. Ahad, T. Newton, T. Elliot, and R. M. Kalin. 2010. Ten year performance evaluation of a field-scale zero-valent iron permeable reactive barrier installed to remediate trichloroethene contaminated groundwater. Environmental Science & Technology 44(10):3861-3869.

Pistocchi, A., and R. Loos. 2009. A map of European emissions and concentrations of PFOS and PFOA. Environmental Science & Technology 43(24):9237-9244.

Probst, K. 2006. Critical Issues Facing the Superfund Program at 9. Washington, DC: Resources for the Future. http://www.rff.org/RFF/Documents/RFF-CTst_06-Probst.pdf.

Rabiet, M., A. Togola, F. Brissaud, J. L. Seidel, H. Budzinski, and F. Elbaz-Poulichet. 2006. Consequences of treated water recycling as regards pharmaceuticals and drugs in surface and ground waters of a medium-sized Mediterranean catchment. Environmental Science & Technology 40(17):5282-5288.

Rak, A., and C. M. Vogel. 2009. Increasing Regulation of Perfluorinated Compounds and the Potential Impacts at Air Force Installation. Poster Presentation. AFCEE EMDQ. http://www.noblis.org/MissionAreas/ees/Documents/AFCEE%20EMDQ%20Poster_2009_FINAL.pdf.

Rayne, S., and K. Forest. 2009. Perfluoroalkyl sulfonic and carboxylic acids: A critical review of physicochemical properties, levels and patterns in waters and wastewaters, and treatment methods. Journal of Environmental Science and Health Part A: Toxic/Hazardous Substances & Environmental Engineering 44(12):1145-1199.

Rhomberg, L. R., L. A. Bailey, and J. E. Goodman. 2010. Hypothesis-based weight of evidence: A tool for evaluating and communicating uncertainties and inconsistencies in the large body of evidence in proposing a carcinogenic mode of action—Naphthalene as an example. Critical Reviews in Toxicology 40:671-696.

Richardson, S. D., and T. A. Ternes. 2005. Water analysis: Emerging contaminants and current issues. Analytical Chemistry 77(12):3807-3838.

Robertson, W. D., C. J. Ptacek, and S. J. Brown. 2007. Geochemical and hydrogeological impacts of a wood particle barrier treating nitrate and perchlorate in ground water. Ground Water Monitoring & Remediation 27(2):85-95.

Rogers, C. 2007. Taking the Next Step in Environmental Due Diligence at 3, American Bar Association, Section of Environment, Energy, and Resources, 36th Conference on Environmental Law, Keystone, CO. www.advancedenvironmentaldimensions.com/documents/ED3.pdf.

Sale, T., and C. Newell. 2010. Guide for Selecting Remedies for Subsurface Releases of Chlorinated Solvents. Draft report, ESTCP Project ER-0530.

Sass, B., A. Gavaskar, W. Yoon, C. Reeter, and E. Drescher. 2002. Geochemical Factors Affecting Performance and Longevity of Permeable Reactive Barriers. Proceedings, Third International Conference on Remediation of Chlorinated and Recalcitrant Compounds, Monterey, California. Columbus, Ohio: Battelle Press.

SEC (Securities and Exchange Commission). 1982. Regulation S-K, 17 C.F.R. § 229.103 (Item 103) and SEC Regulation S-B, Item 103 and Item 103, Instructions 1-4, 17 C.F.R. §228.103. 47 Fed. Reg. 11401, Mar. 16, 1982.

SEC. 2003. Summary by the Division of Corporation Finance of Significant Issues Addressed in the Review of the Periodic Reports of the Fortune 500 Companies: Monitor of the Fortune 500 by the Division of Corporation Finance. http://www.sec.gov/divisions/corpfin/fortune500rep.htm.

SEC. 2011. Staff Accounting Bulletin 114. Release No. SAB 114 at 133-134. http://www.sec.gov/interps/account/sab114.pdf.

Shrout, J. D., A. G. Williams, M. M. Scherer, and G. F. Parkin. 2005. Inhibition of bacterial perchlorate reduction by zero-valent iron. Biodegradation 16(1):23-32.

Shrout, J. D., G. C. Struckhoff, G. F. Parkin, and J. L. Schnoor. 2006. Stimulation and molecular characterization of bacterial perchlorate degradation by plant-produced electron donors. Environmental Science & Technology 40(1):310-317.

Simcik, M. F. 2005. Global transport and fate of perfluorochemicals. Journal of Environmental Monitoring 7(8):759-763.

Singh, A., E. Neuhauser, and A. J. Rabideau. 2011. Dealing with censored vapor concentration data, Case study of former Manufactured Gas Plant sites. 27th Annual Conference on Soil, Sediments, Water, and Energy, Amherst, MA, October 17-21.

Snyder, S. A., J. Leising, P. Westerhoff, Y. Yoon, H. Mash, and B. Vanderford. 2004. Biological and physical attenuation of endocrine disruptors and pharmaceuticals: implications for water reuse. Ground Water Monitoring & Remediation 24(2):108-118.

Soliman, M. A., J. A. Pedersen, H. Park, A. Castaneda-Jimenez, M. K. Stenstrom, and I. H. Suffet. 2007. Human pharmaceuticals, antioxidants, and plasticizers in wastewater treatment plant and water reclamation plant effluents. Water Environment Research 79(2):156-167.

Son, A., J. Lee, P. C. Chiu, B. J. Kim, and D. K. Cha. 2006. Microbial reduction of perchlorate with zero-valent iron. Water Research 40(10):2027-2032.

Spina, P. 2008. State Environmental Easement Statutes as ARARs: A New Option for Enforcing Institutional Controls at Superfund Sites. Columbia Journal of Environmental Law 33:122-129. www.columbiaenvironmentallaw.org/assets/pdfs/33.1/Spina.33.1.pdf.

Standley, L. J., R. A. Rudel, C. H. Swartz, K. R. Attfield, J. Christian, M. Erickson, and J. G. Brody. 2008. Wastewater-contaminated groundwater as a source of endogenous hormones and pharmaceuticals to surface water ecosystems. Environmental Toxicology and Chemistry 27(12):2457-2468.

Steffan, R. J. 2007. Biodegradation of 1,4-Dioxane. Final report for the Strategic Environmental Research and Development Program. ER-1422.

Suh, J. H., and M. Mohseni. 2004. A study on the relationship between biodegradability enhancement and oxidation of 1,4-dioxane using ozone and hydrogen peroxide. Water Research 38(10):2596-2604.

Szecsody, J. E., J. P. McKinley, F. H. Crocker, A. T. Breshears, B. J. Devary, H. L. Fredrickson, and K. T. Thompson. 2009. Abiotic and Biotic Mechanisms Controlling *In Situ* Remediation of NDMA, Final Report. SERDP Project ER-1421. SERDP ER-1421. Richland, WA: Pacific Northwest National Laboratory.

Tanner, R. 2009. Sunburst Sch. Dist. No. 2 V. Texaco, Inc.: Rethinking Restoration Damages as a Remedy for Tortious Injury to Land and Property. http://www.montanalawreview.com/sitebuildercontent/sitebuilderfiles/publication-ready_casenote.pdf.

Technical Review Committee. 1988. Love Canal Emergency Declaration Area Habitability Study, Final Report, Volume I: Introduction and Decision-Making Documentation at 1-2. http://nepis.epa.gov/EPA/html/DLwait.htm?url=/Exe/ZyNET.exe/9100HT40.PDF?ZyActionP=PDF&Client=EPA&Index=1986 Thru 1990&File=D%3A%5CZYFILES%5CINDEX%20DATA%5C86THRU90%5CTXT%5C00000021%5C9100HT40.txt&Query=902488002A%20or%20Love%20or%20Canal%20or%20Emergency%20or%20Declaration%20or%20Area%20or%20%E2%80%93Remediation%20or%20EDA%20and%20%E2%80%93%20or%20Final%20or%20Study%20or%20Report&SearchMethod=1&FuzzyDegree=0&User=ANONYMOUS&Password=anonymous&QField=pubnumber%5E%22902488002A%22&UseQField=pubnumber&IntQFieldOp=1&ExtQFieldOp=1&Docs=.

Thompson, C. M., D. M. Proctor, M. Suh, L. C. Haws, C. D. Hébert, J. F. Mann, H. G. Shertzer, J. G. Hixon, and M. A. Harris. 2012. Comparison of the effects of hexavalent chromium in the alimentary canal of F344 rats and B6C3F1 mice following exposure in drinking water: Implications for carcinogenic modes of action. Toxicological Sciences 125(1):79–90.

University of Buffalo. 2011. Love Canal Collection. http://library.buffalo.edu/specialcollections/lovecanal/about/chronology.php.

USGS. 2003. Ground-Water Depletion Across the Nation. U.S. Geological Survey Fact Sheet 103-03.

Vainberg, S., K. McClay, H. Masuda, D. Root, C. Condee, G. Zylstra, and R. J. Steffan. 2006. Biodegradation of ether pollutants by *Pseudonocardia* sp. strain ENV478. Applied and Environmental Microbiology 72(8):5218-5224.

Vaseashta, A., M. Vaclavikova, S. Vaseasht, G. Gallios, P. Roy, and O. Pummakarnchana. 2007. Nanostructures in environmental pollution detection, monitoring, and remediation. Science and Technology of Advanced Materials 8:47-59.

Vincent, G. K., and V. Velkoff. 2010. The Next Four Decades: The Older Population in the United States: 2010 to 2050. Washington, DC: U.S. Census Bureau.

Waldemer, R. H., and P. G. Tratnyek. 2006. Kinetics of contaminant degradation by permanganate. Environmental Science & Technology 40(3):1055-1061.

Weinfield, N. H. 2007. A Brownfield model collapses under the weight of litigation: City of Wichita v. Trustees of the Apco Oil Corporation Liquidating Trust. Proceedings of the Annual International Conference on Soils, Sediments, Water and Energy: Vol. 12, Article 37. http://scholarworks.umass.edu/soilsproceedings/vol12/iss1/37.

Wilkin, R., and R. Puls. 2003. Capstone Report on the Application, Monitoring, and Performance of Permeable Reactive Barriers for Groundwater Remediation: Volume 1—Performance Evaluation at Two Sites. EPA/600/R-03/045a. Washington, DC: EPA.

Woodhouse, K. S. 2009. Comments on July 2009 Proposed Plan for the VI Pathway. http://www.cpeo.org/pubs/MV-VIPolicy.pdf.

Yabusaki, S., K. Cantrell, B. Sass, and C. Steefel. 2001. Multicomponent reactive transport in an in situ zero-valent iron cell. Environmental Science & Technology 35(7):1493–1503.

Yonkers, T. 2012. Army Urges Hill To Block EPA Policy Changes Midway Through Cleanups. Inside EPA Daily News. March 12, 2012.

Yu, X., C. Amrhein, M. A. Deshusses, and M. R. Matsumoto. 2006. Perchlorate Reduction by Autotrophic Bacteria in the Presence of Zero-Valent Iron. Environmental Science & Technology 40(4):1328-1334.

Zhou, Q., S. McCraven, J. Garcia, M. Gasca, T. A. Johnson, and W. E. Motzer. 2009. Field evidence of biodegradation of N-nitrosodimethylamine (NDMA) in groundwater with incidental and active recycled water recharge. Water Research 43(3):793-805.

6
Technology Development to Support Long-Term Management of Complex Sites

Despite years of characterization and implementation of remedial technologies, many complex federal and private industrial facilities with contaminated groundwater will require long-term management actions that could extend for decades or longer. As discussed in Chapter 2, the Department of Defense (DoD) manages a substantial number of such sites. Chapter 4 concluded that the further application of existing remediation technologies is likely to provide only incremental progress in achieving restoration at the most complex sites. Thus, for these sites the management challenges include optimization of active remedies, reducing mass flux/mass discharge of contaminants from source areas such that natural attenuation may be effective, or ensuring that any active or passive engineered containment system will remain effective over the long term. This chapter discusses technological developments that can aid in addressing these management challenges—in particular, providing the scientific and technical bases for transitioning from active remediation to more passive strategies where applicable.

Optimization of remedial technologies, transitioning to active or passive containment, and improving long-term management can be achieved through (1) better understanding of the spatial distribution of contaminants, exposure pathways, and processes controlling contaminant mass flux and attenuation along exposure pathways; (2) improved spatio-temporal monitoring of groundwater contamination through better application of conventional monitoring techniques, the use of proxy measurements, and development of sensor-based monitoring technologies; and (3) application of emerging diagnostic and modeling tools. In addition to these topics, the chapter explores emerging remediation technologies that have yet to receive extensive field testing and evaluation, and it reviews the state of federal funding for relevant research and development and provides recommendations on research topics relevant to the future management of complex sites where groundwater restoration is unlikely.

SITE CONCEPTUALIZATION

The decision to transition a site from active remediation to long-term management requires a thorough understanding of the geologic framework, history of contamination events, the current location and phase distribution of contaminants, temporal processes that affect groundwater flow and chemical migration, and interactions at hydrogeologic and compliance boundaries. The combined understanding of these factors, referred to here as *site conceptualization*, supports the development of specific management tools such as the conceptual site model (CSM, see Chapter 4) and mathematical models. Typically, the site conceptualization and associated tools are updated as the project progresses from discovery of contamination through closure or transition to long-term management, with the degree of detail dependent on the nature of the contamination and the physical dimensions of the site. The development and enhancement of an accurate and suitably detailed site conceptualization is an important component of addressing future management challenges at these sites including the transition to long-term management.

The current cleanup paradigm distinguishes the *source zone* from the downgradient *plume*, in terms of treating each region differently with respect to characterization and remediation, and it acknowledges the dominant role of geologic heterogeneity in controlling contaminant removal from both regions. In NRC (2005), hydrogeologic heterogeneity was conceptually captured by identifying five generic geologic environments ranging from nearly uniformly homogeneous, unconsolidated porous media (Type I) to fractured rock and carbonate aquifers (Types IV and V). More recently, a 14-compartment model has been proposed (Figure 4-1; Sale and Newell, 2011; ITRC, 2011), in which contaminants can reside in groundwater, sorbed, and vapor phases, either within the source zone or the plume, and which are further subdivided into high- and low-permeability regions. In the high-permeability regions, advective transport will control contaminant migration, while in the low-permeability regions, the dominant transport mechanism is molecular diffusion. The advantage of such multi-compartment conceptual models is the ability to focus on the exchange of contaminant mass between specific compartments that can limit the rate and extent of remediation, recognizing that the controlling processes can change over time.

The 14-compartment framework highlights characterization challenges that significantly influence optimization of remedial actions and the transition to long-term management, including the source/plume distinction, spatial heterogeneity in hydraulic conductivity, and the potential role of the vapor pathway when volatile organic compounds (VOCs) are present. A more comprehensive application of the framework that fully accounts for the relative magnitudes of contaminant mass in each of the compartments and the rates of mass transfer between compartments will require further development to better understand (1) the potential roles of desorption and of back-diffusion from low-permeability compartments to advective zones, (2) the variety of aquifer materials and conditions that comprise the "less transmissive" compartments, (3) the reactive characteristics of the aquifer that control the potential success of long-term strategies such as monitored natural attenuation, and (4) the complex factors that control the fate of volatile contaminants, which can exhibit markedly different behavior at seemingly similar sites because of variability in subsurface conditions, building characteristics at the soil interface, and climate conditions. Each of these issues is further explored below.

Back-Diffusion and Desorption

For many complex sites that have been subject to partial or complete source removal, the transition to long-term management is largely controlled by volatilization into the vapor phase (if applicable) and transport into the aqueous phase plume, as these two phases are the primary media for both off-site contaminant migration and the biotic and abiotic transformation processes associated with natural attenuation. Current conceptualizations of the plume have focused on three potential sources of contaminant mass influx in the groundwater: (1) discharge from undetected mass remaining in the upgradient source zone, (2) aqueous back-diffusion from aquifer materials to the pore water within low-permeability plume material and subsequent diffusive transport to advective zones, and (3) mass transfer (desorption) from aquifer sediments within both transmissive and low-permeability plume materials. For successful transition to long-term management, the contaminant influx from these three processes must be balanced by natural attenuation processes or controlled by physical/hydraulic containment.

The potential loading of dissolved mass from the source zone to the plume has received considerable attention and is straightforward to assess because the mass discharge occurs at the boundary of, rather than within, the plume compartment. However, back-diffusion and desorption of contaminants from materials *within* the plume are much more difficult to analyze because they are spatially nonuniform, dependent on the history of the source and plume migration, and are not easily measurable. In particular, measured groundwater concentrations provide only limited insight into the processes responsible for the persistence of dissolved contaminant plumes because it is difficult to distinguish the relative influence of flow field heterogeneity, back-diffusion, and desorption.

The potential importance of back-diffusion is supported by conceptual and modeling analysis (e.g., MacKay and Cherry, 1989; Wilson, 1997; Parker et al., 2008) and a limited number of field investigations that have directly sampled aquitard material (Ball et al., 1997; Chapman and Parker, 2005). Sorption processes are typically included in contaminant transport models and estimates of time to remediate, although the common use of the retardation factor reflects the optimistic assumptions of a single sorbent and rapid linear partitioning. A considerable body of research over the past two decades has demonstrated that, for many aquifer materials, sorption processes are in fact spatially heterogeneous, nonlinear, and potentially limited by solute diffusion to sorbent material located within the interior of soil particles (e.g., as reviewed by Allen-King et al., 2002). As with back-diffusion, conceptual and modeling analyses have shown that nonlinear and/or rate-limited desorption can potentially contribute to plume persistence over decades (e.g., Ball and Roberts, 1991; Rabideau and Miller, 1994; Rivett et al., 2006). However, at the time of this writing, there is a lack of field data and characterization techniques to distinguish desorption processes from other nonideal effects. A modest step toward better understanding the potential role of sorption processes would be to routinely characterize the organic content of collected soil samples (Simpkin and Norris, 2010), a task that could be accomplished at relatively low cost.

Understanding whether back-diffusion and desorption are occurring at a site is challenging because the relative importance of each process is highly dependent on the site-specific contamination history and the presence and distribution of low-permeability and/or strongly sorbing materials. And yet, current site characterization techniques typically do not fully delineate the structure of these materials, particularly when they are distributed over small spatial scales within the plume interior. Furthermore, there are no proven remedial techniques to preferentially target and accelerate the removal of contaminants from localized sites that are desorption/diffusion limited. Finally, currently used mathematical models are difficult to configure to provide realistic predictions of time to remediation when desorption/diffusion processes are the limiting factor because of the need to assign initial conditions that properly represent the mass located in immobile compartments. Additional research is needed to develop strategies for long-term management that focus on plume zone processes that contribute to plume longevity rather than the processes that occur in the source zone.

Representing Complex Geologic Environments

The 14-compartment model of Sale and Newell (2011) assigns "low-permeability" compartments to both the source and plume domains, highlighting the potential role of back-diffusion in both domains. Such an approach is conceptually similar to the classification scheme proposed by NRC (2005), which included a hierarchy of five geologic environments ranging from nearly uniformly homogeneous, unconsolidated porous media (Type I) to fractured rock and carbonate aquifers (Types IV and V). While both schemes distinguish between contaminants in "mobile" and "immobile" groundwater, the five-region classification recognizes two subtle but potentially significant differences not captured by the 14-compartment model. First, the diffusion rate and storage capacity of contaminants in low-permeability geologic materials can differ substantially among clays, fractures, and/or intrinsic porosity of indurated rock. Second, in addition to providing potential sinks for diffusive exchange of contaminants, some complex domains (highly heterogeneous unconsolidated porous media, fractured rock, karst) are often characterized by large variations in the groundwater velocity. Hence efforts to characterize "complexity" understood in terms of spatial variability must consider both groundwater flow and contaminant transport *within* and *between* discrete compartments, regardless of how such compartments are delineated.

Differences in the diffusion process are relatively straightforward to account for, but require appropriate specification of the geometry and diffusion characteristics of the low-permeability material. In some cases, the necessary information is provided by field characterization, but for many problems of interest, such as diffusion out of thin clay lenses, the relevant diffusion path length is difficult to determine. Similarly, accounting for variation in advective transport pathways typically requires a very detailed conceptualization of the groundwater flow field, particularly the low-permeability features. For example, spatial variations in the hydraulic conductivity of unconsolidated media can lead to preferential pathways in aquifers over significant distances, similar to characteristics associated with fractured rock and karst formations. Such paths of preferential groundwater flow often control the distribution of contaminant mass in both source areas and downgradient plumes, and must be properly considered in the design and implementation of containment and remediation strategies. Chapman et al. (2010) present an example of how information from detailed site characterization can be incorporated into a remedial design that yields good performance despite the presence of preferential flow paths. However, while available modeling tools are increasingly capable of incorporating detailed descriptions of hydraulic conductivity heterogeneity (e.g., see Guilbeault et al., 2005), the requirements for additional site characterization can represent a considerable burden on site management.

Transformation Capacity

As discussed in Chapter 7, monitored natural attenuation (MNA) is the dominant process during long-term management at sites not relying on physical or hydraulic containment. Knowledge of the biogeochemical environment and the identification of potentially important reactive pathways for the target contaminants are necessary prerequisites for initiating MNA after the transition to long-term management has occurred. Relevant considerations include bulk aquifer properties such as mineral composition and pore water chemical constituents, as well as the presence of the necessary microbial consortia. Contaminant transformation during MNA can occur through microbial pathways, abiotic mechanisms, or in many cases a combination of both.

Of critical importance to the aquifer "transformation capacity" for MNA is the spatial pattern of redox zonation. Redox zonation occurs as a result of microbial metabolism where in a homogeneous system terminal electron acceptors with the most favorable free energies are preferably used before the next one can be utilized (termed the "redox ladder" by Borch et al., 2010). Complex sites, however, may have areas of overlapping or patchy redox zonation whereby microbial communities that utilize different terminal electron acceptors can co-exist. Determining whether the site is fully oxic, has extensive zones of anoxia, or is comprised of these patchy suboxic/anoxic regions in conjunction with the target contaminant composition is critical to determining the appropriateness of MNA (Rugge et al., 1998; Hofstetter et al., 1999).

Another important parameter in contaminant transformation is the presence of reactive minerals associated with aquifer solids, such that characterizing these chemical factors can yield clues about the potential effectiveness of MNA. A variety of naturally occurring iron and manganese oxides, iron sulfide minerals, and clays with iron moieties have been shown to be highly reactive and can act as respective reductants and oxidants in abiotic attenuation pathways (Kappler and Straub, 2005; Hofstetter et al., 2003; Neumann et al., 2009; He et al., 2009). Microorganisms play an important role in the controlling both the type and stability of these minerals since many organisms are capable of utilizing mineral oxides as terminal electron acceptors (Lovley, 1993; Tebo et al., 2004). Under some circumstances the microbial population can convert iron oxides to reactive media useful for MNA by producing Fe(II), which can either be chelated by natural ligands, be adsorbed to the remaining iron oxides to create highly potent reductants, or react with sulfides (if sulfate is in abundance as a terminal electron acceptor) to produce potentially reactive iron sulfide minerals (Hakala et al., 2007; Hakala and Chin, 2010). In other cases, however, reduction of manganese oxides (which can mediate oxidation reactions) may result in a decrease in potential MNA. In aquifer pore waters, reactive species such as natural organic matter and reduced sulfur species (bisulfide, polysulfides, and organic thiols) play an important role in MNA by acting as reductants and electron mediators (Kappler and Haderlein, 2003; Hakala and Chin, 2010). Natural organic matter significantly increases the reactivity of reduced sulfur species by acting as an electron mediator, and is an important reductant in sulfur-rich aquifers (Dunnivant et al., 1992).

An example of a well-characterized site with high transformation capacity amenable to MNA is Altus Air Force Base, which has abundant levels of both sulfate and Fe(III) (Kennedy et al., 2006). Microbial metabolic activity at this site produced potent reactive reductants such as reduced sulfur compounds, Fe(II), and iron sulfide minerals, which were capable of abiotically transforming TCE and its derivatives. These investigators reported the absence of sulfate in the area of the TCE plume and the existence of abundant iron sulfide minerals. Further they found no TCE in the area where iron sulfides are abundant and only trace levels of byproducts, suggesting that MNA was occurring.

While much is known about the biological/abiotic conditions necessary to effect contaminant transformation during MNA, there is not yet a complete protocol to determine the extent to which such conditions are present at a site and whether contaminants are being degraded. The tools discussed later in this chapter represent important initial steps toward the development of such a protocol.

Vapor Intrusion Issues

As described in Chapter 5, the vapor intrusion pathway is increasingly considered at complex sites with DNAPL contamination. This pathway can be conceptualized as three distinct zones (Figure 6-1): (1) the source zone where contaminant is immobilized, (2) the subsurface migration pathway, and (3) the influence zone of the building. The management of vapor intrusion requires expanded site characterization, an interpretation of the several types of vapor concentration measurements in the context of site-specific conditions, and, if necessary, development of appropriate mitigation strategies if source removal measures are insufficient to reduce exposure to acceptable levels.

Characterization of the vapor pathway is challenged by the fact that each component is subject to considerable spatial and temporal variability. Fluctuating water table conditions controlled by recharge, pumping, and stream–aquifer interactions can result in transient vapor flux generation at

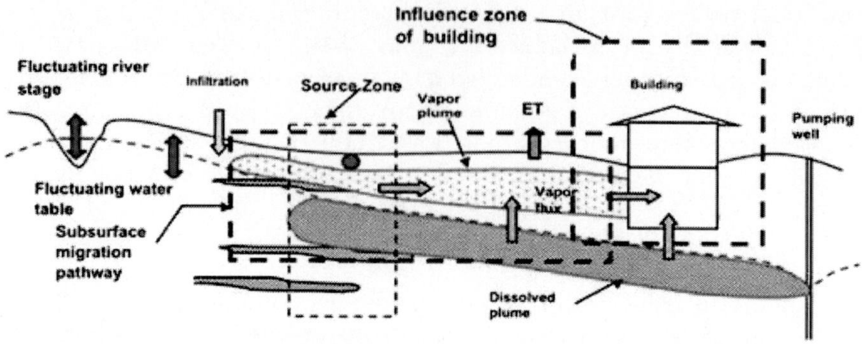

FIGURE 6-1 Vapor intrusion pathways.

the sources. The migration pathway from source to building is significantly affected by changes in soil moisture, temperature, wind, and ambient pressure, and in some cases, biogeochemical transformation processes. Vertical migration is also influenced by changes in building ventilation and heating, ventilation, and air conditioning systems operation. Finally, attempts to characterize the pathway via indoor air sampling can be confounded by indoor sources of contamination.

Among the available guidance for assessing vapor intrusion (e.g., Johnson et al., 1999; Hay-Wilson et al., 2005; McAlary et al., 2005; NYSDOH, 2006; ITRC, 2007), federal guidance is evolving toward an approach based on multiple lines of evidence that involves sampling of indoor air, subslab soil gas, deeper soil gas, groundwater, and soil—in combination with screening-level modeling and empirical assessment (e.g., EPA, 2002, 2011a,b, 2012a,b,c). This reflects experiences with conflicting lines of evidence at some sites, recognition that there will likely be spatial variability in pathway sampling results, low confidence in our ability to correctly interpret the data, and a limited peer-reviewed knowledge base to rely upon. This also suggests that assessment paradigms that rely on too few samples (in space and time) are limited.

Vapor intrusion from groundwater plumes with chlorinated solvents is especially challenging to characterize, partly because such plumes can vary widely in size. Where large plumes encompass an entire neighborhood, assessment of all potentially affected buildings may be impracticable. Furthermore, it is not always the case that the greatest indoor air impacts are found in buildings overlying the highest groundwater concentrations. Groundwater-related vapor intrusion has been documented in some buildings overlying dissolved chlorinated solvent groundwater concentrations as low as 10 μg/L, and no impacts have been observed in other buildings overlying groundwater concentrations as great as 10–100 mg/L (EPA, 2012b).

A number of commercial products can serve as indoor sources of chlorinated solvent vapors, so that interpreting indoor air quality and subslab soil gas data is not always straightforward (Gorder and Dettemmaier, 2011). As a case in point, approximately 3,000 residences overlie chlorinated solvent groundwater plumes originating from Hill Air Force Base, although monitoring has indicated that a very small percentage of the residences have indoor air impacts attributable to groundwater contamination. Detailed study beyond typical pathway assessment monitoring identified numerous indoor air sources of contaminants, including household cleaning products, craft supplies, gun cleaners, and holiday ornaments—leading to a list of 72 household products known to contain TCE and almost another 2,000 products known or suspected of containing chlorinated solvents.

A solid technical basis is lacking for determining which scenarios require indoor sampling and what sampling frequency and duration are appropriate, both over the short term (i.e., daily) and long term (i.e., seasonal). Studies suggest that vapor intrusion emissions into buildings can fluctuate on time scales ranging from days to weeks (Luo, 2009; Luo et al., 2010; Johnson et al., 2012). Research by McHugh et al. (2010) suggests that changes in indoor air concentrations may be different for chemicals emanating from groundwater than those emanating from indoor chemical sources, such that temporal data might be used to distinguish between indoor air impacts from these two sources. However, even with detailed indoor air monitoring data, the issue of temporal variability is further complicated by the dynamics of volatilization from the groundwater plume, which is affected by groundwater table elevation, moisture infiltration rates, moisture profiles, and other climate factors (Sakaki et al., 2013). In general, the temporal changes in the vapor emission rates from groundwater have yet to be studied in great detail and further study is needed to more intelligently design sampling plans.

Because the costs and complexity of vapor intrusion assessment have been increasing without a commensurate increase in the mechanistic understanding of the exposure pathway, the resulting response actions reflect a conservative management approach.

MONITORING

Monitoring of groundwater is conducted over the entire life cycle of a complex site and can represent a significant percentage of life-cycle costs if residual contamination remains after active remediation has been completed, especially when monitoring extends over multiple decades. Traditionally, the monitoring of temporal changes in groundwater contamination relied on conventional well sampling, which is labor intensive and requires costly laboratory analyses. Given that tens to hundreds of monitoring wells are present at most sites, and standard quarterly sampling is often required, estimates of monitoring can exceed $100 million per year at DoD facilities alone, which represents a significant percentage of the financial resources dedicated to remediation efforts. Furthermore, the traditional two-dimensional resolution of monitoring well networks (which produce vertically averaged concentration values) may be insufficient to support an accurate site conceptualization, particularly for highly heterogeneous formations.

Continued development of conventional monitoring techniques has led to more detailed characterization of the distribution of dissolved contaminants, particular in the vertical dimension. However, to support a cost-effective transition to long-term management, additional tools are needed. This section addresses ongoing developments in (1) optimization of conventional monitoring systems, (2) techniques for measuring contaminant flux, (3) sensor technology, and (4) new tools that can be applied to better understand whether MNA is working.

Improved Application of Conventional Monitoring Tools

The deployment of conventional site characterization tools has evolved in a manner that has emphasized greater spatial resolution in regions where contamination is significant. In particular, multi-level monitoring and nested well systems now enable the collection of hydraulic head data and groundwater samples over relatively short vertical intervals (ITRC, 2004; Einarson, 2006; Einarson et al., 2010; Kavanaugh and Deeb, 2011). Although more costly than conventional 2-D monitoring, multi-level monitoring systems can lead to more streamlined and accurate remedial investigations and long-term management.

Formal simulation/optimization techniques have been developed to improve the design of monitoring programs—a process sometimes termed *long-term monitoring optimization* (LTMO). These applications are in a relatively early stage of development and a variety of approaches are available to formulate and solve the optimization problem. For example, one approach might be to analyze the value of information provided by an existing monitoring network to identify monitoring wells that are spatially redundant and could be removed (e.g., Reed et al., 2000, 2001; Babbar-Sebens and Minsker, 2008). Most work to date has focused on monitoring frequency and spatial resolution of well networks, with less attention given to issues such as the number and selection of analytes, sampling analytical techniques, and data processing. In a pilot study comparing two software-driven LTMO systems, the U.S. Environmental Protection Agency (EPA) suggested that annual savings of a few hundred to tens of thousands of dollars might be achievable, particular for sites where more than 50 samples are collected and analyzed annually (EPA, 2004). EPA subsequently issued a "road map" to assist managers with developing a site-specific LTMO program (EPA and USACE, 2005), including user-friendly software tools. Although the underlying concepts are fairly well established, additional documentation of successful case studies would clarify the range of potentially achievable cost savings.

Monitoring of Source Zone Contamination

The successful design of a source zone remediation program depends on sufficiently detailed knowledge of the spatial pattern of immobile source materials. A number of recent reviews have evaluated the variety of tools available to quantify the magnitude and spatial distribution of DNAPL (e.g., NRC, 2005; Mercer et al., 2010). These tools range from low-cost methods to infer the presence of DNAPL (as reviewed by Kram et al., 2001) to more extensive methods designed to delineate the spatial distribution of NAPL saturation to guide source zone remediation (e.g., Saenton and Illangasekare, 2004; Moreno-Barbero and Illangasekare, 2005, 2006). For the latter purpose, the partitioning interwell tracer test (PITT) has proven to be relatively effective (e.g., Annable et al., 1998; Brooks et al., 2002), although its deployment is hindered by high cost and need for relatively sophisticated interpretive tools.

As it is unlikely that complete removal of contaminant source material will be feasible for many complex sites, the transition to long-term management will depend not only on the amount of source mass removed, but on the rate at which mass is transferred between the source and plume compartments during the post-remediation period. One of the most promising recent developments in source zone management is the development of tools for measuring contaminant *mass flux*, either at localized monitoring points or as an integrated *mass discharge* across a control plane. Such knowledge of contaminant discharge is particularly useful in evaluating the potential for downgradient natural attenuation processes.

Conceptually, contaminant discharge is a calculated parameter that reflects both temporal and spatial averaging of the product of groundwater discharge (length per area per time) and contaminant concentration (mass per volume). Field methods include synoptic sampling (e.g., Einarson, 2006), passive flux meters (Annable et al., 2005; Basu et al., 2006), steady-state pumping (e.g., Buschek, 2002), recirculation flux measurements (Goltz et al., 2007), integral pumping tests (Bockelmann et al., 2001; Bauer et al., 2004), and modified integral pumping tests (Brooks et al., 2008). The use of flux measurements as an alternative to concentration-based metrics offers several advantages relevant to long-term management, including less sensitivity to spatial/temporal variability and correspondence with screening models that attempt to correlate source zone mass removal with downgradient plume behavior.

Several recent reviews have explored the relative performance of various techniques for measuring mass flux, which appear to be highly site-specific (EPA, 2009; ITRC, 2010; Kavanaugh and Deeb, 2011). Additional field research is needed to support the more widespread adoption of flux-based performance metrics, including (1) further clarification of the range of uncertainty associated with mass flux and mass discharge measurements, (2) continued refinement of specific aspects of the various techniques, including a better definition of the necessary preliminary site characterization, and (3) new measurement techniques.

Sensor Technology

Because existing monitoring and performance assessment tools are expensive, slow, and consist of point measurements, real-time measurements could provide data for management decisions including optimization of active remedies and assurance that either active or passive containment is effective. Recent advances in microelectronics, wireless communication technologies, and information technologies have produced potentially low-cost techniques to gather and process large amounts of data at very high spatial and temporal resolution. Such wireless sensor networking could be applied to a variety of subsurface systems.

Advances in the development of wireless sensor networks have been successfully applied to problems in infrastructure monitoring, weather and storms, volcanoes, air quality, agriculture, forestry, and ecology (e.g., Culler et al., 2004; Haenggi, 2005; Werner-Allen et al., 2006). Much of this work has highlighted the advantages of deploying a large number of inexpensive sensors to replace of a few highly accurate, but expensive sensors. While environmental monitoring has been considered an ideal application since the field's inception, only a few projects have combined wireless sensing with subsurface monitoring, largely because of the technical difficulty and cost associated with monitoring VOCs in groundwater environments. For example, a study by EPA (2003) concluded that although a sensor might cost as little as $100 to manufacture, a fully developed multiparameter sensor suitable for long-term management applications would cost around $7,500. More recently, an ESTCP-sponsored project (Lieberman, 2007) evaluated sensors for monitoring VOCs, including Halogen-Specific Detector/Membrane Interface Probe systems and laser-induced fluorescence, based on ROST (rapid optical screening tool). However, despite some advances in detection capabilities, the relatively large costs (thousands of dollars per sensor) inhibit deployment in a wireless sensor network setting, and such developments have not advanced to the stage of commercialization. In addition to cost, other outstanding issues must be resolved if wireless sensor network technology is to be adopted for long-term monitoring of groundwater plumes, including the scope and accuracy of contaminant-specific sensors, signal transmission issues in subsurface environments, the mode and frequency of sensor failure, and the availability and efficiency of power sources.

New developments in sensor technology for vapor monitoring could contribute to more effective management of the vapor intrusion pathway. Point-in-time sample collection using Summa Canisters is the standard indoor-air sampling approach. These are limited to time-integration periods of a few days at best, which is likely inadequate for pathway assessment (Luo, 2009; Luo et al., 2010). The ideal vapor intrusion sensing system would be capable of assessing (1) whether conditions exist that can cause unacceptable vapor intrusion (i.e., periods of building under-pressurization, or contaminants present in soil gas at levels above concentrations of concern), (2) the actual impact of vapor intrusion on indoor air quality (i.e., indoor air monitoring), (3) the actual exposure of building occupants to vapor intrusion (i.e., simulated uptake monitoring), and, in the case of mitigation systems, (4) whether the mitigation system is meeting operational goals that eliminate the vapor intrusion pathway (i.e., maintaining a building over-pressurization condition). These sensing systems would need to measure pressure differentials of 0 to 5 Pa without drift for extended periods of time and reliably quantify vapor contaminant concentrations in the 0.1–100 ppb_v range, under a range of humidity conditions and for sampling durations of a few minutes to a few days and over periods of many months. One vision for future sensing systems is something like a household CO monitor with real-time data communication to a home computer, tablet, or PDA to increase occupants' awareness of their indoor air quality.

Evaluating Monitored Natural Attenuation

Chapter 7 discusses the possibility, at many complex sites, of a transition from active source zone remediation to more passive strategies such as MNA or natural attenuation without monitoring. Critical needs in implementing MNA are verification that contaminant transformation is occurring and that the required bacteria are present and active (if biodegradation is the principal attenuation mechanism).

Verification of contaminant transformation can be accomplished by direct groundwater monitoring for the contaminants of concern or alternatively via geophysical techniques, which provide a noninvasive means of identifying changes in biogeochemical conditions in groundwater. Several geophysical parameters are sensitive to redox gradients, microbial activity, and changes in pore-fluid chemistry. For example, changes in *electrical resistivity* can reflect changes in contaminant concentrations, microbial abundances, and the distribution of amendments that promote contaminant degradation; *electrodic potential* is sensitive to the local redox chemistry; *self-potential* can measure natural electrical current sources arising from redox zonation due to contaminant degradation; and *induced polarization* provides evidence of processes near fluid-grain boundaries to infer microbial abundances. Although geophysical measurements can be conducted in existing monitoring wells using removable sensors, the underlying geologic variability in the aquifer may yield subtle variations that do not provide a distinct geophysical signal for particular biogeochemical conditions. Thus, it is usually necessary to monitor geophysical parameters over time and compare parameter values to background conditions.

The additional new technologies reviewed below could enable a more immediate and focused observation of transformation processes relevant to MNA. Not all of these techniques will be required at all sites to document the occurrence of MNA. Laboratories at research universities are able to perform such analyses, and commercial laboratories are beginning to offer such services as well.

Tools from Molecular Biology

New molecular biology tools have facilitated the direct observation of the relevant microbial processes and have enhanced the discovery of new enzymes and biochemical pathways that can be applied to MNA. Determining if site-specific bacteria are capable of degrading the target contaminants can be accomplished using *genomic* tools to detect and quantify gene copies, such as quantitative polymerase chain reaction (qPCR) and PCR-denaturing gradient gel electrophoresis (PCR-DGGE). Similarly, *proteomics* can be used to identify and quantify protein biomarkers that are produced as a stress response during the degradation of contaminants (Nesatyy and Suter, 2007). Box 6-1 describes recent field research using these methods.

Another method for measuring gene expression is *transcriptomics*, which creates complementary DNA from extracted mRNA. Transcriptomics is not as representative of microbial activity as proteomics (Belle et al., 2006), but it does not have the same bias due to database limitations. If new sequences are discovered via transcriptomics, it is then possible to detect them using proteomics.

Metabolomics is the study of the small molecules (e.g., metabolites) produced by cellular processes in response to the environment; monitoring their changes may be a means to verify groundwater contaminant biotransformation (Singh, 2006). For example, metabolic biomarkers for BTEX and

BOX 6-1
"Omics" in the Laboratory and Field

Genomics. Microbial communities at contaminated sites often contain the genes necessary for degradation of BTEX (Hendrickx et al., 2006; Beller et al., 2008; Kao et al., 2010) and chlorinated solvents (Hendrickson et al., 2002; Carreon-Diazconti et al., 2009), and resistance to metals (Waldron et al., 2009). The presence/absence of genes has also been correlated to the rates of reductive dehalogenation of chlorinated ethenes at sites where natural attenuation (Lu et al., 2006; Burgmann et al., 2008) and active bioremediation (Lee et al., 2008) were occurring. Changes in microbial communities due to abiotic treatment schemes (e.g., zero-valent iron) have also been observed (Da Silva et al., 2007), but it is unclear if these changes enhance remediation system performance. The techniques are not foolproof, however, for in studying the biodegradation of RDX, Fuller and Stefan (2008) were unable to find genes associated with RDX degradation in samples in which RDX loss was occurring.

Proteomics. Quantification of specific proteins known to be involved in compound degradation could be used to assess the potential for relevant microbial activity at a site and to verify that bacteria are actively degrading site contaminants. This is promising for *cis*-DCE, which has been shown to lead to the up-regulation in *Polaromonas sp.* Strain JS666 of specific proteins important in *cis*-DCE transformation (Jennings et al., 2009). The proteins associated with bacteria responsible for aerobic biodegradation of vinyl chloride have also been identified (Chuang et al., 2010). Proteomic studies have also focused on bacteria capable of facilitating reductive dehalogenation of groundwater pollutants, such *Dehalococcoides* species that reduce TCE (Werner et al., 2009). Differences in the proteomics of different strains of *Dehalococcoides* can allow evaluation of which dehalogenases are being expressed, which is linked to the capability of the bacteria to degrade specific contaminants (Morris et al., 2007). The proteins associated with the biodegradation of MTBE by a specific bacterial strain have been identified (Eixarch and Constanti, 2010). Similarly, the proteins involved in anaerobic benzene biodegradation have been characterized, and they are different if the bacteria are grown on benzene vs. benzoate (Benndorf et al., 2009).

Proteomics can also be used to evaluate the bioremediation of metal contaminated sites. Wilkins et al. (2009) evaluated the proteins produced by *Geobacter* strains during a biostimulation effort focused on uranium reduction. The proteins associated with metal reduction in *Shewanella onidensis* MR-1 have also been identified (Elias et al., 2007). Use of proteomics could be used to evaluate the success of biostimulation efforts at metal contaminated sites and to monitor changes over time in the microbial consortia responsible for the metal reduction.

PAH include benzylsuccinate for toluene and napthoic acid for naphthalene (Bombach et al, 2010). The use of metabolomics, however, requires knowledge of potential metabolites of the organism under different conditions. One tool to evaluate biodegradation potential and possible metabolites is a database of transformation pathways (e.g., Singh, 2006; Gao et al.,

2010), but substantial effort is required to build the appropriate databases of metabolic profiles for contaminant-degrading organisms.

Isotope Analysis

Another emerging tool for understanding the effectiveness of MNA is compound-specific isotope analysis (CSIA), which is used to monitor the changes in stable isotope ratios of elements within molecules (e.g., $^{13}C/^{12}C$, $^{2}H/^{1}H$) over time. The technique uses isotopes in compounds present at natural abundance—i.e., isotopically labeled compounds are not used. As a transformation reaction proceeds, a molecule of a contaminant containing the lighter isotope (e.g., ^{12}C) will usually react more quickly than the molecule with the heavier one (e.g., ^{13}C) if this atom is included in the bond being broken (i.e., it is at the reactive center). Thus, the remaining parent compound is depleted in the light isotope (and enriched in the heavy isotope) and the reverse is true for reaction products. The change is quantified via the isotope fractionation factor (α), the isotope enrichment factor (e), and/or the apparent kinetic isotope effect. Most studies on contaminant transformation report enrichment factors (in ‰), which can be related to the extent of contaminant transformation. In contrast non-transformative processes such as sorption or dilution result in no fractionation (Pooley et al., 2009; Beller et al., 2008; McKelvie et al., 2007; Amaral et al., 2009).

A guide for the use of CSIA in the assessment of biodegradation of contaminants in groundwater is available from EPA (Hunkeler et al., 2008). Starting with the recommendations of Sherwood Lollar et al. (1999), the EPA report lays out six criteria that must be met to provide evidence for biodegradation of contaminants in groundwater, which would also presumably apply to abiotic reactions. Aelion et al. (2010) also provides detailed information about CSIA and its utility in evaluating biodegradation of contaminants. Several field studies have shown the potential utility of CSIA, particularly in the verification of contaminant degradation during MNA. A summary is given in Box 6-2.

While CSIA is a powerful tool, it has several limitations. As outlined by Blessing et al. (2008), care must be taken in choosing sampling locations and in preserving samples prior to analysis. CSIA is also currently limited to pollutants that have sufficient volatility to be analyzed using gas chromatography, and the isotope ratio mass spectrometer itself often has a limited linear response range and is not particularly sensitive, requiring either large sample sizes or sample preconcentration techniques (Amaral et al., 2009). Furthermore, the relative bioavailability of the contaminant may affect several "masking" processes that alter measured enrichment factors (Elsner et al., 2005; Thullner et al., 2008; Kampara et al., 2008; Aeppli et al. 2009). As outlined in the recommendations, additional research is needed to advance the use of CSIA as a robust and routine measurement for groundwater sampling to support MNA.

MODELING FOR LONG-TERM MANAGEMENT

The implementation of mathematical models to simulate subsurface flow and transport has become an increasingly important component of long-term management. Models can provide insight into the relative importance of the processes that control remediation, although the prediction of the time to meet remediation goals remains an ongoing challenge. Recent advances in computing hardware and computational methods have significantly broadened the scope of available models, including models capable of simulating very complex biogeochemical processes at high resolution, as well as screening models that utilize a simplified representation of site geometry and/or secondary transport processes to provide an approximate representation of processes believed to control remediation and/or attenuation.

Predicting Source Zone Mass Removal during Remediation

Although active source removal typically occurs prior to the transition to long-term management, new developments in remedial technology and/or site characterization might result in scenarios where additional source zone activity is undertaken. Modeling of source zone processes remains challenged by a number of technical constraints, including the need to represent second-order processes such as pore clogging by biofilms and/or precipitated reaction products, reactions with natural organic matter and other non-target compounds, gas production, and other changes in aquifer properties (e.g., Heiderscheidt et al., 2008; Glover et al., 2007). Furthermore, practical applications are almost always constrained by incomplete knowledge of site-specific conditions needed to appropriately assign geologic heterogeneity, NAPL architecture, and/or other contaminant initial conditions (e.g., Illangasekare et al., 1995; Fagerland et al., 2007a,b). In some cases, advanced DNAPL characterization technologies such as PITT could provide the desired initial conditions for modeling (e.g., Basu et al., 2008) and better support the performance assessment of source remediation.

In light of the inherent challenges of modeling source zone processes at field sites, sophisticated process-oriented models have been more commonly applied in the interpretation of laboratory studies, where the relevant input parameters are more straightforward to obtain and assess (e.g., EPA, 2009), and not in the field. In reviewing the current state of remediation technology (Chapter 4), the Committee observed that for many complex sites, the engineering design for source zone remediation is frequently ac-

BOX 6-2
Field Implementation of CSIA Techniques

Biological reductive dechlorination of PCE and TCE was deduced in contaminated groundwater at Dover Air Force base by determining the isotopic enrichment of PCE and TCE in wells downgradient from the source zone and by identifying reaction products (Sherwood Lollar et al., 2001). Imfeld et al. (2008) showed that changes in isotope fractionation of DCEs in a treatment wetland could be related to changes in the hydrogeochemistry of a wetland. Under oxic conditions, the enrichment factor was –1.7‰ but reached –32.6‰ once the wetland became methanogenic. In an aerobic fractured bedrock aquifer contaminated with chlorinated ethenes, Pooley et al. (2009) used a combination of CSIA and reactive transport modeling to verify aerobic biodegradation of TCE and *cis*-DCE and demonstrated the recalcitrance of PCE.

Using two-dimensional CSIA (carbon and hydrogen), Fisher et al. (2007) demonstrated that the degradation process at a BTEX-contaminated site was anaerobic biodegradation. As shown in Figure 6-2, the enrichment factors clearly fall in the region of anaerobic biodegradation. In a controlled field study, Beller et al. (2008) used a combination of CSIA, genomic analysis (qPCR), and metabolite identification to study natural attenuation of BTEX, including the effects of ethanol. Amaral et al. (2010) used CSIA to demonstrate a lack of natural attenuation of TNT and DNT in an oxic contaminated aquifer.

CSIA is also applicable to sites undergoing active groundwater remediation. Song et al. (2002) used isotope effects to verify effectiveness of enhanced *in situ* bioremediation of chlorinated ethenes. After determining the carbon isotope patterns of the reaction of chlorinated ethenes with nano-zero valent iron (nZVI) in the laboratory (Elsner et al., 2008), it was possible to assess the effectiveness of nZVI in the field (Elsner et al., 2010). In the Elsner et al. (2010) study, natural attenuation via biodegradation of TCE and 1,1,1-TCA in the groundwater were confirmed by the extent of isotopic fractionation between the source zone and downgradient wells. After injection of nZVI, an increase of the magnitude of the enrichment factor (and the detection of dechlorination products) was used to confirm the nZVI was reacting with the target contaminants.

,

complished without the use of available process-oriented models, in part because of their mathematical and computational complexity, but in large measure because of the substantial amount of characterization and parameter information required to implement such models with confidence. Indeed, Siegrest et al. (2011), recognizing the potential cost and complexity associated with applying detailed models such as the *in situ* chemical oxidation (ISCO) simulator CORT3D (Heiderscheidt, 2005; Illangasekare

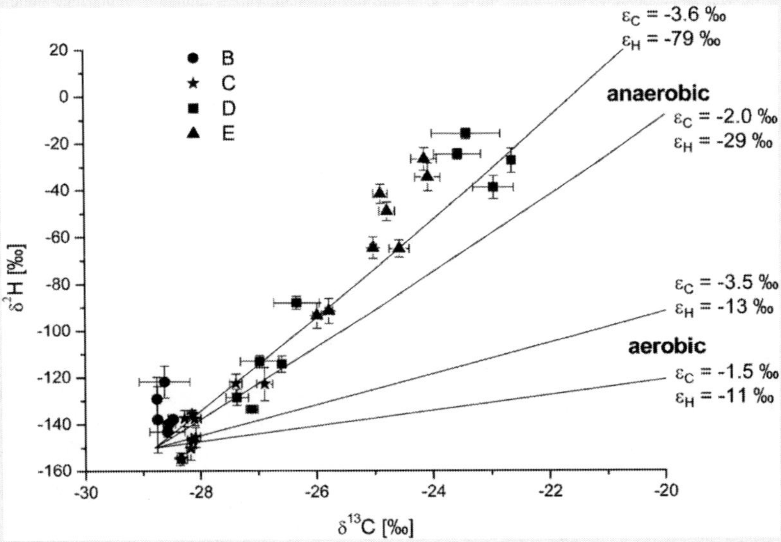

FIGURE 6-2 Concurrent carbon and hydrogen isotope ratios of benzene measure at various sampling depths (B through E) together with isotope patterns for aerobic and anaerobic benzene degradation calculated from published enrichment factors for carbon and hydrogen as well as the isotope signature of the contaminant source using the Rayleigh equation. SOURCE: Reprinted, with permission, from Fisher et al. (2007). © 2007 by American Chemical Society.

et al., 2007), highlighted the need for "medium range" models that could be applied to address basic design questions applicable to most source zone technologies. While research targeting such "intermediate complexity" models could support more effective remedial designs, it is important that model developers work closely with practitioners to develop tools that balance theoretical rigor, mathematical complexity, data requirements, and user-friendliness.

Modeling Plume Processes

Except for the special case of vapor intrusion, modeling for long-term management is primarily concerned with the transport and reaction of dissolved contaminants in groundwater. The most commonly utilized tools for this purpose are numerical models that couple a solution to the groundwater flow equation in saturated media (e.g., MODFLOW, McDonald and Harbaugh, 1988) with a solution to the advective-dispersive-reactive equation (e.g., MT3DMS, Zheng and Wang, 1999). Modern groundwater flow models are capable of representing considerable detail in the flow field, utilizing millions of computational nodes to represent spatially variable aquifer properties, typically over horizontal scales of meters to kilometers (e.g., DOE, 2009) although finer resolution is increasingly feasible. In conjunction with the finer spatial resolution of flow models, reactive transport models are increasingly able to incorporate a wide variety of biogeochemical reactions. Of particular relevance to long-term management, a considerable body of work has addressed the simulation of natural attenuation processes for both petroleum and halogenated organic contaminants, including sequential parent/daughter reactions and multiple electron acceptor/donors. In particular, the modular code structure employed by MT3DMS has facilitated a number of extensions that address reaction scenarios typical of those found at complex sites, including SEAM3D (Waddill and Widdowson, 2003), BioRedox (Carey et al., 1999), and RT3D (Clement, 1997).

Despite the above advances, the prediction of time-to-complete remediation for dissolved plumes remains an elusive goal. The primary reason for this disconnect is that most commonly employed aqueous-phase simulation models lack the capability to represent the various nonideal processes that release contaminants from immobile phases, particularly nonideal desorption and back-diffusion (note: NAPL dissolution is typically not associated with the "plume" compartment). Furthermore, although it would be conceptually straightforward to incorporate such features in reactive transport models, full implementation would require detailed knowledge of the spatial distribution of localized sources, as well as the governing mass transfer processes (e.g., diffusion path lengths, desorption rates). One possible approach is the *dual porosity* or *dual domain* formulation of the advective–dispersive equation (e.g., as in MT3D, Zheng and Wang, 1999; AFCEE, 2007), which can be configured to represent back-diffusion, although this approach has not been widely utilized and would require further refinement to represent highly localized zones of low permeability.

In the short term, credible predictions of time-to-complete remediation based on current modeling tools would be expected only for sites where back-diffusion and desorption are not expected to be significant factors. Such sites, if they exist, would probably be associated with relatively recent contaminant releases. For example, Rivett et al. (2006) generated reasonable modeling predictions of a pump-and-treat field experiment associated with a relatively short (less-than-two-year) plume history. In contrast, Parker et al. (2008) identified back-diffusion as the primary factor responsible for the inability of aqueous transport models to simulate a controlled pump-and-treat study of a more typical (decades-old) plume.

Modeling Hydraulic Containment

While the prediction of time-to-complete remediation remains an elusive goal for mass removal technologies, the common application of pump and treat for hydraulic containment has been supported by ongoing developments to apply groundwater flow/transport models in an optimization framework. A variety of software packages are now available to identify well configurations that provide hydraulic containment while minimizing the overall extraction rates and/or treatment costs, including tools based on the MODFLOW/MT3D simulators. Recent applications of simulation/optimization to P&T design at several DoD facilities were summarized in Chapter 4 (EPA, 1999a,b, 2005). In general, the optimized designs were expected to yield an approximate average life-cycle savings of 10 to 20 percent over trial-and-error designs. However, the models used in these studies were based on conventional advective-dispersive-sorptive transport and did not include the various nonideal processes described above. Thus, while providing design guidance for efficient hydraulic containment, predictions of the time to achieve remedial objectives were likely optimistic.

Although the use of simulation/optimization techniques for hydraulic containment design is relatively mature, continuing developments will provide more realistic cost functions for scenarios in which expected costs are not proportional to the volume of extracted water. Also, because one goal of optimization is to find the least-cost solution to achieve hydraulic containment, optimized designs typically reflect a reduced margin of safety for plume capture, particular if the underlying flow model treats the aquifer as relatively homogeneous. Thus, it is important to account for uncertainty and spatial variability in a robust manner, and there are a number of promising techniques under development that have not yet been widely embraced by practitioners (e.g., Aly and Peralta, 1999; Gopalakrishnan et al., 2003; Guan and Aral, 2004; Bau and Mayer, 2007; Peralta, 2011).

Modeling Natural Attenuation

The decision to switch from an aggressive remediation strategy to MNA or natural attenuation without monitoring requires an estimate of post-remediation plume development, including (1) the amount of plume expansion (if any) that would occur under MNA, (2) the "steady state" plume dimensions, and (3) the rate at which the resulting plume would be depleted. Although complex groundwater flow fields can affect plume development under MNA, for many sites the dominant concerns are biogeochemical reaction processes, which depend on the aquifer transformation capacity and reaction rates for the contaminants of interest. Thus, a simplified one-dimensional steady flow field can often be assumed, which facilitates a closed-form analytical solution to the governing equations for an idealized plane source coupled with multi-solute transport and biodegradation (e.g., BIOCHLOR, Aziz et al., 2000). Although widely used, the success of such screening models is dependent on accurate information concerning reaction rates, appropriate handling of the scale-dependent dispersion process, and accurate assessment of the source term, which is usually represented in terms of known contaminant concentrations or fluxes distributed over a vertical plane source. Recently, the plane source concept has been extended to include the time-dependent mass flux from a DNAPL source zone based on a power function approach, as implemented in the REMCHLOR software.

Screening models such as REMCHLOR represents a significant step forward in the practical application of the 14-compartment model, but its implementation emphasizes mass transfer from the source area to the transmissive plume. The release of contaminants from other potentially problematic compartments, such as low-permeability zones and the sorbed phase, has not yet been studied sufficiently to support implementation of these processes in a screening mode. In particular, the process of back-diffusion from the downgradient plume region is more difficult to conceptualize and approximate mathematically relative to source-zone release, as the geometry and history of the matrix diffusion process is more complex. Furthermore, for both source and plume regions, the potential influence of nonlinear and/or rate-limited desorption has received less attention, although researchers have long recognized the potential influence of nonideal sorption processes (e.g., Brusseau and Rao, 1989; Ball and Roberts, 1991; Rabideau and Miller, 1994; Allen-King et al., 2002; Rivett et al., 2006).

***

Predicting the trajectory of any remediation activity at complex sites (not just MNA) will require further research to clarify the conditions for which back-diffusion and desorption are likely to be contributing factors in the plume zone, development of better characterization tools to establish the necessary initial conditions for modeling, efficient computational power to incorporate the limiting nonideal processes into contaminant transport models, and the careful design of field studies to evaluate the resulting predictive capability of the models.

EMERGING REMEDIATION TECHNOLOGIES

Although the pace of remediation technology development has slowed considerably since the most recent NRC evaluation of source zone strategies (NRC, 2005), a few emerging technologies are in various stages of testing and could eventually provide additional cost-effective tools for managing complex sites. This section provides a snapshot of several emerging technologies, none of which, with exception of nanotechnologies, have received extensive field testing.

The preponderance of research on ***nanoparticles*** used in groundwater remediation has focused on nanoscale zero-valent iron (nZVI) and zero-valent iron doped with a catalytic metal (such as palladium). Contaminants amenable to treatment with nZVI include chlorinated methanes, ethanes, and ethenes (Lien and Zhang, 1999, 2005; Liu et al., 2005; Song and Carraway, 2005, 2006, 2008; Liu and Lowry, 2006), chlorinated phenols (Cheng et al., 2007), PCBs (Wang and Zhang, 1997), hexachlorocyclohexanes (Elliott et al., 2008), TNT (Welch and Riefler, 2008), nitrate (Sohn et al., 2006), perchlorate (Cao et al., 2005), chromate (Xu and Zhao, 2007; Hoch et al., 2008), arsenic (Ramos et al., 2009), and heavy metals (Zn^{II}, Cd^{II}, Pb^{II}, Ni^{II}, Cu^{II}, and Ag^{I}; Li and Zhang, 2007). While the high surface area leads to high reaction rates, the nZVI particles tend to aggregate (Phenrat et al., 2007) when injected into the subsurface, which limits their transport in porous media (Hong et al., 2009). For this reasons, much effort has focused on the development of surface coatings that allow the nZVI to be injected into the subsurface *and* reach the contaminated area.

An EPA compilation of pilot- and full-scale tests with nZVI[1] shows concentration reductions in the target zone of 50 to 90 percent (but sometimes lower), and there was evidence of contaminant rebound once the nZVI is exhausted. The location and amount of nZVI, flow rate, and DNAPL dissolution rate are all critical design parameters, such that emplacement of nZVI downstream of the DNAPL zone provides the best performance (Taghavy et al., 2010; Fagerlund et al., 2012). This latter finding suggests that use of larger-sized iron particles would be more cost effective unless the selectivity/lifetime of nZVI can be improved (Fagerlund et al., 2012). Overall, it appears that nZVI is best applied in limited situations to treat zones where the most contaminant could be removed in a short period of time. It should be kept in mind that while nanoparticles show some promise for remediation of groundwater pollutants, nanoparticles may also present a future environmental risk (Wiesner et al., 2006) and require additional research regarding their fate, transport, and toxicity.

For source zones that contain contaminants amenable to treatment

---

[1] See http://www.clu-in.org/download/remed/nano-site-list.pdf.

with zero-valent iron, the ***in situ mixing of contaminated soil with zero-valent iron and clay*** (ZVI-clay) could afford two advantages: (1) the iron may accelerate the destruction of source zone contaminants, and (2) the clay may reduce contaminant transport from the source zone and redirect upgradient groundwater away from the source zone. While conceptually straightforward, ZVI-clay has received limited field testing (Shackelford et al., 2005; Bozzini et al., 2006; Olson et al., 2012). However, if proven successful, the general approach could be tailored to other contaminants using different reactive media.

The ***self-sustaining treatment for active remediation*** (STAR) technique has been proposed for treatment of creosote, petroleum hydrocarbons, and other combustible NAPLs. STAR is a controlled burning reaction (self-sustaining smoldering) that can be performed *ex situ* or *in situ*, even under fully saturated conditions. A heating element is introduced to the NAPL and heated to the NAPL ignition temperature. Air is then injected to initiate ignition. The heat released then serves to heat NAPL farther away, and the combustion process continues as long as sufficient air is supplied. Although the approach is best suited to readily combustible contaminants, laboratory studies have shown up to 99.9 percent removal of coal tar or crude oil (Switzer et al., 2009; Pironi et al., 2011), with pilot-scale tests ongoing.[2]

An alternative to traditional *in situ* chemical oxidation (ISCO) is the encapsulation of reactive agents in a ***permeable reactive ISCO barrier***. This allows the slow release of the oxidant over time, supporting nearly continuous treatment of the plume until the oxidant is exhausted (Ross et al., 2005; Luster-Teasley et al., 2010; Liang et al., 2011). The release rate is dependent upon whether the design is based upon oxidant diffusion or erosion of the encapsulating polymer. In some cases, both processes control the release rate. The amount of oxidant added to the polymer matrix coupled with the release rate will dictate both its life cycle and effectiveness as an oxidant. The major challenge in utilizing this technology is optimizing the dose of oxidant needed to degrade the target contaminant with a slow enough release rate to minimize frequent media replacement.

In situ electrodes can be deployed for the sequential reduction and oxidation of contaminants (Wani et al., 2005, 2006) or for the generation of ozone (Vera et al., 2009). The electrode approach allows one to change the potentials of a cathode and anode such that the process responsible for the degradation of the contaminant of concern can be set to either reduction (e.g., production of hydrogen) or oxidation (e.g., generation of ozone or other reactive species). The mode of operation is dependent upon the target contaminants; many halogenated substances are more amenable to reduction processes, while BTEX and PAHs are better suited for oxidation.

[2]See http://star.siremlab.com.

This approach was shown to be effective for the reduction of both RDX and TNT (Wani et al., 2006) and does not require the use of chemical additives. The use of electrodes to produce ozone *in situ* (Vera et al., 2009) has an advantage over on-site *ex situ* ozone generators in that it is passive in nature, which circumvents the logistical difficulties of pumping ozone-saturated water into the contaminated zone, and it can continuously generate ozone at the site, which maintains a constant level of oxidant. None of these *in situ* electrode methods, however, has been tested at field sites, and both cost and scaling issues may be important.

RESEARCH FUNDING

As the focus at complex sites shifts from active remediation to long-term management, the development and effective deployment of appropriate concepts and tools may require a redirection of research efforts. The majority of support for research applicable to groundwater remediation has been provided by federal agencies, particularly EPA, the DoD, the National Science Foundation (NSF), the Department of Energy (DOE), and the National Institute for Environmental Health Sciences (NIEHS). The funding estimates shown in Table 6-1 were obtained primarily by searching

TABLE 6-1 Federal Research Funding for Groundwater Remediation

Federal Program	Estimated Cumulative Funding ($M) (1996–2011)[a]	Number of Projects (1996–2011)[a]
Department of Defense	315	250–300
National Science Foundation (HS)	2–4	< 10
National Science Foundation (CBET)	25	80–110
Department of Energy (including actinide research)	138	200
Environmental Protection Agency (including ten Hazardous Substance Research Centers)	14	85 projects plus centers
National Institute for Environmental Health Sciences	500–800[b]	Unknown

[a] Estimated by the Committee through searches of public databases.

[b] The $500-$800 million in research funding for NIEHS was primarily for research on human health impacts of contaminant exposure. Although relevant to groundwater remediation, the funding was not specifically directed to groundwater-specific contamination issues, and is thus not directly comparable to the other federal programs.

public databases using appropriate keywords (e.g., groundwater, remediation, etc.) but, in some cases, considerable judgment was required to distinguish relevant remediation research from more general environmental programs. Furthermore, some relevant projects were primarily focused on non-groundwater issues such as sediment remediation. It should be noted that other federal agencies not shown in Table 6-1, such as the U.S. Geological Survey (USGS), occasionally fund projects that have relevance to groundwater remediation (e.g., the USGS Water Resources Institutes). Although the Committee found it difficult to quantify historic research funding specifically applicable to groundwater remediation, its opinion is that such funding has generally declined over the past decade, with the single exception of the DoD. Additional details related to the primary agencies are given below.

DoD's primary research mechanism is its Strategic Environmental Research and Development Program (SERDP) and Environmental Security and Technology Certification Program (ESTCP). SERDP, in partnership with EPA, supports a wide range of projects related to DoD-generated environmental issues, including diverse topics such as ecosystem effects to innovative subsurface remediation strategies. ESTCP constitutes DoD's environmental technology demonstration program and involves no other federal partners.

Relevant NSF programs include the Division of Geosciences program in Hydrologic Sciences (HS), and several programs within the Division of Chemical Bioengineering, Environmental, and Transport Systems (CBET). Much of the Hydrologic Sciences research is focused on nano- to global-scale hydrologic and/or chemical processes. A review of this database yielded only a handful of projects that deal directly or indirectly with groundwater remediation over the past decade. Significantly more projects related to subsurface remediation/characterization have been funded through the NSF Division of Chemical Bioengineering, Environmental, and Transport Systems to address the development of tools for understanding contaminant behavior in the subsurface and remediation strategies. While the cumulative funding is significantly more than for the Hydrologic Sciences, the annual average is on the order of $1.4 million/year for remediation-focused research, with some projects addressing contaminants that are not the primary drivers for complex hazardous sites.

DOE funding for research on subsurface contamination is currently administered through the Office of Biological and Environmental Research. While a large number of proposals have been funded through this office, many of the projects are focused on actinide and inorganic contaminants, with fewer projects focused on chlorinated solvents and hydrocarbons, although a large number of funded projects have addressed microbial activity in the subsurface. Most of the sponsored research is conducted at the

National Labs (e.g., Oak Ridge, Pacific Northwest), often in conjunction with academic partnerships.

EPA has funded research on groundwater remediation through a variety of mechanisms, including its Science to Achieve Results (STAR) basic research program, several agency research laboratories, and externally funded Hazardous Substance Research Centers (which have been discontinued). The available database records were insufficient to provide more than an approximate estimate of overall funding levels across the external programs and it was difficult to distinguish in-house research from those projects awarded to universities.

Since 1987, NIEHS has operated the Superfund (Basic) Research Program, which is similar to the EPA Hazardous Substance Research Centers program in that it supports the operation of stand-alone research centers. Remediation-oriented projects have received a very small fraction of the overall funding, which has been largely directed toward human health rather than technology research.

In addition to federal agencies, some research funding for subsurface science and technology has been provided by state agencies and the private sector. In general, individual corporations fund both internal and external programs to address specific challenges important to fulfilling their obligations to protect human health and the environment via remediation. Contracting strategies vary from site-specific feasibility experiments to university contracts of sufficient duration to support full Ph.D. programs, often targeting basic "first-principles" research. As U.S. government research funding has declined, corporate research programs have sought international partners to provide matching funds, such as the Source Area *in situ* BioREmediation (SABRE) program centered in the United Kingdom.

It is important to note that research investments by government agencies and the private sector have yielded several innovative approaches to site remediation such as ZVI, ISCO, and thermal methods. Nonetheless, given budget constraints facing both government agencies and companies that fund remediation research, the development of more effective treatment technologies is likely to occur at a much reduced pace. Some private sector organizations are working with universities to pursue targeted and applied research on new solutions to legacy site issues. In addition, companies selling products and services in the remediation business continue to develop innovative technical strategies and products to improve all components of groundwater remediation. Whether this level of funding and other market-driven technical innovations will be sufficient to address the challenges of long-term management is uncertain. Other consequences of the lack of government funding are a reduction in support of graduate programs and the migration of students away from the remediation field. This may lead to a shortage of qualified personnel over the next decade to respond to the long-term management issues of these complex sites.

CONCLUSIONS AND RECOMMENDATIONS

Many complex federal and private industrial facilities with contaminated groundwater will require long-term management actions that could extend for decades or longer. Technological developments can aid in the transition from active remediation to more passive strategies and provide more cost-effective and protective long-term management of complex sites. Further improvements in long-term site management are likely to emphasize more cost-effective containment, new diagnostic tools for performance and compliance monitoring, and modeling strategies that can be used for decision making. The following conclusions and recommendations are offered.

Long-term management of complex sites requires an appropriately detailed understanding of geologic complexity and the potential distribution of contaminants among the aqueous, vapor, sorbed, and NAPL phases, as well as the unique biogeochemical dynamics associated with both the source area and the downgradient plume. Recent improvements to the understanding of subsurface biogeochemical processes have not been accompanied by cost-effective site characterization methods capable of fully distinguishing the distribution of contaminants between different subsurface zones or compartments (as described by the recently proposed conceptual 14-compartment model). Management of residual contamination to reduce the exposure risks via the vapor intrusion pathway is challenged by the highly variable nature of exposure, as well as uncertain interactions between subsurface sources and indoor background contamination.

Existing protocols for assessing monitored natural attenuation and other remediation technologies should be expanded to integrate compound-specific isotope analysis and molecular biological methods with more conventional biogeochemical characterization and groundwater dating methods. The development of molecular and isotopic diagnostic tools has significantly enhanced the ability to evaluate the performance of degradation technologies and monitored natural attenuation at complex sites.

Mathematical models are increasingly important tools for key decision points in the management of complex sites, despite the inherent difficulty in predicting the time-to-complete remediation. In particular, the use of more realistic models that can account for processes that significantly decrease the rate of mass removal during remediation is critical for deciding whether to transition to active or passive long-term management. The implementation of site-specific models is often constrained by the lack of spatially detailed information about contaminant distributions and/or reaction processes. While modeling predictions will always be subject to uncertainty due to the inherent limitations in site characterization and the high degree of heterogeneity, accurate models can bound the likely timeframes for restoration and provide another line of evidence needed to make decisions on the ultimate disposition of complex sites.

Although the Committee did not attempt a comprehensive assessment of research needs, research in the following areas would help address technical challenges associated with long-term management at complex contaminated sites:

- **Remediation Technology Development.** Additional work is needed to advance the development of emerging and novel remediation technologies, improve their performance, and understand any potential broader environmental impacts. A few developing remediation techniques could provide more cost-effective remediation for particular combinations of contaminants and site conditions at complex sites, but they are in the early stages of development.
- **Tools for Characterizing Complex Subsurface Conditions.** More refined concepts and tools are needed to better delineate and manage localized contamination associated with the processes of back-diffusion and desorption, complex geologic environments, and aquifer transformation capacity. These include better characterization of subsurface media to assess the magnitude of back-diffusion and desorption processes and to identify the type and abundances of naturally occurring reactive chemical species and microorganisms that are involved in natural biodegradation processes.
- **Tools to Assess Vapor Intrusion.** Further research and development should identify, test, and demonstrate tools and paradigms that are practicable for assessing the significance of vapor intrusion, especially for multi-building sites and preferably through short-term diagnostic tests. Development of real-time unobtrusive and low-cost air quality sensors would allow verification of those short-term results over longer times at buildings not needing immediate mitigation.
- **Molecular Biological Tools and Databases.** Robust databases are necessary to conduct molecular biological analyses for various contaminant degradation pathways, in conjunction with further refinement of field sampling protocols. There is also a need for better methods of protein extraction from environmental matrices, as well as more cost effective methods to detect specific peptides that do

not require knowledge of the exact gene sequence. For CSIA, data analysis tools for enrichment factors should be expanded to address a wider range of groundwater pollutants at low concentrations and should consider the effects of bioavailability and mass transfer limitations.

- **Modeling.** Additional targeted modeling research and software development that will benefit the transition of sites from active remediation to long-term management should be initiated. Particular needs include concepts and algorithms for including the processes of back-diffusion and desorption in screening and plume models, and the development of a larger suite of intermediate-complexity modeling tools to support engineering design for source remediation.

Overall, research and development have been unable to keep pace with the needs of practitioners trying to conduct remediation on complex sites. Currently, a national strategy for technology development to support long-term management of complex sites is lacking. It is not clear that the pertinent federal agencies will be capable of providing the funding and other support for the fundamental research and development that is necessary to meet the challenges facing complex sites. A comprehensive assessment of future research needs, undertaken at the federal level and involving coordination between federal agencies, would allow research funding to be allocated in an efficient and targeted manner.

REFERENCES

Aelion, C. M., R. Aravena, D. Hunkeler, and P. Hoehnener (Eds.). 2010. Environmental Isotopes in Biodegradation and Bioremediation. Boca Raton, FL: CRC Press.

Aeppli, C., M. Berg, O. A. Cirpka, C. Holliger, R. P. Schwarzenbach, and T. B. Hofstetter. 2009. Influence of mass-transfer limitations on carbon isotope fractionation during microbial dechlorination of trichloroethene. Environmental Science & Technology 43(23):8813-8820.

AFCEE (Air Force Center for Engineering and the Environmental). 2007. AFCEE Source Zone Initiative, Final Report, Colorado State University, Boulder CO, 234 pp.

Allen-King, R. M., P. Grathwohl, and W. P. Ball. 2002. New modeling paradigms for the sorption of hydrophobic organic chemicals to heterogeneous carbonaceous matter in soils, sediments, and rocks. Advances in Water Resources 25:985-1016.

Aly, A. H., and R. C. Peralta. 1999. Optimal design of aquifer cleanup systems under uncertainty using a neural network and genetic algorithm. Water Resources Research 35(8):2523-2532.

Amaral, H. I. F., J. Fernades, M. Berg, R. P. Schwarzenbach, and R. Kipfer. 2009. Assessing TNT and DNT groundwater contamination by compound-specific isotope analysis and ^{3}H–^{3}He groundwater dating: A case study in Portugal. Chemosphere 77:805-812.

Amaral, H. I. F., M. Berg, M. S. Brennwald, M. Hofer, and R. Kipfer. 2010. $^{13}C/^{12}C$ analysis of ultra-trace amounts of volatile organic contaminants in groundwater by vacuum extraction. Environmental Science & Technology 44(3):1023-1029.

Annable, M. D., P. S. C. Rao, K. Hatfield, W. D. Graham, A. L. Wood, and C. G. Enfield. 1998. Partitioning tracers for measuring residual NAPL: Field-scale test results. Journal of Environmental Engineering 124:498-503.

Annable, M. D., K. Hatfield, J. Cho, H. Klammler, B. L. Parker, J. A. Cherry, and P. S. C. Rao. 2005. Field-scale evaluation of the passive flux meter for simultaneous measurement of groundwater and contaminant fluxes. Environmental Science & Technology 39(18):7194-7201.

Aziz, C. E., C. J. Newell, J. R. Gonzales, P. Haas, T. P. Clement, and Y. Sun. 2000. BIOCHLOR: Natural Attenuation Decision Support System Version 1.0, User's Manual. EPA/600/R-00/008. Ada, OK: EPA ORD. http://www.epa.gov/ada/csmos/models/biochlor.html.

Babbar-Sebens, M., and B. Minsker. 2008. Standard interactive genetic algorithm—comprehensive optimization framework for groundwater monitoring design. Journal of Water Resources Planning and Management 134(6):538-547.

Ball, W. P., and P. V. Roberts. 1991. Long-term sorption of halogenated organic-chemicals by aquifer material. 1. Equilibrium. Environmental Science & Technology 25(7):1223-1237.

Ball, W., C. Liu, G. Xia, and D. F. Young. 1997. A diffusion-based interpretation of tetrachloroethene and trichloroethene concentration profiles in a groundwater aquitard. Water Resources Research 33(12):2741-2757.

Basu, N. B., P. S. C. Rao, I. C. Poyer, M. D. Annable, and K. Hatfield. 2006. Flux-based assessment at a manufacturing site contaminated with trichloroethylene. Journal of Contaminant Hydrology 86(1-2):105-127.

Basu, N. B., A. D. Fure, and J. W. Jawitz. 2008. Predicting DNAPL dissolution using a simplified source depletion model parameterized with partitioning tracers. Water Resources Research 44(7):W07414.1-13, doi:10.1029/2007WR006008.

Bau, D. A., and A. S. Mayer. 2007. Data-worth analysis for multiobjective optimal design of pump-and-treat remediation systems. Advances in Water Resources 30(12):1815-1830.

Bauer, S., M. Bayer-Raich, T. Holder, C. Kolesar, D. Muller, and T. Ptak. 2004. Quantification of groundwater contamination in an urban area using integral pump tests. Journal of Contaminant Hydrology 75(3-4):183-213.

Belle, A., A. Tanay, L. Bitincka, R. Shamir, and E. K. O'Shea. 2006. Quantification of protein half-lives in the budding yeast proteome. Proceedings of the National Academy of Sciences 103(35):13004-13009.

Beller, H. R., S. R. Kane, T. C. Legler, J. R. McKelvie, B. Sherwood Lollar, F. Pearson, L. Balser, and D. M. MacKay. 2008. Comparative assessments of benzene, toluene, and xylene natural attenuation by quantitative polymerase chain reaction analysis of a catabolic gene, signature metabolites, and compound-specific isotope analysis. Environmental Science & Technology 42(16):6065-6072.

Benndorf, D., C. Vogt, N. Jehmlich, Y. Schmidt, H. Thomas, G. Woffendin, A. Shevchenko, H. H. Richnow, and M. von Bergen. 2009. Improving protein extraction and separation methods for investigating the metaproteome of anaerobic benzene communities within sediments. Biodegradation 20(6):737-750.

Blessing, M., M. A. Jochmann, and T. C. Schmidt. 2008. Pitfalls in compound-specific isotope analysis of environmental samples. Analytical and Bioanalytical Chemistry 390(2):591-603.

Bockelmann, A., T. Ptak, and G. Teutsch. 2001. An analytical quantification of mass fluxes and natural attenuation rate constants at a former gasworks site. Journal of Contaminant Hydrology 53(3-4):429-453.

Bombach, P., H. H. Richnow, M. Kastner, and A. Fischer. 2010. Current approaches for the assessment of in situ biodegradation. Applied Microbiology and Biotechnology 86:839-852.

Borch, T., R. Kretzschmar, A. Kappler, P. Van Cappellen, M. Ginder-Vogel, A. Voegelin, and K. Campbell. 2010. Biogeochemical redox processes and their impact on contaminant dynamics. Environmental Science & Technology 44 (1):15-23.

Bozzini, C., T. Simpkin, T. Sale, D. Hood, and B. Lowder. 2006. DNAPL remediation at Camp Lejeune using ZVI-clay soil mixing. Proceedings of the International Conference on Remediation of Chlorinated and Recalcitrant Compounds, Monterey, CA, May 22-25.

Brooks, M. C., M. D. Annable, P. S. C. Rao, K. Hatfield, J. W. Jawitz, W. R. Wise, A. L. Wood, and C. G. Enfield. 2002. Controlled release, blind tests of DNAPL characterization using partitioning tracers. Journal of Contaminant Hydrology 59:187–210.

Brooks, M. C., A. L. Wood, M. D. Annable, K. Hatfield, J. Cho, C. Holbert, P. S. C. Rao, C. G. Enfield, K. Lynch, and R. E. Smith. 2008. Changes in contaminant mass discharge from DNAPL source mass depletion: Evaluation at two field sites. Journal of Contaminant Hydrology 102(1):140-153.

Brusseau, M. L., and P. S. C. Rao. 1989. Sorption nonideality during organic contaminant transport in porous media. CRC Critical Reviews in Environmental Control 19(1):33-99.

Burgmann, H., J. Kleikemper, L. Duc, M. Bunge, M. H. Schroth, and J. Zeyer. 2008. Detection and quantification of dehalococcoides-related bacteria in a chlorinated ethene-contaminated aquifer undergoing natural attenuation. Bioremediation Journal 12(4):193-209.

Buscheck, T. 2002. Mass Flux Estimates to Assist Decision-Making: Technical Bulletin. Version 1.0. ChevronTexaco Energy Research and Technology Company, June 2002.

Cao, J., D. Elliott, and W.-X. Zhang. 2005. Perchlorate reduction by nanoscale iron particles. Journal of Nanoparticle Research 7(4-5):499-506.

Carey, G. R., P. J. Van Geel, and J. R. Murphy. 1999. BIOREDOX-MT3DMS V2.0: A Coupled Biodegradation-Redox Model for Simulating Natural and Enhanced Bioremediation of Organic Pollutants – User's Guide. Waterloo, Ontario: Conestoga-Rovers & Associates.

Carreon-Diazconti, C., J. Santamaria, J. Berkompas, J. A. Field, and M. L. Brusseau. 2009. Assessment of *in situ* reductive dechlorination using compound-specific stable isotopes, functional gene PCR, and geochemical data. Environmental Science & Technology 43(12):4301-4307.

Chapman, S. W., and B. L. Parker. 2005. Plume persistence due to aquitard back diffusion following dense nonaqueous phase liquid source removal or isolation. Water Resources Research 41. doi:10.1029/2005WR004224.

Chapman, S. W., B. Parker, J. Cherry, J. Langenbach, and S. Thotapalli. 2010. Performance of an Innovative Hydraulic Capture System for DNAPL Source Cutoff. 2010 Battelle Conference, Remediation of Chlorinated and Recalcitrant Compounds. May 25, 2010.

Cheng, R., J. L. Wang, and W.-X. Zhang. 2007. Comparison of reductive dechlorination of p-chlorophenol using Fe0 and nanosized Fe0. Journal of Hazardous Materials 144(1-2):334-339.

Chuang, A. S., Y. O. Jin, L. S. Schmidt, Y. Li, S. Fogel, D. Smoler, and T. E. Mattes. 2010. Proteomic analysis of ethene-enriched groundwater microcosms from a vinyl chloride-contaminated site. Environmental Science & Technology 44(5):1594-1601.

Clement, T. P. 1997. RT3D - A modular computer code for simulating reactive multi-species transport in 3-dimensional groundwater aquifers. Battelle Pacific Northwest National Laboratory Research Report, PNNL-SA-28967. http://bioprocess.pnl.gov/rt3d.htm.

Culler, D., D. Estrin, and M. Sirivastava. 2004. Overview of sensor networks. IEEE Computer Society 37(8):41-49.

Da Silva, M. L. B., R. L. Johnson, and P. J. J. Alvarez. 2007. Microbial characterization of groundwater undergoing treatment with a permeable reactive iron barrier. Environmental Engineering Science 24(8):1122-1127.

Davey, N. G., E. T. Krogh and C. G. Gill. 2011. Membrane-introduction mass spectrometry (MIMS). TrAC Trends in Analytical Chemistry 30(9):1477-1485.

DOE (Department of Energy). 2009. Draft Tank Closure and Waste Management Environmental Impact Statement for the Hanford Site, Richland, Washington. DOE/EIS-0391. Richland, WA: Department of Energy.

Dunnivant, F. M., R. P. Schwarzenbach, and D. L. Macalady. 1992. Reduction of substituted nitrobenzenes in aqueous solutions containing natural organic matter. Environmental Science & Technology 26(11):2133-2141.

Einarson, M. 2006. Multilevel ground-water monitoring. Pp. 808-845 In D.M. Nielsen (Ed.), Practical Handbook of Environmental Site Characterization and Ground-Water Monitoring (2nd ed.). Boca Raton, FL: CRC Press.

Einarson, M. D., D. M. Mackay, and P. J. Bennett. 2010. Sampling transects for affordable, high-resolution plume characterization and monitoring. Ground Water 48(6):799-808.

Eixarch, H., and M. Constanti. 2010. Biodegradation of MTBE by *Achromobacter xylosoxidans* MCM1/1 induces synthesis of proteins that may be related to cell survival. Process Biochemistry 45(5):794-798.

Elias, D. A., F. Yang, H. M. Mottaz, A. S. Beliaev, and M. S. Lipton. 2007. Enrichment of functional redox reactive proteins and identification of mass spectrometry results in several terminal Fe(III)-reducing candidate proteins in *Shewanella oneidensis* MR-1. Journal of Microbiological Methods 68(2):367-375.

Elliott, D. W., H.-L. Lien, and W.-X. Zhang. 2008. Zerovalent iron nanoparticles for treatment of ground water contaminated by hexachlorocyclohexanes. Journal of Environmental Quality 37(6):2192-2201.

Elsner, M., L. Zwank, D. Hunkeler, and R. P. Schwarzenbach. 2005. A new concept linking observable stable isotope fractionation to transformation pathways of organic pollutants. Environmental Science & Technology 39:6896-6916.

Elsner, M., M. Chartrand, N. VanStone, G. Lacrampe Couloume, and B. Sherwood Lollar. 2008. Identifying abiotic chlorinated ethene degradation: Characteristic isotope patterns in reaction products with nanoscale zero-valent iron. Environmental Science & Technology 42(16):5963-5970.

Elsner, M., G. Lacrampe-Couloume, S. Mancini, L. Burns, and B. Sherwood Lollar. 2010. Carbon isotope analysis to evaluate nanonscale Fe(0) treatment at a chlorohydrocarbon contaminated site. Groundwater Monitoring & Remediation 30(3):79-95.

EPA (U.S. Environmental Protection Agency). 1999a. Hydraulic optimization demonstration for groundwater pump and treat systems, volume I: Pre-optimization screening (method and demonstration). EPA/542/R 99/011A. Washington, DC: Office of Research and Development and Office of Solid Waste and Emergency Response.

EPA. 1999b. Hydraulic optimization demonstration for groundwater pump and treat systems, volume II: Application of hydraulic optimization. EPA/542/R-99/011B. Washington, DC: Office of Research and Development and Office of Solid Waste and Emergency Response.

EPA. 2002. Draft Guidance for Evaluating the Vapor Intrusion to Indoor Air Pathway from Groundwater and Soils (Subsurface Vapor Intrusion Guidance). Washington, DC: EPA OSWER. http://www.epa.gov/correctiveaction/eis/vapor/guidance.pdf.

EPA. 2003. A review of emerging sensor technologies for facilitating long-term ground water monitoring of volatile organic compounds. EPA/542/R-03/007. Washington, DC: EPA Office of Solid Waste and Emergency Response.

EPA. 2004. Demonstration of two long-term groundwater monitoring approaches. EPA/542/R-04/001A. Washington, DC: EPA Office of Solid Waste and Emergency Response.

EPA. 2005. O&M Report Template for Ground Water Remedies (with Emphasis on Pump and Treat Systems). EPA 542-R-05-010. Washington, DC: EPA Office of Solid Waste and Emergency Response.

EPA. 2009. Impacts of DNAPL Source Treatment: Experimental and Modeling Assessment of the Benefits of Partial DNAPL Source Removal. 600-09-R-096. Ada, OK: EPA ORD.

EPA. 2011a. Background Indoor Air Concentrations of Volatile Organic Compounds in North American Residences (1990–2005): A Compilation of Statistics for Assessing Vapor Intrusion. EPA 530-R-10-001. Washington, DC: EPA Office of Solid Waste and Emergency Response.

EPA. 2011b. Petroleum Hydrocarbons and Chlorinated Hydrocarbons Differ in Their Potential for Vapor Intrusion. Washington, DC: EPA Office of Underground Storage Tanks.

EPA. 2012a. Conceptual Model Scenarios for the Vapor Intrusion Pathway. EPA 530-R-10-003. Washington, DC: EPA Office of Solid Waste and Emergency Response.

EPA. 2012b. EPA's Vapor Intrusion Database: Evaluation and Characterization of Attenuation Factors for Chlorinated Volatile Organic Compounds and Residential Buildings. EPA 530-R-10-003. EPA 530-R-10-002. Washington, DC: EPA Office of Solid Waste and Emergency Response.

EPA. 2012c. Superfund Vapor Intrusion FAQs. Washington, DC: EPA Office of Solid Waste and Emergency Response.

EPA and USACE (U.S. Army Corps of Engineers). 2005. Road map to long-term monitoring optimization. EPA/542/R-05/003. Washington, DC: EPA Office of Research and Development.

Fagerlund, F., T. H. Illangasekare, and A. Niemi. 2007a. Nonaqueous-Phase Liquid Infiltration and Immobilization in Heterogeneous Media: 1. Experimental Methods and Two-Layered Reference. Vadose Zone Journal 6:471-482.

Fagerlund, F, T. H. Illangasekare, and A. Niemi. 2007b. Nonaqueous-phase liquid infiltration and immobilization in heterogeneous media: 2. Application to stochastically heterogeneous formations. Vadose Zone Journal 6:483-495.

Fagerlund, F., T. H. Illangaserkare, T. Phenrat, H.-J. Kim, and G. V. Lowry. 2012. PCE dissolution and simultaneous dechlorination by nanoscale zero-valent iron particles in a DNAPL source zone. Journal of Contaminant Hydrology 131:9-28.

Fisher, A., K. Theuerkorn, N. Stelzer, M. Gehre, M. Thullner, and H. H. Richnow. 2007. Applicability of stable isotope fractionation analysis for the characterization of benzene biodegradation in a BTEX-contaminated aquifer. Environmental Science & Technology 41(10):3689-3969.

Fuller, M. E., and R. J. Stefan. 2008. ER-1378 groundwater chemistry and microbial ecology effects on explosives biodegradation. Final report for the Strategic Environmental Research and Development Program.

Gao, J., L. B. M. Ellis, and L. P. Wackett. 2010. The University of Minnesota biocatalysis/biodegradation database: Improving public access. Nucleic Acids Research 38:D488-D491.

Glover, K., J. Munakata-Marr, and T. H. Illangasekare. 2007. Biologically-enhanced mass transfer of tetrachloroethene from DNAPL in source zones: Experimental evaluation and influence of pool morphology. Environmental Science & Technology 41(4):1384-1389.

Goltz, M. N., Kim, S., Yoon, H., and Park J. 2007. Review of groundwater contaminant mass flux measurement. Environmental Engineering Research 12(4):176-193.

Gopalakrishnan, G., B. S. Minsker, and D. E. Goldberg. 2003. Optimal sampling in a noisy genetic algorithm for risk-based remediation design. Journal of Hydroinformatics 5:11-25.

Gorder, K. A., and E. M. Dettemaier. 2011. Portable GC/MS methods to evaluate sources of cVOC contamination in indoor air. Ground Water Monitoring & Remediation 31(4):113-119.

Guan, J., and M. M. Aral. 2004. Optimal design of groundwater remediation systems using fuzzy set theory. Water Resources Research 40(1):W01518.

Guilbeault, M. A., B. L. Parker, and J. A. Cherry. 2005. Mass and flux distributions from DNAPL zones in sandy aquifers. Ground Water 43(1):70-86.

Haenggi, M. 2005. Opportunities and challenges in Wireless Sensor Networks. *In:* Handbook of Sensor Networks: Compact Wireless and Wired Sensing Systems. M. Ilyas and I. Mahgoub (eds.). Boca Raton, FL: CRC Press.

Hakala, J. A., and Y.-P. Chin. 2010. Abiotic reduction of pendimethalin and trifluralin in controlled and natural systems containing Fe(II) and dissolved organic matter. Journal of Agriculture and Food Chemistry 58(24):12840-12846.

Hakala J. A., Y.-P. Chin, and E. J. Weber. 2007. Influence of dissolved organic matter and Fe(II) on the abiotic reduction of pentachloronitrobenzene. Environmental Science & Technology 41:7337-7342.

Hay-Wilson, L., P. C. Johnson, and J. Rocco. 2005. Collecting and interpreting soil gas samples from the vadose zone: A practical strategy for assessing the subsurface-vapor-to-indoor-air migration pathway at petroleum hydrocarbon sites. American Petroleum Institute. Publication Number 4741.

He, Y. T., C. Su, J. T. Wilson, R. T. Wilkin, C. Adair, T. Lee, P. Bradley, and M. Ferrey. 2009. Identification and characterization methods for reactive minerals responsible for natural attenuation of chlorinated organic compounds in ground water. EPA 600/R-09/115. Ada, OK: EPA Office of Research and Development.

Heiderscheidt, J. L. 2005. DNAPL Source Zone Depletion During In Situ Chemical Oxidation (ISCO): Experimental and Modeling Studies. PhD Dissertation, Environmental Science and Engineering Division, Colorado School of Mines, Golden, CO.

Heiderscheidt, J., R. L. Siegrist, and T. H. Illangasekare. 2008. Intermediate-scale 2D experimental investigation of *in situ* chemical oxidation using potassium permanganate for remediation of complex DNAPL source zones. Journal of Contaminant Hydrology 102(1-2):3-16.

Hendrickson, E. R., J. Payne, R. M. Young, M. G. Starr, M. P. Perry, S. Fahnestock, D. E. Ellis, R. C. Ebersole. 2002. Molecular analysis of dehalococcoides 16S ribosomal DNA from chloroethene-contaminated sites throughout North America and Europe. Applied and Environmental Microbiology 68(2):485-495.

Hendrickx, B., W. Dejonghe, F. Faber, W. Boenne, L. Bastiaens, W. Verstraete, E. M. Top, and D. Springael. 2006. PCR-DGGE method to assess the diversity of BTEX mono-oxygenase genes at contaminated sites. FEMS Microbiology Ecology 55(2):262-273.

Hoch, L. B., E. J. Mack, B. W. Hydutsky, J. M. Hershman, J. M. Skluzacek, and T. E. Mallouk. 2008. Carbothermal synthesis of carbon-supported nanoscale zero-valent iron particles for the remediation of hexavalent chromium. Environmental Science & Technology 42(7):2600-2605.

Hofstetter, T. B., C. G. Heijman, S. B. Haderlein, C. R. Holliger, and R. P. Schwarzenbach. 1999. Complete reduction of TNT and other (poly)nitroaromatic compounds under iron-reducing subsurface conditions. Environmental Science & Technology 33:1479-1487.

Hofstetter, T. B., R. P. Schwarzenbach, and S. B. Haderlein. 2003. Reactivity of Fe(II) species associated with clay minerals. Environmental Science & Technology 37:519-528.

Hong, Y., R. J. Honda, N. V. Myung, and S. L. Walker. 2009. Transport of iron-based nanoparticles: Role of magnetic properties. Environmental Science & Technology 43(23):8834-8839.

Hunkeler, D., R. U. Meckenstock, B. Sherwood Lollar, T. C. Schmidt, and J. T. Wilson. 2008. A Guide for Assessing Biodegradation and Source Identification of Organic Ground Water Contaminants using Compound Specific Isotope Analysis. EPA report 600/R-08/148.

Illangasekare, T. H., J. L. Ramsey, K. H. Jensen, and M. Butts. 1995. Experimental study of movement and distribution of dense organic contaminants in heterogeneous aquifers. Journal of Contaminant Hydrology 20:1-25.

Illangasekare, T. H., J. Munakata Marr, R. L. Siegrist, K. Soga, K. C. Glover, E. Moreno-Barbero, J. L. Heiderscheidt, S. Saenton, M. Matthew, A. R. Kaplan, Y. Kim, D. Dai, J. L. Gago, and J. W. E. Page. 2007. Mass Transfer from Entrapped DNAPL Sources Undergoing Remediation: Characterization Methods and Prediction Tools, SERDP Project CU-1294.

Imfeld, G., C. Estop Aragones, S. Zeiger, C. V. von Eckstadt, H. Paschke, R. Trabitzsch, H. Weiss, and H. H. Richnow. 2008. Tracking in situ biodegradation of 1,2-dichloroethenes in a model wetland. Environmental Science & Technology 42(21):7924-7930.

ITRC (Interstate Technology & Regulatory Council). 2004. Strategies for Monitoring the Performance of DNAPL Source Zone Remedies. http://www.itrcweb.org/Documents/DNAPLs-5.pdf.

ITRC. 2007. Vapor Intrusion Pathway: A Practical Guideline. Washington, DC: Interstate Technology Regulatory Council. http://www.itrcweb.org/gd_VI.asp.

ITRC. 2010. Use and Measurement of Mass Flux and Mass Discharge. MASSFLUX-1. Washington, DC: ITRC Integrated DNAPL Site Strategy Team.

ITRC. 2011. Technical and Regulatory Guidance: Integrated Strategies for Chlorinated Solvent Sites; Interstate Technology & Regulatory Council: Washington, DC.

Jennings, L. K., M. M. G. Chartrand, G. Lacrampe-Couloume, B. Sherwood Lollar, J. C. Spain, and J. M. Gossett. 2009. Proteomic and transcriptomic analyses reveal genes upregulated by cis-dichloroethene in *Polaromonas* sp. strain JS666. Applied and Environmental Microbiology 75(11):3733-3744.

Johnson, P. C., M. W. Kemblowski, and R. L. Johnson. 1999. Assessing the significance of subsurface contaminant vapor migration to enclosed spaces: Site-specific alternatives to generic estimates. Journal of Soil Contamination 8(3):389-421.

Johnson, P. C., H. Luo, C. Holton, P. Dahlen, and Y. Guo. 2012. Vapor intrusion above a dilute CHC plume: Lessons-learned from two years of monitoring. EPA-AEHS Workshop, Recent Advances to VI Application & Implementation, 20 March 2012, San Diego. https://iavi.rti.org/WorkshopsAndConferences.cfm.

Kampara, M., M. Thullner, H. H. Richnow, H. Harms, and L. Y. Wick. 2008. Impact of bioavailability restrictions on microbially induced stable isotope fractionation. 2. Experimental evidence. Environmental Science & Technology 42(17):6552-6558.

Kao, C. M., H. Y. Chien, R. Y. Surampalli, C. C. Chien, and C. Y. Chen. 2010. Assessing of natural attenuation and intrinsic bioremediation rates at a petroleum-hydrocarbon spill site: Laboratory and field studies. Journal of Environmental Engineering 136(1):54-67.

Kappler, A., and S. B. Haderlein. 2003. Natural organic matter as reductant for chlorinated aliphatic pollutants. Environmental Science & Technology 37:2707-2713.

Kappler A., and K. L. Straub. 2005. Geomicrobiological cycling of iron. Reviews in Mineralogy and Geochemistry 59:85-108.

Kavanaugh, M., and R. Deeb. 2011. Diagnostic Tools for Performance Evaluation of Innovative In-Situ Remediation Technologies at Chlorinated Solvent-Contaminated Sites. ER-200318. Washington, DC: ESTCP.

Kennedy, L. G., J. W. Everett, and J. Gonzales. 2006. Assessment of biogeochemical natural attenuation and treatment of chlorinated solvents, Altus Air Force Base, Altus, Oklahoma. Journal of Contaminant Hydrology 83(3-4):221-236.

Kram, L. K., Keller, A. L, Rossabi, J., and Everett, L. G. 2001. DNAPL characterization methods and approaches, Part 1: Performance comparisons. Ground Water Monitoring and Remediation 21(4):109-123.

Lee, P. K. H., T. W. Macbeth, K. S. Sorenson Jr, R. A. Deeb, and L. Alvarez-Cohen. 2008. Quantifying genes and transcripts to assess the in situ physiology of *Dehalococcoides* spp. in a trichloroethene-contaminated groundwater site. Applied and Environmental Microbiology 74(9):2728-2739.

Li, X., and W.-X. Zhang. 2007. Sequestration of metal cations with zerovalent iron nanoparticles—A study with high resolution x-ray photoelectron spectroscopy (HR-XPS). Journal of Physical Chemistry C 111(19):6939-6946.

Liang, H., R. Falta, C. Newell, S. Farhat, S. Rao, and N. Basu. 2011. Decision and Management Tools for DNAPL Sites: Optimization of Chlorinated Solvent Source and Plume Remediation Considering Uncertainty. ESTCP Project ER-200704.

Lieberman, S. 2007. Direct Push Chemical Sensors for DNAPL. ER-0109. Washington, DC: ESTCP.

Lien, H. L., and W.-X. Zhang. 1999. Transformation of chlorinated methanes by nanoscale iron particles. Journal of Environmental Engineering 125(11):1042-1047.

Lien, H. L., and W.-X. Zhang. 2005. Hydrodechlorination of chlorinated ethanes by nanoscale Pd/Fe bimetallic particles. Journal of Environmental Engineering 131(1):4-10.

Liu, Y., and G. V. Lowry. 2006. Effect of particle age (Fe0 Content) and solution pH on NZVI reactivity: H2 evolution and TCE dechlorination. Environmental Science & Technology 40(19):6085-6090.

Liu, Y., S. A. Majetich, R. D. Tilton, D. S. Sholl, and G. V. Lowry. 2005. TCE dechlorination rates, pathways, and efficiency of nanoscale iron particles with different properties. Environmental Science & Technology 39(5):1338-1345.

Lovley, D. R., E. E. Roden, E. J. P. Phillips, and J. C. Woodward. 1993. Enzymatic iron and uranium reduction by sulfate-reducing bacteria. Marine Geology 113(1-2):41-53.

Lu, X., J. T. Wilson, and D. H. Kampbell. 2006. Relationship between dehalococcoides DNA in ground water and rates of reductive dechlorination at field scale. Water Research 40(16):3131-3140.

Luo, H., P. Dahlen, P. C. Johnson, T. Peargin, and T. Creamer. 2009. Spatial variability of soil-gas concentrations near and beneath a building overlying shallow petroleum hydrocarbon–impacted soils. Ground Water Monitoring & Remediation 29(1):81-91.

Luo, H., P. Dahlen, and P. C. Johnson. 2010. Hydrocarbon and oxygen transport in the vicinity of a building overlying a NAPL source zone. Air and Waste Management Association: Vapor Intrusion 2010:155-185.

Luster-Teasley, S., P. Onochie, and V. Shirley. 2010. Encapsulation of Potassium Permanganate Oxidant in Biodegradable Polymers to Develop a Novel Form of Controlled Release Remediation. *In:* Emerging Environmental Technologies, Volume 2. Vishal Shah (ed.) Springer.

MacKay, D. M., and J. A. Cherry. 1989. Groundwater contamination: Pump-and-treat remediation. Environmental Science & Technology 23(6):630-636.

MacMillan, D. K., and D. E. Splichal. 2005. A review of field technologies for long-term monitoring of ordnance-related compounds in groundwater. ERDC/EL TR-05-14. Vicksburg, MS: U.S. Army Engineer Research and Development Center.

McAlary, T., R. A. Ettinger, and P. C. Johnson. 2005. Reference Handbook for Site-Specific Assessment of Subsurface Vapor Intrusion to Indoor Air. EPRI Technical Report 1008492. Palo Alto, CA.

McDonald, M. G., and A. W. Harbaugh. 1988. A modular three-dimensional finite-difference ground-water flow model. Techniques of Water-Resources Investigations. Book 6. U.S. Geological Survey.

McHugh, T., R. Davis, G. DeVaull, H. Hopkins, J. Menatti, and T. Peargin. 2010. Evaluation of vapor attenuation at petroleum hydrocarbon sites. Soil and Sediment Contamination: An International Journal 19(6):725-745.

McKelvie, J. R., S. K. Hirschorn, G. Lacrampe-Couloume, J. Lindstrom, J. Braddock, K. Finneran, D. Trego, and B. Sherwood Lollar. 2007. Evaluation of TCE and MTBE in situ biodegradation: Integrating stable isotope, metabolic intermediate, and microbial lines of evidence. Ground Water Monitoring & Remediation 27(4):63-73.

Mercer, J. W., R. M. Cohen, and M. R. Noel. 2010. DNAPL site characterization issues at chlorinated solvent sites. Pp. 217-280 in In Situ Remediation of Chlorinated Solvent Plumes. Stroo, H. F., Ward, C. H., Eds. New York: Springer.

Moreno-Barbero, E., and T. H. Illangasekare. 2005. Simulation and performance assessment of partitioning tracer tests in heterogeneous aquifers. Environmental & Engineering Geoscience XI(4):395-404.

Moreno-Barbero, E., and T. H. Illangasekare. 2006. Influence of pool morphology on the performance of partitioning tracer tests: evaluation of the equilibrium assumption. Water Resources Research 42:11. doi:10.1029/2005WR004074.

Morris, R. M., J. M. Fung, B. G. Rahm, S. Zhang, D. L. Freedman, S. H. Zinder, and R. E. Richardson. 2007. Comparative proteomics of *Dehalococcoides* spp. reveals strain-specific peptides associated with activity. Applied and Environmental Microbiology 73(1):320-326.

Nesatyy, V. J., and M. J. F. Suter. 2007. Proteomics for the analysis of environmental stress responses in organisms. Environmental Science & Technology 41(20):6891-6900.

Neumann, A., T. B. Hofstetter, M. Skarpeli-Liati, and R. R. Schwarzenbach. 2009. Reduction of polychlorinated ethanes and carbon tetrachloride by structural Fe(II) in smectites. Environmental Science & Technology 43(11):4082-4089.

NRC (National Research Council). 2005. Contaminants in the Subsurface. Washington, DC: National Academies Press.

NYSDOH (New York State Department of Health). 2006. Final NYSDOH CEH BEEI Soil Vapor Intrusion Guidance. http://www.health.state.ny.us/environmental/investigations/soil_gas/svi_guidance/.

Olson, M. R., T. C. Sale, C. D. Shackelford, C. Bozzini, and J. Skeean. 2012. Chlorinated solvent source-zone remediation via ZVI-clay soil mixing: 1-year results. Ground Water Monitoring & Remediation 32:63–74. doi:10.1111/j.1745-6592.2011.01391.x.

Ouyang, G., and J. Pawliszyn. 2006. Recent developments in SPME for on-site analysis and monitoring. Trends in Analytical Chemistry 25(7):692-703.

Parker, B. L., S. W. Chapman, and M. A. Guilbeault. 2008. Plume persistence caused by back diffusion from thin clay layers in a sand aquifer following TCE source-zone hydraulic isolation. Journal of Contaminant Hydrology 102:86-104.

Peralta, R. C. 2011. Simulation/Optimization Modeling for Groundwater and Conjunctive Management. Boca Raton, FL: CRC Press.

Phenrat, T., N. Saleh, K. Sirk, R. D. Tilton, and G. V. Lowry. 2007. Aggregation and sedimentation of aqueous nanoscale zerovalent iron dispersions. Environmental Science & Technology 41(1):284-290.

Pironi, P., C. Switzer, J. I. Gerhard, G. Rein, and J. L. Torero. 2011. Self-sustaining smoldering combustion for NAPL remediation: Laboratory evaluation of process sensitivity to key parameters. Environmental Science & Technology 45:2980-2986.

Pooley, K. E., M. Blessing, T. C. Schmidt, S. B. Haderlein, K. T. B. MacQuarrie, and H. Prommer. 2009. Aerobic biodegradation of chlorinated ethenes in a fractured bedrock aquifer: Quantitative assessment by compound-specific isotope analysis (CSIA) and reactive transport modeling. Environmental Science & Technology 43(19):7458-7464.

Rabideau, A. J., and C. T. Miller. 1994. 2-Dimensional modeling of aquifer remediation influenced by sorption nonequilibrium and hydraulic conductivity heterogeneity. Water Resources Research 30(5):1457-1470.

Ramos, M. A. V., W. Yan, X. Q. Li, B. E. Koel, and W.-X. Zhang. 2009. Simultaneous oxidation and reduction of arsenic by zero-valent iron nanoparticles: Understanding the significance of the core-shell structure. Journal of Physical Chemistry C 113(33):14591-14594.

Reed, P., B. S. Minsker, and D. Goldberg. 2001. A multiobjective approach to cost effective long-term groundwater monitoring using an elitist nondominated sorted genetic algorithm with historical data. Journal of Hydroinformatics 3(2):71-90.

Reed, P. M., B. S. Minsker, and A. J. Valocchi. 2000. Cost-effective long-term groundwater monitoring design using a genetic algorithm and global mass interpolation. Water Resources Research 36(12):3731-3741.

Rivett, M. O., S. W. Chapman, R. M. Allen-King, S. Feenstra, and J. A. Cherry. 2006. Pump-and treat remediation of chlorinated solvent contamination at a controlled field-experiment site. Environmental Science & Technology 40(21):6770-6781.

Ross, C., L. C. Murdoch, D. L. Freedman, and R. L. Siegrist. 2005. Characteristics of potassium permanganate encapsulated in polymer. Journal of Environmental Engineering 131:1203-1211.

Rügge, K., T. B. Hofstetter, S. B. Haderlein, P. L. Bjerg, S. Knudsen, C. Zraunig, H. Mosbæk, and T. H. Christensen. 1998. Characterization of predominant reductants in an anaerobic leachate-contaminated aquifer by nitroaromatic probe compounds. Environmental Science & Technology 32(1):23-31.

Saenton, S., and T. H. Illangasekare. 2004. Determination of DNAPL entrapment architecture using experimentally validated numerical codes and inverse modeling. Proceedings of XVth Computational Methods in Water Resources 2004 International Conference (ed. C. T. Miller, M. W. Farthing and W. G. Gray), Elsevier.

Sakaki, T., P. E. Schulte, A. Cihan, J. A. Christ, and T. H. Illangasekare. 2013. Airflow pathway development as affected by soil moisture variability in heterogeneous soils. Vadose Zone Journal. In Press.

Sale, T. C., and C. Newell. 2011. Guide for Selecting Remedies for Subsurface Releases of Chlorinated Solvents. ESTCP Project ER-200530. Washington, DC: ESTCP.

Shackelford, C. D., T. C. Sale, and M. R. Liberati. 2005. In-situ remediation of chlorinated solvents using zero valent iron and clay mixtures: A case history. Proceedings of the Sessions of the Geo-Frontiers 2005 Congress. Geotechnical Special Publication. 142:5996.

Sherwood Lollar, B., G. F. Slater, J. M. E. Ahad, B. Sleep, J. Spivak, M. Brennan, and P. MacKenzie. 1999. Contrasting carbon isotope fractionation during biodegradation of trichloroethylene and toluene: Implications for intrinsic biodegradation. Organic Geochemistry 30(8):813-820.

Sherwood Lollar, B., G. F. Slater, B. Sleep, M. Witt, G. M. Klecka, M. Harkness, and J. Spivack. 2001. Stable carbon isotope evidence for intrinsic bioremediation of tetrachloroethene and trichloroethene at Area 6, Dover Air Force Base. Environmental Science & Technology 35(2):261-269.

Siegrist, R. L., M. Crimi, and R. A. Brown. 2011. *In situ* chemical oxidation: technology description and status. In *In Situ* Chemical Oxidation for Groundwater Remediation. Siegrist, R. L., Crimi, M., Simpkin, T. J., Eds. New York: Springer.

Simpkin, T. J., and R. D. Norris. 2010. Engineering and implementation challenges for chlorinated solvent remediation. Pp. 109-144 In *In Situ* Remediation of Chlorinated Solvent Plumes. Stroo, H. F., Ward, C. H., Eds. New York: Springer.

Singh, O. V. 2006. Proteomic and metabolomics: The molecular make-up of toxic aromatic pollutant bioremediation. Proteomics 6:5481-5492.

Sohn, K., S. W. Kang, S. Ahn, M. Woo, and S. K. Yang. 2006. Fe(0) nanoparticles for nitrate reduction: Stability, reactivity, and transformation. Environmental Science & Technology 40(17):5514-5519.

Song, D. L., M. E. Conrad, K. S. Sorenson, and L. Alvarez-Cohen. 2002. Stable carbon isotope fractionation during enhanced in situ bioremediation of trichloroethene. Environmental Science & Technology 36(10):2262-2268.

Song, H., and E. R. Carraway. 2005. Reduction of chlorinated ethanes by nanosized zero-valent iron: Kinetics, pathways, and effects of reaction conditions. Environmental Science & Technology 39:6237-6245.

Song, H., and E. R. Carraway. 2006. Reduction of chlorinated methanes by nanosized zero-valent iron: Kinetics, pathways, and effect of reaction conditions. Environmental Engineering Science 23:272-284.

Song, H., and E. R. Carraway. 2008. Catalytic hydrodechlorination of chlorinated ethenes by nanoscale zero-valent iron. Applied Catalysis B: Environmental. 78:53-60.

Switzer, C., P. Pironi, J. I. Gerhard, G. Rein, and J. L. Torero. 2009. Self-sustaining smoldering combustion: A novel remediation process for non-aqueous phase liquids in porous media. Environmental Science & Technology 43:5871-5877.

Taghavy, A., J. Costanza, K. D. Pennell, and L. M. Abriola. 2010. Effectivness of nanoscale zero-valent iron for treatment of a PCE-DNAPL source zone. Journal of Contaminant Hydrology 118:128-142.

Tebo, B. M., J. R. Bargar, B. G. Clement, G. J. Dick, K. J. Murray, D. Parker, R. Verity, and S. M. Webb. 2004. Biogenic manganese oxides: Properties and mechanisms of formation. Annual Review of Earth and Planetary Sciences 32:287-328.

Thullner, M., M. Kampara, H. H, Richnow, H. Harms, and L. Y. Wick. 2008. Impact of bioavailability restrictions on microbially induced stable isotope fractionation. 1. Theoretical calculation. Environmental Science & Technology 42(17):6544-6551.

Vera, Y. M., R. J. de Carvalho, M. L. Torem, and B. A. Calfa. 2009. Atrazine degradation by *in situ* electrochemically generated ozone. Chemical Engineering Journal 155(3):691-697.

Waddill, D. W., and M. A. Widdowson. 2003. SEAM3D: A numerical model for three-dimensional solute transport and sequential electron acceptor-based biodegradation in groundwater. ERDC/EL TR-00-18. Vicksburg, MS: U.S. Army Research and Development Center.

Waldron, P. J., L. Wu, J. D. Van Nostrand, C. W. Schadt, Z. He, D. B. Watson, P. M. Jardine, A. V. Palumbo, T. C. Hazen, and J. Zhou. 2009. Functional gene array-based analysis of microbial community structure in groundwaters with a gradient of contaminant levels. Environmental Science & Technology 43(10):3529-3534.

Wang, C. B., and W.-X. Zhang. 1997. Synthesizing nanoscale iron particles for rapid and complete dechlorination of TCE and PCBs. Environmental Science & Technology 31(7):2154-2156.

Wani, A. H., B. R. O'Neal, D. M. Gilbert, D. B. Gent, and J. L. Davis. 2005. Electrolytic Transformation of Hexahydro-1,3,5-Trinitro-1,3,5-Triazine (RDX) and 2,4,6-Trinitrotoluene (TNT) in Aqueous Solutions. U.S. Army Corps of Engineers. ERDC/EL TR-05-10. Washington, DC.

Wani, A. H., B. R. O'Neal, D. M. Gilbert, G. B. Gent, and J. L. Davis. 2006. Electrolytic transformation of ordnance related compounds in groundwater: Laboratory mass balance studies. Chemosphere 62:689-698.

Welch, R., and R. G. Riefler. 2008. Estimating treatment capacity of nanoscale zero-valent iron reducing 2,4,6-trinitrotoluene. Environmental Engineering Science 25(9):1255-1262.

Werner, J. J., A. C. Ptak, B. G. Rahm, S. Zhang, and R. E. Richardon. 2009. Absolute quantification of *Dehalococcoides* proteins: Enzyme bioindicators of chlorinated ethene dehalorespiration. Environmental Microbiology 11:2687-2697.

Werner-Allen, G., K. Lorincz, M. Welsh, O. Marcillo, J. Johnson, M. Ruiz, and J. Lees. 2006. Deploying a wireless sensor network on an active volcano. IEEE Internet Computing 10(2):18-25.

Wiesner, M. R., G. V. Lowry, P. Alvarez, D. Dionysiou, and P. Biswas. 2006. Assessing the risks of manufactured nanomaterials. Environmental Science & Technology 40(14):4336-4345.

Wilkins, M. J., N. C. VerBerkmoes, K. H. Williams, S. J. Callister, P. J. Mouser, H. Elifantz, A. L. N'Guessan, B. C. Thomas, C. D. Nicora, M. B. Shah, P. Abraham, M. S. Lipton, D. R. Lovley, R. L. Hettich, P. E. Long, and J. F. Banfield. 2009. Proteogenomic monitoring of *Geobacter* physiology during stimulated uranium bioremediation. Applied and Environmental Microbiology 75(20):6591-6599.

Wilson, J. L. 1997. Removal of aqueous phase dissolved contamination: Non-chemically-enhanced pump and treat. In Subsurface Remediation Handbook. Ward, C. H., Cherry, J., Scalf, M., eds. Chelsea, MI: Ann Arbor Press.

Xu, Y., and D. Zhao. 2007. Reductive immobilization of chromate in water and soil using stabilized iron nanoparticles. Water Research 41(10):2101-2108.

Zheng, C., and P. P. Wang. 1999. MT3DMS, A Modular Three-Dimensional Multi-Species Transport Model for Simulation of Advection, Dispersion and Chemical Reactions of Contaminants in Groundwater Systems: Documentation and User's Guide. SERDP-99-1. Vicksburg, MS: U.S. Army Corps of Engineers Engineer Research and Development Center.

7

Better Decision Making During the Long-Term Management of Complex Groundwater Contamination Sites

This chapter builds on the lessons learned in the earlier chapters of this report and focuses on recommendations that encourage better decision making *during the post-remedy-selection phase of remediation* at complex groundwater sites (as opposed to Chapter 3, which focuses primarily on the original remedy selection process). This focus is critical because the Committee has concluded that regardless of the remedial technologies applied at complex sites, removal of sufficient mass to reduce contaminant concentrations in groundwater to levels that allow for unlimited use and unrestricted exposure is unlikely for many decades. Furthermore, no transformational remedial technology or combination of technologies appears capable of overcoming the inherent technical challenges to restoration at these complex sites. Rather, the nation's cleanup programs are transitioning from remedy selection into remedy operation and long-term management (LTM), potentially over long timeframes. The implications of the limitations of existing technologies to attain unlimited use and unrestricted exposure levels throughout the impacted aquifer should be more fully reflected in the decision-making process used in existing cleanup programs and should be recognized earlier in the regulatory cycle of these complex sites.

Better decision making is needed at key points in the life cycle of a complex groundwater contamination site to address issues that frequently arise, including (1) what is a "reasonable time frame;" (2) what is the definition of contaminant removal "to the maximum extent practicable;" (3) when should active remediation at a complex groundwater site be transitioned to a passive remedy, such as monitored natural attenuation (MNA) or natural attenuation (NA); and (4) can consensus be reached on a "diminishing returns" concept applied to the performance of active remedies for groundwater cleanup at complex sites?

This chapter provides a series of recommendations that, in the judgment of the Committee, will accelerate the transition of a site to one of three possible "end states," where this term simply means a state where long-term management will be implemented if required. These "end states" are (1) *closure* in which unlimited use and unrestricted exposure levels have been attained (presumably no long-term management will be required at such sites, although even these sites can be subject to reopeners should conditions change); (2) *long-term passive management* (e.g., using MNA, NA, physical containment, PRBs, institutional controls, or some combination thereof), and (3) *long-term active management* (e.g., indefinite hydraulic containment using pump and treat or other active remedies requiring continuous operation). Complex sites under both passive and active long-term management could eventually transition to the *closure* end state, but the time frame extends many decades into the future.

The acceleration of this transition to one of three end states is premised on using remedies that are fully protective of human health and the environment in combination with more rapid acceptance of alternative end states other than clean closure, taking risk reduction, life-cycle costs, and technical feasibility into account. The transition of a site to either passive or active long-term management must be accomplished in a manner that is transparent, reduces long-term risks to an acceptable level, and is practical and cost effective, among other goals.

SETTING THE STAGE

All complex contaminated groundwater sites will ultimately transition from investigation, through remedy selection, implementation, and operations, to long-term management and ultimately (without any time frame constraint) to attainment of unlimited use and unrestricted exposure goals. If these concentration levels have not been achieved after some reasonable time period, a site will require long-term monitoring and management under either passive remedies such as MNA or an active remedy such as pump and treat. The number of complex sites where achieving unlimited use and unrestricted exposure goals within a reasonable timeframe is unlikely is not known precisely, but as discussed in Chapter 2 is estimated to be in the range of 12,000 sites and may be much higher.

Over more than 30 years of remediation experience at contaminated groundwater sites following the passage of federal statutes (CERCLA and RCRA) and state regulations have shown that the duration of the cleanup process—from initial discovery of contamination to installation of "final" remedies (as opposed to interim actions designed to eliminate imminent threats)—can take 10 to 25 years at complex sites. If during this time interim actions have not halted or substantially slowed the migration of contamination, contaminants can spread both laterally and vertically, making the site even more difficult and costly to address. A desire to transition more quickly from remedy operation to some final end state was expressed frequently to the Committee over the last two years. In particular, many of the DoD personnel expressed frustration over the long timeframes and financial demands resulting from continued delays in decision making at key points in the remedial process. A reduction in these timeframes is desired during many phases of cleanup, not just between remedy operation and the transition to long-term management or actual closure, but also between initial site characterization and remedy selection and in the period subsequent to remedy implementation during which the effectiveness of the remedy is assessed.

Accelerating decisions throughout the cleanup process is difficult for a number of reasons, as many initiatives undertaken in the past have made clear (Clean Sites, 1990; DOE, 1998; EPA, 2011a; various ITRC documents). First, the investigatory process is inherently difficult, expensive, and may not result in an accurate conceptual model of the site, at least initially. There are uncertainties and subjectivity in selecting the appropriate technologies for site characterization and remediation, including differing perspectives over the intensity of data collection and its timing and cost effectiveness. Second, there can be disagreements among potentially responsible parties, the states and even within the ten EPA Regional offices over the appropriate risk management approach to take at an individual site (i.e., how much aggressive source removal is warranted when considering the time to reach groundwater cleanup goals and the high costs of such actions). Third, when residual contamination is left in place, stakeholders have increasingly argued for remedy modification to accelerate risk reduction, to consider more recently developed sustainability metrics, or to address previously unknown risks such as exposure to volatile chemicals via vapor intrusion (see Chapters 5 and 6).

In addition to reducing the time to reach cleanup goals, potentially responsible parties (PRPs) such as the DoD strive to minimize life-cycle costs at these complex sites. Some estimates of future costs for site cleanup exceed $300 billion (in 2004 U.S. dollars) (EPA, 2004). As discussed in Chapter 2, the Committee's upper estimate of $127 billion (see Chapter 2) is likely an underestimate of future liabilities. At the same time, severe budget constraints at the federal and state levels have led to the need for prioritization of resource expenditures to ensure that the greatest risks at contaminated sites are mitigated and that long-term containment of contamination is achieved to ensure no unacceptable risks to human health and the environment. In this context, the DoD in particular has established aggressive goals to reduce costs at the majority of their sites (Conger, 2011; Yonkers, 2011). Based on the experience of some Committee members, private sector PRPs with large portfolios of complex groundwater sites have also established strategies to accelerate cleanup, with a goal of reducing life-cycle costs. Strategies have included setting up separate companies responsible for remediation of legacy sites, or outsourcing the management of complex sites through the use of experienced program management companies that oversee remediation.

Innovation would be particularly helpful in three areas of decision making. The first is making decisions in a more prompt manner, as delays result in higher transaction costs and increase the risks of ineffective interim remedies. Second, better decision making requires better metrics for demonstrating progress—based not upon regulatory milestones, but upon quantifiable, transparent metrics of remedial performance and human health risk reduction. Third, the decision on when to transition to long-term management should be formally recognized as the point where further active remediation results in little or no decreases in contaminant concentration, and the unit cost of the remedy increases much faster than the reduction in contaminant concentrations. The benefits of improved decision making at complex sites may include (1) reduction in the duration between decisions in the cleanup process; (2) cost savings at particular sites; (3) more rapid restoration of impaired groundwater resources, thereby allowing unrestricted use of at least some portions of a site; (4) more rapid mitigation of exposure pathways while long-term strategies are being considered; and (5) minimizing the long-term risks and financial burdens, to the public and PRPs, associated with groundwater sites where residual contamination is likely to persist for long timeframes. Of course, not all of these benefits can be simultaneously realized.

EPA's Existing Site Remediation Process for Groundwater

EPA recently summarized its existing guidance for restoring contaminated groundwater (EPA, 2011b). While its groundwater restoration Road Map focuses on CERCLA, EPA has stated elsewhere that the methodology also applies to groundwater remediation under the RCRA Corrective Action program (see Chapter 3). Figure 7-1 shows the current EPA decision framework, which includes five key decision points (diamonds in the figure): (1) determine if selected remedy is viable, (2) post operations, determine if operational data are sufficient to evaluate performance of the remedy, (3) determine if achievement of the remedial action objectives (RAOs) in the record of decision (ROD)—usually defined as restoration where groundwater is a potential source of drinking water—can be achieved or not, (4) if restoration is likely, determine whether in fact RAOs have been achieved,

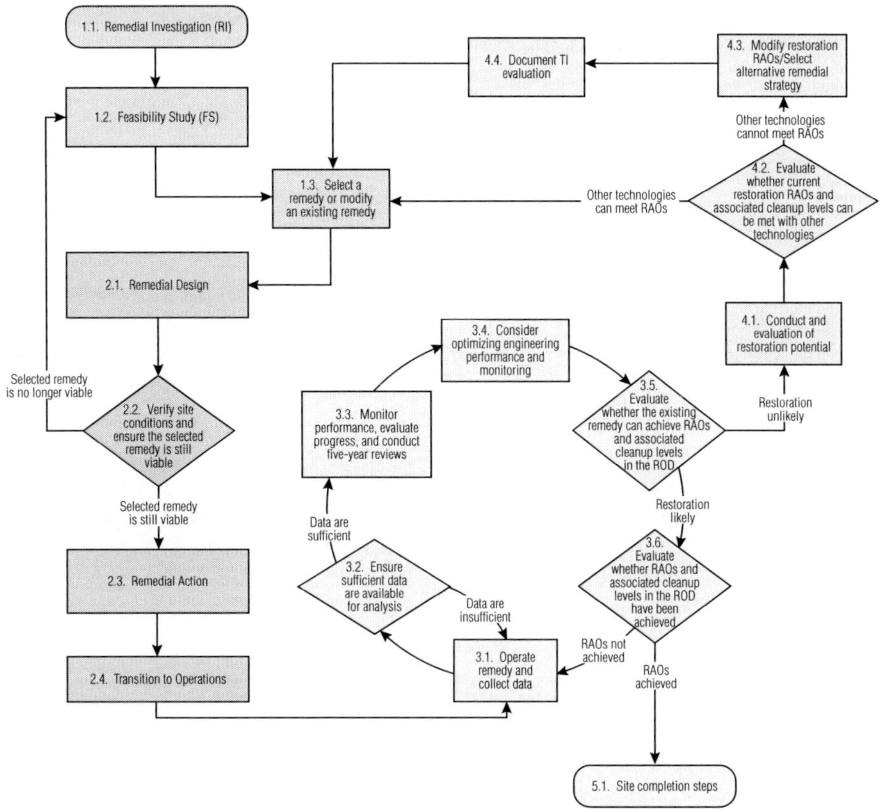

FIGURE 7-1 EPA recommended process for restoring contaminated groundwater at Superfund sites.
SOURCE: EPA (2011b).

and (5) if restoration is unlikely, evaluate whether alternative technologies can overcome the restoration limitations of the technologies included in the ROD. In the event that available and demonstrated technologies cannot meet RAOs, a Technical Impracticability (TI) waiver can be granted. The transition to site completion thus requires modification of the RAOs, possibly modification to the remedy, and continued evaluation and assessment of an active remedy. There is, however, no discussion of a transition to passive or active long-term management. In the opinion of the Committee, lack of guidance on this transition is unfortunate given the likelihood that many contaminated groundwater sites pose technical challenges that will not allow for achievement of unlimited use and unrestricted exposure throughout the entire contaminated aquifer for many decades.

AN ALTERNATIVE DECISION PROCESS FOR CONTAMINATED GROUNDWATER

The EPA Road Map includes a process to monitor performance of the remedy and conduct regular (every five years) reviews of the adequacy and protectiveness of the remedy, considering new information if available. Similarly, under the RCRA corrective action program continued operation of the industrial facility requires approval of a permit that will include necessary monitoring and reporting requirements where contamination remains on site, as well as verification that remedies are meeting such requirements as eliminating off-site contaminant migration.[1] Many Superfund facilities have now been through multiple five-year reviews, although statistics on the average number of five-year reviews or the number of non-Superfund sites that have received similar regulatory reviews are not readily available. Given the lengthy life cycle of these complex sites (and associated costs), the development of an alternative process for addressing groundwater contamination is warranted that accounts more explicitly for the likelihood of residual contamination and provides sufficient engineering and legal controls to ensure that exposure risks are below acceptable limits. An alternative decision process is shown in Figure 7-2.

Initial Steps of the Alternative Decision Process

Figure 7-2 includes the processes currently followed at all sites regulated under CERCLA and at many complex sites regulated under other federal or state programs (RCRA Corrective Action or state Superfund sites), but it provides more detailed guidance for sites where recalcitrant contamination remains in place at levels above those allowing for unlimited use and unrestricted exposure. The first two decisions in Figure 7-2, shown as diamonds, come after a remedy is in place as directed in the ROD. These diamonds represent two key questions that must be answered (usually during the five-year review): (1) is the remedy protective of human health and the environment (considering all exposure pathways, including vapor intrusion), and (2) have maximum contaminant levels (MCLs) or other RAOs been met at the designated point(s) of compliance (POCs).

With respect to remedy protectiveness, EPA's five-year review guidance (EPA, 2001a) states that "if the risk associated with the cleanup levels currently being achieved by the remedy are within EPA's acceptable risk range, the remedy generally should be considered protective." Protectiveness may be demonstrated through a variety of means including a human health risk assessment that demonstrates site risks for all exposure pathways fall

[1] See http://www.epa.gov/epawaste/hazard/correctiveaction/index.htm.

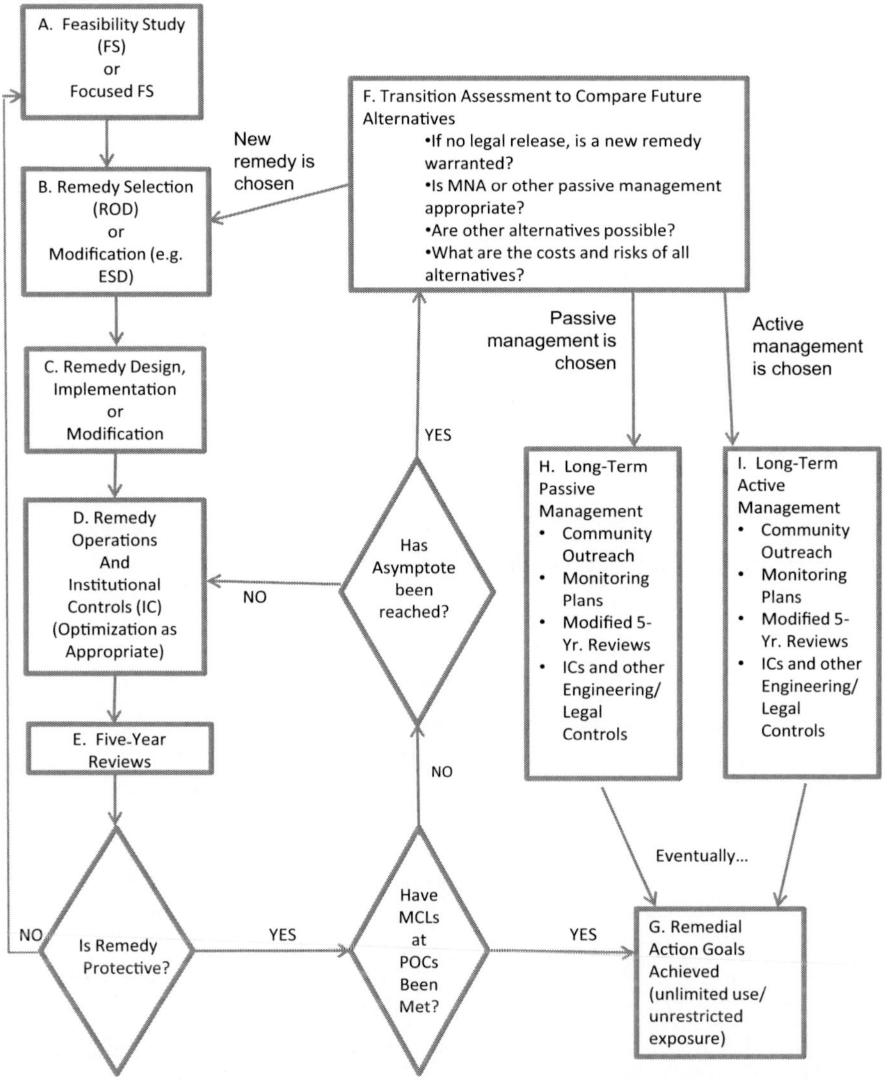

FIGURE 7-2 Key decisions for complex sites with groundwater contamination.

within EPA's acceptable risk range of 10^{-6} to 10^{-4}, interruption of exposure pathways, and other measures. Based on the experience of the Committee, designation of a CERCLA remedy as "not protective" during the five-year review is relatively infrequent and when it does occur, action is eventually taken. The MEW Superfund facility in Mountain View, California, is an example of a site that was considered "protective" through one five-year review cycle until 2009, when the remedy was no longer considered protective because of newly identified human health risks from vapor intrusion. Similar cases may arise elsewhere as the potential risks from vapor intrusion become better understood.

With respect to the second decision, there are certainly instances where sites regulated under CERCLA have reached the level of unlimited use/unrestricted exposure at the points of compliance, but as discussed in Chapter 2 many of these sites were listed early in the program's history and likely did not have significant contamination to begin with. A larger number of these delisted sites have persistent groundwater contamination (see Chapter 2). For complex DoD sites, few have achieved unlimited use and unrestricted exposure levels in groundwater. Thus, at many groundwater sites the answer to this second question will be "no," leading to the third decision.

Has an Asymptote Been Reached?

Figure 7-2 begins to diverge from the EPA Road Map for remediating groundwater at complex sites at the third decision point, which asks whether an asymptote, or a point of diminishing returns, has been reached in the benefits gained by applying the current remedy, considering both time and cost. The benefits metrics considered here are derived from the cleanup objectives for the site and might include reductions in groundwater or indoor air concentrations, shrinking of the dissolved plume footprint, reductions in source zone mass, or an increase in the potential beneficial uses of the site and any affected resources (e.g., aquifers and surface water bodies).

Remedy performance is typically reported on a quarterly schedule and may include data collected at daily or weekly intervals. This provides sufficient data to adjust and optimize systems on monthly to quarterly timeframes and to assess, on annual timeframes, if operation of the remedy continues to provide significant benefits. Given this context of typical data monitoring and reporting cycles, the question of whether an asymptote has been reached can be rewritten as "given their current trends, will there be significant changes in the benefits metrics (e.g., groundwater concentrations) over the next 12 months and after another five years of remedy operation?" The 12-month and five-year periods are relevant and practical because they represent a reasonable minimum future projection period given typical data sets and the mandated CERCLA five-year review cycle, respectively. The Committee's experience with complex sites suggests that there will be few cases where optimal remedy implementation reaches asymptotic conditions in less than 12 months.

Answering the question of "will there be significant changes in the benefits metrics over the next 12 months and after another five years" will involve professional technical judgment, input from all stakeholders, and consideration of the legal framework. Discussion will likely center on the definition of "significant changes" for the benefits metrics. It should be noted that this question is only relevant in cases where there is groundwater contamination above levels allowing for unlimited use and unrestricted exposure and only in cases for which remedy optimization has already occurred.

Some of the monitoring tools mentioned in Chapter 6 could be particularly useful for determining whether a site has reached an asymptote, such as compound specific isotope analysis. For example, if contaminant concentrations were found to be steady (or even decreasing), and isotopic signatures were not changing along the treatment gradient, it would suggest that no further degradation is occurring (i.e., concentrations are too low for biodegradation to operate or the transformation capacity of the aquifer has been exhausted). In that case, it would suggest that the remedy is no longer having an effect and that the point of diminishing returns had been reached. Other examples discussed in Chapter 6 include the application of diagnostic tools to measure reductions in mass flux/mass discharge from source areas, which could also be used to assess whether asymptotic limits have been reached.

It should be noted that while a clear, purely technical trigger for determining when the asymptote has been reached would be desirable, this decision must be made in the context of the legal framework within which the site is managed and the sometimes competing stakeholder interests. Regulators and impacted communities are often unwilling to accept that a remedy will not reach its objectives in a certain timeframe. However, it is generally the experience of the Committee that stakeholders will understand the technical limitations of the remedy provided that such discussions are conducted in a transparent manner (see further discussion of stakeholder interactions later in this chapter). Furthermore, the Committee has observed that members of affected communities do not support the indefinite expenditure of resources on activities that do little or nothing to reduce risk. Thus, the asymptote analysis is not just a valuable decision-making tool, it is also a vital part of communicating the challenge of groundwater remediation to the public.

The paradigm embodied in Figure 7-2 strikes a balance between deciding not to undertake any risk reduction efforts as part of the original remedy because of the impracticability of attaining drinking water standards at complex sites versus continuing to expend extensive resources when further remedial work provides no substantial incremental risk reduction (as measured by decreases in contaminant concentration or other appropriate benefits metrics).

The Transition Assessment

For sites that have reached an asymptote, Figure 7-2 leads to a transition assessment. The transition assessment is envisioned as an analysis similar to a focused feasibility study that considers alternatives for site management—choosing a new remedy or transitioning to long-term management (such as monitored natural attenuation) or the other alternative approaches outlined in Chapter 3 and ESTCP (2011). For each of the possible alternatives, the transition assessment considers the nine remedy selection criteria of CERCLA or similar criteria established under other regulatory regimes, particularly the risk of any residual contamination expected to remain in aquifers that are not likely to be restored; costs such as life-cycle costs and the marginal costs of remediation compared to the level of risk reduction achieved; and state and community acceptance. Although the actual decision making occurs at the same time through comparison of the alternatives, for simplicity we begin first with a discussion of whether a new remedy is warranted, followed by the transition to long-term management. Risk and cost considerations are also detailed here, while stakeholder concerns are presented in a subsequent section.

The transition assessment is partly a response to the ongoing debates regarding the cost effectiveness of remedies at complex sites that have been operating for several years. Although some additional costs will be incurred to conduct the transition assessment, it is anticipated that if the paradigm in Figure 7-2 is followed, overall life-cycle costs will be reduced while maintaining the goal of protecting human health.

Is a New Remedy Warranted?

When asymptotic conditions have occurred and concentrations remain above MCLs or other restoration goals, an evaluation is required of whether modifications to or replacement of the existing remedy are warranted. This decision must be consistent with the legal requirements in existing consent orders or settlements (as well as existing remedy selection criteria). Where a non-federal PRP has entered into a legally binding agreement, this agreement generally provides a legal release from future cleanup liability if the PRP implements the original remedy selected for the site and the remedy remains protective. Thus, at such sites the decision would proceed to the question about passive management without consideration of whether a new remedy is warranted.

Regulators and PRPs may agree to modify the mandates in existing regulatory instruments (e.g., an administrative order or settlement) to include consideration of alternative remedies. This may be desirable where a private company is trying to follow a new internal policy (e.g., a sustainability policy), or where the application of a new remedy is likely to attain drinking water standards and the incremental cost is less than the existing remedy. Similarly, if attaining drinking water standards will contribute to making the area more likely to be redeveloped, a company may agree to perform a new remedy even if the cost of implementing the new remedy exceeds the cost of the existing remedy. There is no legal release provided in the federal facilities agreements that govern cleanup at DoD sites. Governmental PRPs may decide as a matter of agency policy, or be ordered by states, to perform additional remedial work in the appropriate circumstances.

In cases where the site does not have a legal release, one question to be asked is, "is a new remedy warranted based on the original CERCLA or State remedy selection criteria?" The answer to this question could be "yes" if a new technology has been recently developed that could address the contamination more effectively (i.e., significantly reduces the timeframe to achieve restoration compared to the remedies considered during the initial feasibility study), *and* is cost effective and practical (see Chapter 3 for a more complete discussion of remedy selection criteria). The Committee expects that a new remedy might be warranted where its implementation would achieve drinking water standards in a significantly shorter period of time, the technology is clearly feasible, and the total incremental life-cycle costs of a new remedy are less than the continued costs of the existing remedy.

The answer to the question could be "no" if it is determined, based on balancing all the remedy selection criteria, that aquifer restoration using an active remedy is unlikely within a certain timeframe (say 100 years). For example, if a new remedy were predicted to reduce contaminant concentrations in groundwater to drinking water standards in 1,000 years versus 10,000 years, the answer to the question would be "no." Predictive numerical models of the type described in Chapter 6 are critical to answering such questions. Furthermore, this approach must be tailored by the relevant regulatory agencies to conform to existing statutes and regulations. If no new remedy is warranted, the transition assessment for such a site would then determine which of the two long-term management end states would be preferable.

The Southeast Industrial Area (SIA) on the Anniston Army Depot is an example of a site where the questions in Figure 7-2 have been variously tackled, but the physical complexities in the subsurface at the SIA

BOX 7-1
Anniston Army Depot, Anniston, Alabama

The Anniston Army Depot (ANAD) is an active military facility that occupies 15,200 acres in northeastern Alabama. The storage, maintenance, and industrial functions of ANAD have generated solid and liquid wastes including metals, pesticides and herbicides, chlorinated and petroleum hydrocarbons, and solvents, among others. From the 1940s through the late 1970s, wastes were disposed of on-site in trenches, lagoons, landfills, and other surface impoundments. Since 1976, investigations have focused on contamination of shallow groundwater at the facility and off-site groundwater by chlorinated solvents and metals. The contaminant source areas are the Landfill Area, Trench Area, Northeast Area, and Southeast Industrial Area (OU1), which are underlain by a residuum layer and bedrock. The total mass of TCE in groundwater is estimated to range from 3.6 to 27.1 million pounds, 87 percent of which is in the residuum lithologic unit (Malcolm Pirnie, 2006; SAIC, 2005). In addition, over 99 percent of the total TCE mass is present as DNAPL. TCE and other contaminants continue to migrate vertically and horizontally from the source areas, impacting groundwater. The primary receptor is Coldwater Springs, a prolific natural spring supplying potable water to the City of Anniston, Alabama.

The 1991 interim ROD required a groundwater extraction and treatment system known as the Groundwater Interceptor System (GWIS). An air stripping system is removing TCE to levels below drinking water standards (i.e., less than 5 μg/L) at the springs. However, in the most recent five-year review (EPA, 2010a), EPA determined that the interim remedy at OU1 is not protective because the on-site pump and treat system is not significantly reducing the extent or mobility of TCE contamination in the groundwater. Possible exposure to contaminants in Coldwater Springs water is unlikely if the treatment system is operating, and regular monitoring continues to show that TCE is effectively removed by the air stripper (EPA, 2010a).

Given the large amount of TCE DNAPL present at this site, the inability to

and disagreements over the applicability of a TI waiver have resulted in lengthy delays in agreeing on a final remedy for the site. As discussed in Box 7-1, human health exposure due to TCE releases from the site has been eliminated by the use of wellhead treatment at the point of exposure (water supply source), and yet migration of the TCE in groundwater is not under control and thus the remedy is not fully protective as noted in a recent five-year review (EPA, 2010a). A focused feasibility study was recently conducted to compare alternative remedies, including combinations of enhanced groundwater extraction, additional use of *in situ* oxidation technologies, enhanced bioremediation, and land use controls, among oth-

locate the contaminant transport pathways between OU1 and Coldwater Springs, the technical difficulties in restoring the groundwater quality beneath the site, and the complex lithology in the subsurface, the Army applied for a TI waiver to replace the RAO of achieving the TCE MCL throughout the aquifer system. However, because the current interim remedy is considered "not protective," the waiver was denied and additional efforts have been made by the Army, under direction from EPA and the State of Alabama, to remove greater quantities of TCE and other compounds from the subsurface.

The Army's 2011 Focused Feasibility Study (Tetra Tech, 2011) evaluated enhanced pump and treat and bioremediation for their estimated time to achieve RAOs in the groundwater compared to the present remedy, using solute flow and transport models (Modflow, Remchlor, and MT3D). The predicted time for the TCE to reach the MCL ranged from 1,233 years to over 10,000 years for the various alternatives. Estimated net present values costs for applying the two alternative strategies are approximately $17 million and $21 million (in 2011 dollars), respectively. Another interim ROD is expected soon.

ANAD exhibits many characteristics of complex sites—complex hydrogeology, large quantities of DNAPL, continued uncertainties in establishing the exact pathways of contaminant transport, and very long timeframes to achieve restoration. In addition, the site is affecting a critical water supply, but these impacts have been mitigated through the installation of air stripping to meet drinking water standards. The continued use of the air stripping system does pose risks to the consumers, but these risks can be mitigated through proper monitoring and operation of the treatment system. Finally, because the interim remedy was not protective, alternatives were evaluated that appear to be capable of reducing the mass discharge of TCE to the major receptors. However, the estimated time of remediation greatly exceeded 100 years, and it is thus difficult to determine the benefits of choosing any alternative compared to continued operation of the pump-and-treat system at Coldwater Springs. As long as the interim remedy is deemed "not protective," additional remedial efforts will be required to meet revised RAOs for the site.

ers. Modeling of the potential performance of the alternatives predicted timeframes for reaching groundwater cleanup goals ranging from 1,233 to 10,000 years (neither end of which is likely to be considered "reasonable"). Furthermore, attempts to negotiate a TI waiver for TCE in a portion of the site were rejected by regulators. The current status of the site following completion of the focused feasibility study appears to be preparation of an interim ROD, which will require implementation of an *in situ* bioremediation remedy to achieve revised remedial action objectives for the site. The path to a final ROD is uncertain at this time (Laurie Haines-Eklund, AEC, personal communication, 2012).

Consideration of Costs Curing the Transition Assessment

The issues discussed in Chapter 3 concerning the consideration of cost in the remedy selection phase are likely to arise during the transition assessment, particularly the question of whether to use a broader range of discount rates in calculating costs. For example, the current present value of costs could be recomputed using the opportunity cost of capital discount rate that is appropriate at that time. (As in Chapter 3, the Committee is not advocating for any particular discount rate; the choice of discount rate is based on balancing various policy considerations including statutory requirements). Costs for the various future alternatives—MNA, other passive management, continued or new active management—could then be compared to one another on a common economic basis.

With respect to the methods used to estimate the costs of the various alternatives, probabilistic cost models could be considered for estimating remedial and operation/maintenance costs and potential environmental liabilities (e.g., see Hayes et al., 1996). This approach identifies a range of statistical probabilities to address the uncertainty of critical issues faced during remediation, including not only the extent of contamination, but also construction, operation and maintenance, and legal/regulatory risks that may occur. The approach generates ranges for the median costs, fair value, and other parameters as a probability-weighted average for several possible scenarios.

This reevaluation of costs during the transition assessment is relevant for several reasons. First, existing costs represent a sunk investment and the relevant questions deal with the best path forward from the current point in time. Second, initial remedy selection is based on projected risk reductions and projected costs, both of which could be wrong. Third, economic conditions change and might make remedy implementation more or less costly. Finally, failing to do this assumes that the present value of costs is fixed at the time of the ROD. Indeed, by allowing the cost analyses to adapt through time, one can better inform *any* remedial actions that might be undertaken. Such updating of costs would occur during a transition assessment or during the five-year review if the remedy was found not to be protective.

Risk Assessment as Part of the Transition Assessment

The transition assessment provides an opportunity to reevaluate site risks above and beyond what would normally occur during a five-year review. When new information becomes available that would indicate the remedy may no longer be protective, such as new toxicity information that indicates greater chemical toxicity, the development of new health-based criteria (e.g., MCLs), or the identification of new exposure pathways (see Chapter 5 and the MEW case study), this is supposed to be reflected in five-year review reports. Under RCRA corrective action, such changes might trigger modifications to the existing Part B permit for operating industrial facilities. Such considerations can indicate the need for additional actions if a remedy is no longer protective. However, because the five-year review focuses primarily (if not solely) on lack of protectiveness, the risk analyses that occur during a typical five-year review provide an incomplete understanding of risks posed at a site presently and in the future. A transition assessment would instead include more comprehensive risk assessment, including

- Better understanding of risk reductions as predicted in the ROD compared with actual risks reduced from ongoing remedial activities. Evaluation of both individual risks and population risks could be conducted (see Chapter 3).
- Risks associated with various future alternatives—passive management, MNA, continued or new active management. These risks could be compared to one another, in terms of the magnitude of the relative risk reduction and the time scale to achieve the projected risk reduction.
- More explicit consideration of uncertainty that would reveal not only the confidence in existing risk estimates, but also the potential impact of other factors (e.g., new dose-response studies, future groundwater and land uses) on risk estimates in the future.

Is Passive Long-Term Management Acceptable?

The transition assessment considers whether or not a passive long-term management option (such as MNA, NA, a permeable reactive barrier, or physical containment) is appropriate. The current regulatory approach to the use of MNA is to require that the timeframe for restoration using MNA is "reasonable," usually considered to be less than 100 years (EPA, 1999, 2007, 2011c; ITRC, 2010; USGS, 2007). However, for most non-petroleum-based contaminants, MNA will often require longer timeframes for restoration, and there have been instances where MNA has been allowed even though restoration will likely not occur in 100 years. Where site-specific risk analysis indicates that residual contamination will not pose unacceptable future risks, the use of a longer timeframe criterion for transition to MNA is critical to avoid inefficient use of limited resources. This may require the definition of a zone within the aquifer where alternative RAOs would be established (equivalent to a "TI Zone" in the current EPA Road Map).

An example of the transition to passive long-term management is the MEW case study, where regional pump-and-treat systems have been operating for about 15 years. The site will likely be transitioning to MNA once a certain target concentration has been met. Another example is the Solvents Recovery Service of New England Superfund facility in Southington, Connecticut, that had solvents in bedrock (EPA Region 1, 2010). Modeling indicated that the solvent plumes would not reach MCLs for 500 years under baseline conditions. Even if 95 to 99 percent of the overburden mass could be removed, it would take approximately 250 years to clean up the bedrock. Thus, the chosen approach is to use pump-and-treat, MNA, and institutional controls.

The end state of long-term passive management (see Figure 7-2) requires that an acceptable and cost-effective monitoring program be established to ensure that the passive management alternative is protective of human health and the environment. The necessary engineering and legal controls would be established as part of the management plan. Finally, transparent communication with community representatives would be an essential component of the plan for any site in this "end state." It should be stressed that sites in this end state may eventually achieve restoration throughout the aquifer, but it is expected that the timeframe for this to occur would exceed many decades (see the dashed lines in Figure 7-2).

If MNA or passive containment is not appropriate based on a balancing of all of the remedy selection factors, a long-term active management program would be developed. This would include a comprehensive outreach program to the affected community, a thorough plan for monitoring contaminant concentrations in groundwater, five-year reviews for Superfund facilities, and maintenance of institutional and engineering controls.

The Role of the Five-Year Review in More Effective Long-Term Management

Many of the above sections have mentioned doing certain tasks as part of the five-year review process. Indeed, although the five-year review is shown as a single box in Figure 7-2, it would occur at regular intervals at all sites with residual contamination, under either active or passive long-term management. Thus, the questions posed in Figure 7-2, including those in the transition assessment, could theoretically all occur under the umbrella of an improved five-review process. This suggests the need to comment on the current status of five-year reviews.

In 2011, EPA's Office of the Inspector General (OIG) found that 84 percent of the five-year reviews performed since 2006 were overdue as of April 28, 2009 (EPA, 2011d). Also, the EPA OIG noted that EPA regional staff did not consistently follow CERCLA policies for updating the Comprehensive Environmental Response, Compensation, and Liability Information System (CERCLIS) database on the status and actions taken to address prior review issues. In addition, recommendations from prior reviews were not always followed through. The report concluded that "EPA does not have effective management controls to monitor the completion of five-year review recommendations at federal government Superfund sites."

In response to the Inspector General's report, EPA recently issued priorities for the five-year review program that would require more stringent management controls to ensure that recommendations in the five-year review reports are tracked, monitored, and implemented by EPA and the federal agencies (EPA, 2011e). The Inspector General's report specifies the actions that remedial project managers must undertake to ensure that the reviews are completed on time and are performed in an independent manner, particularly with respect to the protectiveness determination. Most importantly for federal facilities, given the historic disagreements between EPA and DoD, it confirms that five-year reviews are enforceable under the Federal Facility Agreements (EPA, 2011e).

This Committee did not perform a review of the adequacy of five-year review reports, although it read many such reports as part of its evaluation of groundwater sites removed from the National Priority List (see Appendix C). The Committee also considered other evaluations of five-year reviews (EPA, 2011e; GAO, 2009; Pham, 2010; Probst, 1999; Schiller, 2009). In general, these evaluations indicate a need to improve the quality and comprehensiveness of these reviews, particularly because of the importance of the "protectiveness" decision. Given the opportunity to improve decision making as outlined above, both revision of the five-year review guidance and issuance of separate guidance for non-Superfund sites to incorporate and institutionalize the transition assessment is the recommended path forward for implementing the process shown in Figure 7-2. For each new component, the five-year review reports should have enough documentation for stakeholders to replicate the calculations and understand the support for the conclusions. This suggests that five-year review reports will need to become much more comprehensive than they have been in the past.

Because EPA guidance argues that cleanups at similar types of sites should follow the same process and generally should result in the same or very similar remedial actions (see Chapter 3), the transition assessment should apply to RCRA sites undergoing corrective action for groundwater cleanup and other cleanup sites. However, the RCRA corrective action program is implemented through corrective action permits while a facility is still in operation. As a practical matter, while a plant is operating the primary focus of the corrective action plan is to protect human health and prevent expansion of the plume. As a result, typically RCRA corrective actions early in the process do not involve the selection of a "final" remedy.

Once the final RCRA corrective action remedy is selected and implemented, then the transition assessment applies and should be used. It should also be noted that Figure 7-2 applies primarily to complex sites. A transition assessment is not necessary at sites where the remedy is likely to attain drinking water standards within a reasonable period of time.

THE ROLE OF COMMUNITY INVOLVEMENT IN TRANSITION ASSESSMENT AND LONG-TERM MANAGEMENT

At complex groundwater sites that are the focus of Figure 7-2 where achievement of long-term remedial objectives is difficult, complex community engagement strategies are essential because the potential risk from the contaminated groundwater is likely to persist for decades and in some cases many human generations. Informed community input can improve cleanup decisions at these complex sites, and communities that have had the opportunity to participate in the decision-making process are more likely to support, rather than challenge, key decisions. Furthermore, tools such as asymptote analysis that give communities a clear sense of the options for long-term site management can enable community members to both understand and participate in the making of highly technical decisions.

Three decades of experience at hazardous waste cleanup sites have demonstrated that the affected public responds positively when it is viewed as a partner with a stake in the outcome of cleanup activities. This partnership goes beyond simply communicating to the public what decision-makers want to do, but it must not go as far as always doing what site neighbors propose. Rather, it involves listening to stakeholder viewpoints in such a way that they feel empowered to provide constructive input into the decision-making process. The trust inherent in this interaction is the primary reason the public will support cleanup decisions (EPA, 2001b; Siegel, 2007). At complex groundwater sites where future predictions are inherently uncertain, the public tends to trust decision makers when they believe that a serious effort has been made to achieve cleanup goals and that systems are in place to ensure that any residual contamination can be effectively and safely managed.

Most regulatory programs provide for some level of community involvement, but CERCLA is the most explicit, with community acceptance being the ninth criterion for remedy selection under the National Contingency Plan. It does not give members of affected communities—even if they all agree—direct authority over the selection of remedies, but it legally recognizes that they have a significant role to play. At Superfund facilities and at projects addressed under other federal and state authorities, community members are routinely invited to comment on proposed plans, in writing and sometimes at public meetings, or to indicate their preferences among remedial alternatives. Federal agencies—notably EPA, DoD, and the Department of Energy—have gone further and promoted the formation of site-specific advisory boards. The Defense Department currently sponsors 265 Restoration Advisory Boards at active, closed, and former installations, while DOE hosts local boards at eight major cleanup sites (DoD, 2011; DOE, 2011). Similarly, 68 EPA Community Advisory Groups facilitate information exchange among the local community, EPA, state regulators, and other federal agencies regarding the remediation of Superfund facilities (EPA, 2011f). Other similar advisory groups have been formed under state and local auspices. These community-based organizations are advisory and do not hold statutory decision-making authority. Yet where they function properly, they provide site neighbors with the opportunity to exert real influence over cleanup decisions.

Under the classic model of public input, agencies develop a small number of remedial options, selected a preferred alternative, and ask the public what it prefers. In most cases, key decisions, such as the establishment of remedial objectives, have already been made. Judgments on future land uses, which often influence remedial options, have also been developed. It is difficult for the public to do anything more than push for the expenditure of more money. On the other hand, with advisory groups engaged community members can help make the early decisions—about remedial objectives or anticipated land use—that influence future decisions (EPA, 2011g). EPA has long-standing guidance (EPA, 2011g) to encourage community involvement at the site-characterization stage of the process even though the statute does not include explicit requirements for community involvement activities at this stage.

For community involvement to be successful, a great deal of effort needs to go into building trust among the parties, and community members need to be educated on the technical and process aspects of cleanup. Ideally, they have access to their own, independent technical consultants. The most effective of these consultants not only translate official documents and the underlying science into a form more easily understood by the public, but they are in a position to focus public concerns on decisions where they can make a difference. A number of programs provide independent technical assistance to community groups, the largest and oldest of which is EPA's Technical Assistance Grant (TAG) program. Since 1988, 323 TAGs have been awarded (205 providing $50,000 or less and 15 providing more than $250,000) (Catalogue of Federal Domestic Assistance, 2011).

Once initial remedial decisions are made, advisory board members are likely to show less interest in meetings, and agencies may be unwilling to support regular meetings with no clear agenda. There are few examples of effective public participation in the reopening of remedial decisions. All this may suggest a reduced role for the public once a site has reached the stages found in Figure 7-2. The Committee believes that just the opposite is needed, because at these complex sites if remedies fail to achieve objectives or become unprotective many important decisions will have to be made. Reconvening dormant advisory groups just prior to the initial five-year review would allow the public to gather and digest new information about the remedy, understand site complexities that are preventing the attainment of remedial objectives, and frame sophisticated responses to the questions posed in Figure 7-2.

This reengagement of the public is supported by CERCLA, which states that EPA must provide an opportunity for a public meeting prior to the "adoption of *any* plan for remedial action" (underlining added) (42 U.S.C. § 9617(a)). EPA policy supports "going beyond the letter of the law" and recommends the implementation of additional community involvement activities not required by the NCP (EPA, 2005). In light of this policy and the benefits, it is reasonable to interpret the statute as allowing, if not requiring, a public meeting whenever there is a major change to the remedial plan, e.g., a decision that additional work is needed to ensure the protectiveness of the remedy as a result of a five-year review, or a decision to transition to more passive long-term management.

One example of where a community advisory body has been engaged over the long term is at the Moffett-MEW Regional Groundwater Plume in Mountain View, California (see Appendix B). At this site, EPA is leading a focused feasibility study to consider ways to accelerate groundwater remediation in response to a five-year review that found a long-standing pump-and-treat remedy to be unprotective. The local Community Advisory Board developed a list of remedial priorities and offered its own screening of remedial technologies, and EPA has incorporated the community's ideas into its study. One of the reasons that community involvement was sustained long after the 1998 installation of the regional groundwater remedy is that other issues became pressing, such as wetlands cleanup, the emergence of vapor intrusion from the regional plume into an adjacent residential area, and the preservation of historic Hangar One at Moffett Field.

Another very important aspect of community involvement in the process found in Figure 7-2 is that communities should have a role to play in establishing and overseeing institutional controls. Because institutional controls may impede a community's future groundwater, building, and land uses, EPA Regions are already directed to "ensure communities have meaningful opportunity to review proposals for site remedies and provide adequate information to allow informed public comment regarding the choices between cleanup alternatives that either achieves levels that allow for unrestricted use, or leave levels that lead to restricted uses and rely on institutional controls" (EPA, 2010b, p. 9). Once engineering and institutional controls are in place, informed neighbors can be in a position to provide valuable information on possible failures that might otherwise go unnoticed by regulators who seldom visit the site (EPA, 2010b). In fact, EPA already supports a larger role for local residents, community associations, and interested organizations (representing those who live or work near the site and thus have a vested interest in ensuring compliance with institutional controls) (EPA, 2010b).

EPA's five-year review guidance (EPA, 2001a) recognizes the need for some level of community notification at the beginning and end of the five-year review process, with optional additional engagement at high-profile sites. The guidance suggests contacting local health and other government officials, community organizations, members of advisory groups, and technical assistance grant recipients at these stages. Historically, there have been few instances where the public has played an active role in the five-year review process, particularly at sites or facilities where remedies are all in place. The challenge is to devise mechanisms for long-term public involvement at sites where people think they are "done." First, it is essential that PRPs or regulators retain and regularly update lists of interested members of the public, and local media should be encouraged to publish annual site updates. Where advisory boards have been meeting, it may be useful to convene annual reunions to review site progress. At sites where most engaged community members frequently use the Internet or social networks, discussion forums should be encouraged. At some sites, it may prove helpful to move the locus of public involvement to permanent local institutions such as public health departments. All of these vehicles could be "re-activated" at the beginning of each five-year review cycle.

CONCLUSIONS AND RECOMMENDATIONS

At many complex sites, contaminant concentrations in the plume remain stalled at levels above cleanup goals despite continued operation of remedial systems. At such sites, which may number in the tens of thousands, there is no clear path forward to a final end state, such that money continues to be spent with no concomitant reduction in the risks posed by the sites. **If the effectiveness of site remediation reaches a point of diminishing returns prior to reaching cleanup goals and optimization has been exhausted, the transition to MNA or some other active or passive management should be considered using a formal evaluation.** This transition assessment, which is akin to a focused feasibility study, would determine whether a new remedy is warranted at the site or whether long-term management is appropriate (see Figure 7-2). This analysis, particularly the graphical demonstration that contaminant concentrations at a site have reached an asymptote, is likely to enhance public involvement in long-term decision making and thus strengthen public confidence in the decisions made.

As part of the transition assessment, the costs of remedial options should be updated, ideally using probabilistic methods, to reflect the current present value of costs in order to more effectively compare the costs of various future alternatives. To the extent possible, the marginal cost of remediation through time should be quantified. Also as part of the transition assessment, formal post-remedy risk assessment should be conducted to compare alternatives, including long-term active or passive management. This risk analysis should explicitly consider uncertainty and take into account both individual and population risks, where appropriate, such that the relative risk reduction of the alternatives can be determined.

Five-year reviews are an extremely valuable source of field data for evaluating the performance of remedial strategies that have been implemented at CERCLA facilities and could be improved. To increase transparency and allow EPA, the public, and other researchers to assess lessons learned, more should be done, on a national basis, to analyze the results of five-year reviews in order to evaluate the current performance of implemented technologies. **EPA's technical guidance for five-year reviews should be updated to provide a uniform protocol for analyzing the data collected during the reviews, reporting their results, and improving their quality.**

Informed public involvement characterized by two-way communications and that builds trust is a key element of good remedial decision making, particularly at complex sites. Federal agencies generally go beyond the minimum level of public involvement by forming or recognizing advisory groups and providing technical assistance, such that members of affected communities have the opportunity to influence remedial decision making. **Nonetheless, public involvement tends to diminish once remedies at a site or facility are in place. No agency has a clear policy for sustaining public involvement during long-term management.**

Regulators and federal responsible parties should work with members of existing advisory groups and technical assistance recipients to devise models for ongoing public oversight once remedies are in place. Such mechanisms may include annual meetings, Internet communications, or the shifting of the locus of public involvement to permanent local institutions such as public health departments. Such involvement is critical to the success of the processes found in Figure 7-2.

The cost of new remedial actions may decrease at complex sites if more of them undergo a transition to passive long-term management. However, long-term management at complex sites will still demand substantial long-term funding obligations. Failure to fund adequately the long-term management of complex sites may result in unacceptable risks to the public due to unintended exposure to site contaminants.

REFERENCES

Catalogue of Federal Domestic Assistance. 2011. Superfund Technical Assistance Grants (TAG) for Community Groups at National Priority List (NPL) Sites. https://www.cfda.gov/index?s=program&mode=form&tab=step1&id=a83b10849f1dc74453b499c5d9e0370f.

Clean Sites. 1990. Improving Remedy Selection: An Explicit and Interactive Process for the Superfund Program.

Conger, J. 2011. Memo: New goals for DERP. July 18, 2011.

DoD (Department of Defense). 2011. Defense Environmental Programs Annual Report to Congress, FY 2010, p. 54. http://www.denix.osd.mil/arc/upload/FY10DEPARC_06_Restoration_Final_Report-2.pdf.

DOE (Department of Energy). 1998. Accelerating Cleanup: Paths to Closure. DOE/EM-0362.

DOE. 2011. About the EM Site-Specific Advisory Board. http://www.em.doe.gov/stakepages/ssababout.aspx.

EPA (U.S. Environmental Protection Agency). 1999. Use of Monitored Natural Attenuation at Superfund, RCRA Corrective Action, and Underground Storage Tank Sites. EPA/540/R-99/009. Washington, DC: EPA Office of Solid Waste and Emergency Response.

EPA. 2001a. Comprehensive Five-Year Review Guidance. Washington, DC: EPA Office of Solid Waste and Emergency Response.

EPA. 2001b. Stakeholder Involvement & Public Participation at the U.S. EPA: Lessons Learned, Barriers, & Innovative Approaches. EPA-100-R-00-040. http://www.epa.gov/publicinvolvement/pdf/sipp.pdf.

EPA. 2004. Cleaning up the nation's waste sites: Markets and technology trends. Washington, DC: EPA Office of Solid Waste and Emergency Response.

EPA. 2005. Superfund Community Involvement Handbook. EPA 540-K-05-003. Washington, DC: EPA Office of Solid Waste and Emergency Response. http://www.epa.gov/superfund/community/cag/pdfs/ci_handbook.pdf.

EPA. 2007. Monitored Natural Attenuation of Inorganic Contaminants in Ground Water Volume 1—Technical Basis for Assessment. EPA/600/R-04/027. Washington, DC: EPA Office of Solid Waste and Emergency Response.

EPA. 2010a. Five Year Review for Anniston Army Depot, SIA OU1.

EPA. 2010b. Institutional Controls: A Guide to Planning, Implementing, Maintaining, and Enforcing Institutional Controls at Contaminated Sites. EPA-540-R-09-001. Washington, DC: EPA Office of Solid Waste and Emergency Response. http://www.epa.gov/superfund/policy/ic/pdfs/PIME-IC-Guidance-Interim.pdf.

EPA. 2011a. Presumptive Remedy for CERCLA Municipal Landfill Sites. http://www.epa.gov/superfund/policy/remedy/presump/clms.htm.

EPA. 2011b. Groundwater Road Map Recommended Process for Restoring Contaminated Groundwater at Superfund Sites. OSWER 9283.1-34. Washington, DC: EPA OSWER.

EPA. 2011c. An Approach for Evaluating the Progress of Natural Attenuation in Groundwater. EPA 600/R-11/204. Ada, OK: EPA Office of Research and Development.

EPA. 2011d. Stronger Management Controls Will Improve EPA Five-Year Reviews of Superfund Sites Report No. 12-P-0251. Washington, DC: EPA Office of Inspector General.

EPA. 2011e. Memorandum from Mathy Stanislaus, Assistant Administrator for Solid Waste and Emergency Response to Carolyn Copper, Director for Program Evaluation, EPA Office of Inspector General, Re: Response to the Office of Inspector General Final Evaluation Report, "Stronger Management Controls Will Improve EPA Five Year Reviews of Superfund Sites (Report No. 12-P-0251) (April 18, 2012).

EPA. 2011f. Technical Assistance Services for Communities. EPA 540-K-08-001. http://www.epa.gov/superfund/community/tasc/pdfs/tasc_brochure_english.pdf.

EPA. 2011g. Community Involvement Initiative, Action 1: Draft Work Plans, Superfund Remedial Process Improvements at 4. http://www.epa.gov/oswer/engagementinitiative/superfund.pdf.

EPA Region 1. 2010. First Five-Year Review Report for the Solvents Recovery Service of New England, Inc. Superfund Site. Boston, MA: EPA Region 1.

ESTCP (Environmental Security Technology Certification Program). 2011. Alternative Endpoints and Approaches Selected for the Remediation of Contaminated Groundwater.

GAO (Government Accountability Office). 2009. Hazardous Waste Sites: Improved Effectiveness of Controls at Sites Could Better Protect the Public. http://www.gao.gov/products/GAO-05-163.

Hayes, T. D., D. G. Linz, D. V. Nakles, and A. P. Leuschner, eds. 1996. Management of Manufactured Gas Plant Sites, Vol. 2. Amherst, MA: Amherst Scientific Publishers.

ITRC (Interstate Technology and Regulatory Council). 2010. Technical/Regulatory Guidance: A Decision Framework for Applying Monitored Natural Attenuation Processes to Metals and Radionuclides in Groundwater. December 2010.

Malcolm Pirnie. 2006. Technical Impracticability Evaluation (Draft), Southeast Industrial Area, Anniston, Army Depot, Anniston, Alabama.

Pham, T. 2010. Assessment of Failures in Long-Term Management of Contaminated Sites. Worchester Polytechnical Institute. March 10, 2010.

Probst, K. 1999. Institutional Controls: The Next Frontier. Washington, DC: Resources for the Future. http://www.wpi.edu/Pubs/E-project/Available/E-project-032410-230908/unrestricted/Final_-_Failures_at_Remediated_Sites.pdf.

SAIC (Science Applications International Corporation). 2005. Comprehensive Groundwater Remedial Investigation, Southeast Industrial Area, Phase III, at Anniston Army Depot, Anniston, Alabama (Draft).

Siegel, L. 2007. Communities and Chemical Warfare Material Disposal. Mountain View, CA: Center for Public Environmental Oversight. http://www.cpeo.org/pubs/C&CWMD.pdf.

Tetra Tech. 2011. Focused Feasibility Study for Southeast Industrial Area OU1 of ANAD, Draft, May 2011.

USGS (U.S. Geological Survey). 2007. A Framework for Assessing the Sustainability of Monitored Natural Attenuation. Circular 1303. 2007.

Yonkers, T. 2011. Policy for Refocusing the Air Force Environmental Restoration Program. February 24, 2011.

Acronyms

AAP	Army ammunition plant
ACL	Alternate concentration limit
ANAD	Anniston Army depot
AEC	Army Environmental Command
AFB	Air Force Base
ARAR	Applicable or relevant and appropriate requirement
ASM	Adaptive site management
ASR	Aquifer storage and recovery
AS	Air sparging
BRAC	Base realignment and closure
BTEX	benzene, toluene, ethylbenzene, and xylene
CCL	Contaminant Candidate List
CERCLA	Comprehensive Environmental Response, Compensation, and Liability Act
CERCLIS	Comprehensive Environmental Response, Compensation, and Liability Information System
CSIA	Compound-specific isotope analysis
CSM	Conceptual site model
CVOC	Chlorinated volatile organic compound
DCA	Dichloroethane
DCE	Dichloroethene
DERP	Defense Environmental Restoration Program

DNAPL	Dense nonaqueous phase liquid
DoD	U.S. Department of Defense
DOE	U.S. Department of Energy
DOI	U.S. Department of the Interior
ERH	Electrical resistance heating
EPA	U.S. Environmental Protection Agency
ESD	Explanation of Significant Differences
ESTCP	Environmental Security Technology Certification Program
FUDS	Formerly used defense sites
FUSRAP	Formerly Used Sites Remedial Action Program
GAC	Granular activated carbon
GAO	Government Accountability Office
GW	Groundwater
HRS	Hazard Ranking System
HVOC	Halogenated volatile organic compound
IC	Institutional control
IRP	Installation Restoration Program
ISB	*In situ* bioremediation
ISCO	*In situ* chemical oxidation
ISTD	*In situ* thermal desorption
ITRC	Interstate Technology and Regulatory Council
LNAPL	Light nonaqueous phase liquid
LSRP	Licensed Site Remediation Professional
LTM	Long-term monitoring
LTMO	Long-term monitoring optimization
LUST	Leaking underground storage tank
MCL	Maximum contaminant level
MCLG	Maximum contaminant level goal
MEP	Maximum extent practicable
MMRP	Military Munitions Response Program
MNA	Monitored natural attenuation
MTBE	Methyl tert-butyl ether
NA	Natural attenuation
NAPL	Nonaqueous phase liquid
NAS	Naval Air Station

NASA	National Aeronautics and Space Administration
NCP	National Contingency Plan
NFESC	Naval Facilities Engineering Service Center
NPL	National Priorities List
NRC	National Research Council
NRD	Natural resource damage
NSF	National Science Foundation
OMB	Office of Management and Budget
O&M	Operation and maintenance
OSWER	EPA Office of Solid Waste and Emergency Response
OU	Operational unit
OUST	Office of Underground Storage Tanks
PAH	Polycyclic aromatic hydrocarbon
PCB	Polychlorinated biphenyl
PCE	Perchloroethylene or tetrachloroethene
PCR	Polymerase chain reaction
POC	Point of compliance
PRB	Permeable reactive barrier
PRP	Potentially responsible party
P&T	Pump and treat
RAB	Restoration Advisory Board
RAGS	Risk Assessment Guidance for Superfund
RAO	Remedial action objective
RC	Response complete
RCRA	Resource Conservation and Recovery Act
R&D	Research and development
RD/RA	Remedial Design/Remedial Action
RfC	Reference concentration
RFH	Radio frequency heating
RIP	Remedy in place
RI/FS	Remedial Investigation/Feasibility Study
ROD	Record of Decision
RPM	Remedial project manager
RSE	Remedial system evaluation
SARA	Superfund Amendments and Reauthorization Act
SCM	Site conceptual model
SCRD	State Coalition for Remediation of Drycleaners
SDWA	Safe Drinking Water Act
SEAR	Surfactant-enhanced aquifer remediation

SEE	Steam enhanced extraction
SERDP	Strategic Environmental Research and Development Program
STAR	Self-sustaining treatment for active remediation
SVE	Soil vapor extraction
SVOC	Semivolatile organic compound
SWMU	Solid waste management unit
TAG	Technical assistance grant
TCA	Trichloroethane
TCE	Trichloroethene
TEA	Terminal electron acceptor
TI	Technical Impracticability
TSCA	Toxic Substances Control Act
USACE	U.S. Army Corps of Engineers
USDA	U.S. Department of Agriculture
USGS	U.S. Geological Survey
UST	Underground storage tank
UU/UE	Unlimited use and unrestricted exposure
VC	Vinyl chloride
VCP	Voluntary cleanup program
VI	Vapor intrusion
VOC	Volatile organic compound
ZVI	Zero-valent iron

Appendix A

Biographical Sketches of Committee Members and Staff

COMMITTEE MEMBERS

MICHAEL KAVANAUGH, Chair, is a principal at Geosyntec. Before Geosyntec, Dr. Kavanaugh served as Vice President and a Global Science and Technology Leader at Malcolm Pirnie, Inc. Dr. Kavanaugh's primary areas of expertise include hazardous waste management with a particular focus on groundwater remediation, risk and decision analysis, water quality, water treatment, potable and non-potable water reuse, and fate and transport of chemical contaminants in the environment. Dr. Kavanaugh has served on numerous NRC boards and committees, chairing both the Water Science and Technology Board and the Board on Radioactive Waste Management. He is also a Consulting Professor in the Department of Civil and Environmental Engineering at Stanford University. A member of the National Academy of Engineering, Dr. Kavanaugh received a B.S. in chemical engineering from Stanford University, and an M.S. in chemical engineering and a Ph.D. in civil/sanitary engineering from University of California, Berkeley.

WILLIAM A. ARNOLD is the Joseph T. and Rose S. Ling Professor at the University of Minnesota's Department of Civil Engineering. He specializes in the fate and transport of anthropogenic organic chemicals (solvents, pesticides, and pharmaceuticals) in natural and engineered aquatic systems. In particular, he studies diffusion, mass transfer, and partitioning processes and how knowledge of these processes can be used to develop containment/remediation schemes. Dr Arnold is familiar with subsurface remediation techniques such as zero-valent metals, phytoremediation, surfactants, reactive membranes, and sediment capping. He received a B.S. in chemical engineering from MIT, an M.S. in chemical engineering from Yale, and a Ph.D. in environmental engineering from the Johns Hopkins University.

BARBARA D. BECK is a principal at Gradient. She is an expert in toxicology and in human health risk assessment for environmental chemicals, especially metals and air pollutants. Dr. Beck directs Gradient's toxicology and risk assessment practice and has performed numerous site-specific and chemical-specific risk assessments, as well having developed exposure and risk assessment methodologies. She is a Diplomate of the American Board of Toxicology and a Fellow and past President of the Academy of Toxicological Sciences. Dr. Beck is also a Visiting Scientist in the Molecular and Integrative Physiological Sciences Program in the Department of Environmental Health at the Harvard School of Public Health. She has previously held the position of Chief of the Air Toxics Staff for U.S. EPA Region I. Dr. Beck received her A.B. in biology from Bryn Mawr College and Ph.D. in molecular biology and microbiology from Tufts University.

YU-PING CHIN is a Professor in the School of Earth Sciences at The Ohio State University. Prior to joining The Ohio State University, Dr. Chin conducted research at the Ralph M. Parsons Laboratory at the Massachusetts Institute of Technology on the properties of organic humic materials in marine and lacustrine porewaters and on the fluxes of particle reactive contaminants across the sediment/water interface. He is a specialist in natural attenuation, redox processes, and surfactant and cosolvent behavior in the subsurface. He is a current member of the Water Science and Technology Board. Dr. Chin received his A.B in geology from Columbia University and his M.S. and Ph.D. in aquatic chemistry from the University of Michigan.

ZAID CHOWDHURY is a vice president of Malcolm Pirnie, Inc., in Phoenix, Arizona. He has an extensive background in water treatment processes and is involved in all advanced drinking water process evaluations performed by the company. Dr. Chowdhury has managed many high profile projects including the development of the water treatment plant simulation model which he also incorporated into the Surface Water Analytical Tool (SWAT) for the regulatory development of the D/DBP Rule by the EPA. During his tenure at Malcolm Pirnie, he has managed numerous water quality evaluations involving bench- and pilot-scale studies for alternative treatment processes and the development of information needs for capital improvement programs for water utilities. He received his B.S. in civil engineering from the Bangladesh University of Engineering & Technology and his M.S. and Ph.D. in civil engineering from the University of Arizona.

DAVID E. ELLIS is a Principal Consultant in DuPont's Corporate Remediation Group in Wilmington, Delaware. He has 30 years experience in the science of subsurface cleanup. Dr. Ellis is Chair of the Sustainable Remediation Forum (SuRF) and Chair of the UK's SABRE consortium on bioremediation of chlorinated solvent source areas. He has served on two previous NRC committees—those investigating natural attenuation and source removal. Dr. Ellis is very active in the ITRC, serving both on the Board of Advisors and a lead instructor in several ITRC classes. He received his B.S. in geology from Allegheny College and his M.Phil. and Ph.D. in geology and geophysics from Yale University.

TISSA H. ILLANGASEKARE is a Professor and AMAX Chair of Civil and Environmental Engineering at the Colorado School of Mines and the Director of the Center for the Experimental Study of Subsurface Environmental Processes. His research interests include subsurface hydrology, numerical and physical modeling, subsurface chemical transport and multiphase flow land-atmospheric interaction and remediation engineering. He is a licensed Professional Engineer, Board Certified Environmental Engineer, and Diplomate of the American Academy of Water Resources Engineers. He has served on several previous NRC committees including the Committee on Subsurface Contamination at Department of Energy Complex Sites: Research Needs and Opportunities. He was the European Geosciences Union's 2012 Henry Darcy medalist. Dr. Illangasekare received a Ph.D. in civil engineering from Colorado State University, a M.E. from the Asian Institute of Technology, and a B.S. from University of Ceylon, Sri Lanka.

PAUL C. JOHNSON is a Professor in the Department of Civil and Environmental Engineering and is also the Executive Dean of the Ira A. Fulton School of Engineering at Arizona State University. Prior to joining ASU he was a Senior Research Engineer at the Shell Oil Westhollow Technology Center. His teaching, research, and professional activities focus on the application of contaminant fate and transport fundamentals to subsurface remediation and risk assessment problems. Dr. Johnson is recognized for contributions to the fields of soil and groundwater remediation and risk assessment; more specifically, the design, monitoring, and optimization of soil and groundwater remediation systems and the monitoring and modeling of exposure pathways, including vapor intrusion. Dr. Johnson is also the editor-in-chief for the journal *Ground Water Monitoring & Remediation*. He received a B.S. from the University of California, Davis, and his M.A. and Ph.D. in chemical engineering from Princeton University.

MOHSEN MEHRAN is a principal hydrologist and chief executive officer of Rubicon Engineering Corporation. In the last 40 years, he has been the principal investigator and manager for Remedial Investigation/Feasibility Studies; RCRA Facility Investigations, risk assessment; and design, installation, and operation of remediation systems. Dr. Mehran has developed and applied numerous computer models to solve groundwater flow problems and investigate the migration of various chemical compounds in fractured/porous media. He has applied this technical specialty to site characterization, evaluation of remedial alternatives, development of cleanup criteria, and allocation of cost among potentially responsible parties for the aerospace, petroleum, electronics, chemical, wood preserving, and communications industries. Dr. Mehran holds a B.S. from Tehran University and M.S. and Ph.D. degrees from the University of California, Davis.

JAMES W. MERCER spent eight years with the U.S. Geological Survey in the Northeastern Research Group working on water resources and contaminant and heat transport issues, including multiphase flow. In 1979 he co-founded GeoTrans (now Tetra Tech GEO) and in 1980 began working on dense nonaqueous phase liquid issues at Love Canal (for which he received the Wesley W. Horner Award of the American Society of Civil Engineers). He continues to work on DNAPL issues and co-authored a book on *DNAPL Site Evaluation* in 1993. In 1994, Dr. Mercer received the American Institute of Hydrology's Theis Award for contributions to groundwater hydrology. He has served on the NRC's Water Science and Technology Board and several committees and was a member of the U.S. EPA Science Advisory Board. He is currently on the Scientific Advisory Board of the Strategic Environmental Research and Development Program. Dr. Mercer received his Ph.D. in geology from the University of Illinois.

KURT D. PENNELL is a Professor and Chair at the Department of Civil and Environmental Engineering at Tufts University. His research interests include the fate and transport of engineered nanomaterials and nonaqueous phase liquids in the subsurface; development and testing of *in situ* remediation technologies including thermal treatment, surfactant flushing and bioremediation; and the link between chronic exposure to persistent organic pollutants, oxidative stress and neurodegenerative disease. He is a licensed professional engineer, board-certified environmental engineer, and currently serves as an Associate Editor of the *Journal of Contaminant Hydrology*. Dr. Pennell received a B.S. from the University of Maine, an M.S. from North Carolina State University, and a Ph.D. from the University of Florida.

ALAN J. RABIDEAU is a Professor of Civil, Structural, and Environmental Engineering at the State University of New York at Buffalo, where he also administers the University's interdisciplinary doctoral program in ecosystem restoration. He has served on the editorial boards for *Advances in Water*

Resources and the *Journal of Environmental Engineering*. Dr. Rabideau's primary research interests include mathematical modeling of flow and reactive contaminant transport in groundwater, subsurface remediation, and decision and risk analysis for environmental systems. Past awards include the American Society of Civil Engineers' Rudolf Hering medal for the best paper in environmental engineering and the Brigham Award for outstanding service from the New York Water Environment Association. In 2011, he participated in the project team that was awarded the National Ground Water Association Remediation Project of the Year. Dr. Rabideau received his B.S. in civil engineering from the University of Notre Dame, an M.E. in civil engineering from the State University of New York at Buffalo, and a Ph.D. in environmental sciences and engineering from the University of North Carolina at Chapel Hill.

ALLEN M. SHAPIRO is a Senior Research Hydrologist with the National Research Program of the U.S. Geological Survey in Reston, VA. His research focuses on the development of field techniques and methods of integrating and interpreting geologic, geophysical, hydraulic, and geochemical information in order to characterize fluid and chemical transport in fractured rock over dimensions from meters to kilometers. His research has been applied to issues of water supply, geotechnical engineering, waste isolation, and groundwater contamination and restoration, including the fate of DNAPLs and transport of pathogens in fractured rock. Dr. Shapiro is the Principal Investigator investigating the fate of DNAPLs in fractured sedimentary rock at the former Naval Air Warfare Center in West Trenton, NJ. Dr. Shapiro serves as an Associate Editor of Ground Water. In 2004, the National Ground Water Association selected Dr. Shapiro as the 2004 Distinguished Darcy Lecturer. He received a B.S. in civil engineering from Lafayette College, and his M.S., M.A., and Ph.D. degrees in civil and geological engineering from Princeton University.

LEONARD M. SIEGEL is director of the Center for Public Environmental Oversight (CPEO), a project of the Pacific Studies Center that facilitates public participation in the oversight of military environmental programs, federal facilities cleanup, and Brownfields revitalization. He is one of the environmental movement's leading experts on military facility contamination, community oversight of cleanup, and the vapor intrusion pathway. For his organization he runs two Internet newsgroups: the Military Environmental Forum and the Brownfields Internet Forum. He is a member of the Interstate Technology and Regulatory Council's Munitions Response Work Team, the Department of Toxic Substances Control (California) External Advisory Group, and the Moffett Field (formerly Moffett Naval Air Station) Restoration Advisory Board. He has served on several NRC committees, most recently as a member of the Committee to Review Possible Toxic Effects from Past Environmental Contamination at Fort Detrick.

WILLIAM J. WALSH is an attorney in the Washington, D.C., office of Pepper Hamilton LLP. Prior to joining Pepper, he was section chief in the EPA Office of Enforcement. His legal experience includes environmental regulatory advice and advocacy and defense of environmental injury litigation involving a broad spectrum of issues pursuant to a variety of environmental statutes, including the Resources Conservation and Recovery Act (RCRA) and the Toxic Substances Control Act (TSCA). He represents trade associations, including the Rubber Manufacturers Association and the American Dental Association, in rule-making and other public policy advocacy. He has negotiated protective, yet cost-effective remedies in pollution cases involving water, air, and hazardous waste; and advised technology developers and users on taking advantage of the incentives for, and eliminating the regulatory barriers to, the use of innovative environmental technologies. Mr. Walsh has served on several NRC committees, including the Committee to Review Operations for Remediation of Recovered Chemical Warfare Materiel from Burial Sites. He holds a J.D. from George Washington University Law School and a B.S. in physics from Manhattan College.

STAFF

LAURA J. EHLERS is a senior staff officer for the Water Science and Technology Board of the National Research Council. Since joining the NRC in 1997, she has served as the study director for 18 committees, including the Committee to Review the New York City Watershed Management Strategy, the Committee on Bioavailability of Contaminants in Soils and Sediment, the Committee on Assessment of Water Resources Research, the Committee on Reducing Stormwater Discharge Contributions to Water Pollution, and the Committee to Review EPA's Economic Analysis of Final Water Quality Standards for Nutrients for Lakes and Flowing Waters in Florida. Ehlers has periodically consulted for EPA's Office of Research Development regarding their water quality research programs. She received her B.S. from the California Institute of Technology, majoring in biology and engineering and applied science. She earned both an M.S.E. and a Ph.D. in environmental engineering at the Johns Hopkins University.

Appendix B

Complex Site List

In order to better understand the relevant issues faced at complex sites of groundwater contamination, the Committee compiled the following information on a number of both public and private hazardous waste sites:

- Background
- Hydrogeology and source zone architecture
- Whether there is drinking water or indoor air exposure
- Regulatory information including remedial goals
- How site-specific risk assessment was taken into account
- Remedial action
- Current status
- Cost information (if available)

Some of the sites are highlighted in the body of the report, either in a text box or in the main text to illustrate a point. The rest are listed in Table B-1 along with relevant references. Most of these sites are still under active remediation.

TABLE B-1 Complex Sites Studied by the Committee

Site Name	**Location of Relevant Information**
Anniston Army Depot, Anniston, AL	*Chapter 7*
Bachman Road Dry Cleaners, Oscoda, MI	*Chapter 4*
Bethpage Navy Base, Long Island, NY	Siegel, L. 2011. The Limitations of Wellhead Treatment: Bethpage and Massapequa, Long Island, New York. http://www.cpeo.org/pubs/Bethpage.pdf.
CTS Asheville Site, Skyland, NC	EPA Region 4. 2002. Request for a Removal Action at Mills Gap Road Site in Asheville, NC. Memo from James Webster, Emergency Response and Removal Branch, to Richard Green, Waste Management Division. EPA Office of the Inspector General. 2010. EPA Activities Provide Limited Assurance of the Extent of Contamination and Risk at a North Carolina Hazardous Waste Site. Report No. 10-P-0130. http://www.mountainx.com/news/2007/cts_contanimation_report_says_no_new_dangers_residents_say_the_study_is_fla/ http://www.mountainx.com/news/2007/cts_health_assessment_released_declares_no_increased_cancer_little_risk_of_/ http://www.mountainx.com/news/2007/buncombe_residents_petition_gov_purdue_denr_officials_for_cts_cleanup/
Del Amo/Montrose Superfund Sites, Los Angeles County, CA	http://yosemite.epa.gov/r9/sfund/r9sfdocw.nsf/vwsoalphabetic/Del+Amo+Facility?OpenDocument http://yosemite.epa.gov/r9/sfund/r9sfdocw.nsf/vwsoalphabetic/Montrose%20Chemical%20Corp?OpenDocument EPA Region 9. 1999. Record of Decision for Dual Site Groundwater Operable Unit, Montrose Chemical and Del Amo Superfund Sites. EPA Region 9. 2010. Notice of Public Meeting; Groundwater Cleanup Project at the Montrose and Del Amo Superfund Sites in Los Angeles County, CA.
Former Koppers Company Wood Treating Plant, Salisbury, MD	*Chapter 4*
Hardage/Criner Site, McClain County, OK	EPA. 1986. Superfund Record of Decision, Hardage/Criner, OK. EPA/ROD/RO6-87/017. EPA Office of Emergency and Remedial Response. EPA. 2002. First Five-Year Review Report for the Hardage/Criner Superfund Site, McClain County, OK.
Hill Air Force Base, Salt Lake City, UT	*Chapter 5*

Letterkenny Army Depot, Chambers, PA	EPA. 2009a. Letterkenny Army Depot Property Disposal Office fact sheet, last updated December 2009. http://www.epa.gov/reg3hwmd/npl/PA2210090054.htm. EPA. 2009b. Letterkenny Army Depot Southeast fact sheet, last updated December 2009. http://www.epa.gov/reg3hwmd/npl/PA6213820503.htm
Machias Gravel Pit, Machias, NY	Gnat, R., Loch, M. et al. 1996. Machias Gravel Pit -Assessment through remediation in under three years. Proceedings of the Twenty-Eighth Mid-Atlantic Industrial and Hazardous Waste Conference. Technomic Publishing Company, Inc., Lancaster, PA. Rabideau, A. J., J. M. Blayden, and C. Ganguly. 1999. Field performance of air sparging for removing TCE from groundwater. Environmental Science & Technology 33(1):157-162.
MEW "Regional" VOC Plume, Mountain View, CA	EPA Region 9. 2009. Second Five-Year Review for Middlefield-Ellis-Whisman (MEW) Superfund Study Area Mountain View And Moffett Field, California.
Mission Valley Terminal, San Diego, CA	http://www.swrcb.ca.gov/rwqcb9/board_info/agendas/2009/aug/aug_09.shtml http://www.swrcb.ca.gov/rwqcb9/water_issues/programs/tsmc/mvt.shtml
Nano-Zero-Valent Iron Demonstration, Trenton, NJ	http://www.clu-in.org/conf/tio/nano-iron_121410/
Orange County, CA water supply	*Chapter 5*
Orica Botany Bay, Australia	http://www.oricabotanygroundwater.com/PDFs/Notice_of_Clean_Up_Action.pdf http://www.oricabotanygroundwater.com/Clean%20up%20Plan%20Documents/GCP_Final_31102003%5B1%5D.pdf http://www.environment.nsw.gov.au/resources/clm/docs/pdf/n20101714.pdf
Palm Beach Gardens, FL, chlorinated solvent plume	Mercer, J.W., G.C. Frederickson, D. Burnell, S. Dublin, J.E. Donahue and R.M. Ferris. 2006. Successful Remediation of Chlorinated Solvents Using Source Treatment and Natural Attenuation, The Fifth International Conference on Remediation of Chlorinated and Recalcitrant Compounds (Battelle), Monterey, CA, May 22-25. Metcalf & Eddy. 1997. Records Review / Work Plan, Lilac Street Wellfield, Vinyl Chloride Study Project, Palm Beach Gardens, Florida, FDEP Site No. 298. Metcalf & Eddy. 1998. Groundwater Investigation, Lilac Street Wellfield Chlorinated Solvent Study Project, Palm Beach Gardens, Florida FDEP Site No. 298.

Sages Dry Cleaners, Jacksonville, FL	ITRC. 2003. Technical and Regulatory Guidance for Surfactant/Cosolvent Flushing of DNAPL Source Zones, ITRC. Jawitz, J., R. K. Sillan, M. D. Annable, P. S. C. Rao and K. Warner. 2000. In-situ alcohol flushing of a DNAPL source zone at a dry cleaner site. Environmental Science & Technology 34: 3722-3729. Mravik, S., R. K. Sillan, A. L. Wood, and G. W. Sewell. 2003. Field evaluation of the solvent extraction residual biotreatment (SERB) technology. Environmental Science & Technology 37:5040-5049. Sewell, G. W., et al. 2005. Chlorinated Solvent Contaminated Soil and Ground Water: Field Application of the Solvent Extraction Residual Biotreatment Technology. Chapter 5 *In* Bioremediation of Recalcitrant Compounds. Taylor & Francis Publishers, Boca Raton, pp 59-149.
San Fernando Valley Ground Water Basin, Burbank and Glendale, CA	*Chapter 5*
San Gabriel, CA water supply	*Chapter 5*
Santa Monica, CA water supply	*Chapter 5*
Schofield Barracks, HI	*Chapter 2*
Twin Cities Army Ammunition Plant, New Brighton/Arden Hills, MN	*Chapter 4*
Visalia Pole Yard Superfund Site, Visalia, CA	EPA. 2009. DNAPL Remediation: Selected Projects Where Regulatory Closure Goals Have Been Achieved, EPA 542/R-09/008. Southern California Edison. 2008. Remedial Action Completion Report, 9 pp.
West Side Corporation, Queens, NY	http://www.dec.ny.gov/chemical/45221.html Sundquist, J. A., and Chiusano, D. J. 2008. Electrical Resistance Heating remediation of tetrachloroethene DNAPL and groundwater contamination. *In* Proceedings of the Sixth International Conference on Remediation of Chlorinated and Recalcitrant Compounds, Monterey, CA, May 2008, Battelle, Columbus, OH.

Young-Rainey STAR Center, Largo, FL	DOE. 2009. Pinellas Environmental Restoration Project, Sitewide Environmental Monitoring Semiannual Progress Report for the Young-Rainey STAR Center, June through November 2009, LMS/PIN/N01439. Heron, G., S. Carroll, and S. G. Nielsen. 2005. Full-scale removal of DNAPL constituents using steam-enhanced extraction and electrical resistance heating. Ground Water Monitoring & Remediation 25(4):92-107.

Appendix C

Analysis of 80 Facilities with Contaminated Groundwater Deleted from the National Priorities List

80 Superfund facilities with groundwater contamination were identified by EPA personnel as having met cleanup goals and being delisted from the National Priorities List (NPL). Information on the 80 facilities found in this appendix was compiled by the Committee based on Site Deletion documents, Records of Decision, EPA Fact Sheets, and other information available from the CERCLIS website (http://cumulis.epa.gov/supercpad/cursites/srchsites.cfm). The Committee strongly advises anyone using these data to review the primary documents concerning these sites.

For the 80 facilities, Tables C-1 through C-4 provide information on the site types, the contaminants of concern (COCs) present, the remedial action objectives (RAOs), and the extent to which contaminant concentrations in groundwater met maximum contaminant levels (MCLs) prior to facility delisting, respectively. Contaminants were divided into categories of volatile organic compounds (VOCs), semivolatile organic compounds (SVOCs) and metals. Figures C-1 through C-4 provide the same information in graphical form. Further explanation of these tables and graphs can be found in Chapter 2.

NOTE: PAGES 321-408 OF THIS APPENDIX ARE AVAILABLE ONLINE AT WWW.NAP.EDU/CATALOG.PHP?RECORD_ID=14668.

FACILITY INDEX

Region 1

1. Cheshire Groundwater Contamination
2. Pinette's Salvage Yard
3. SACO Tannery Waste Pits
4. Tansitor Electronics, Inc.

Region 2

5. Anchor Chemicals
6. Jones Sanitation
7. Katonah Municipal Well
8. Krysowaty Farm
9. Love Canal
10. Mannheim Avenue Dump
11. Pfohl Brothers Landfill
12. Pijak Farm
13. Spence Farm
14. Suffern Village Well Field
15. Tabernacle Drum Dump
16. Tronic Plating Co., Inc
17. Upper Deerfield Township Sanitary Landfill
18. V&M/Albaladejo
19. Vestal Water Supply Well 4-2

Region 3

20. Aladdin Plating
21. Berks Landfill
22. Brodhead Creek
23. Matthews Electroplating
24. Mid-Atlantic Wood Preservers, Inc.
25. New Castle Spill
26. Southern Maryland Wood Treating
27. Sussex County Landfill No. 5

Region 4

28. BMI-Textron
29. Cedartown Industries, Inc.
30. Cedartown Municipal Landfill
31. Davie Landfill
32. Dubose Oil Products Co.
33. Gold Coast Oil Corp.
34. Independent Nail Co.
35. Lee's Lane Landfill
36. Lewisburg Dump
37. Palmetto Recycling, Inc.
38. Rochester Property
39. Sixty-Second Street Dump
40. Standard Auto Bumper Corp.
41. Woodbury Chemical Co. (Princeton Plant)

Region 5

42. Agate Lake Scrapyard
43. Alsco Anaconda
44. Avenue "E" Groundwater Contamination
45. Berlin & Farro
46. Dakhue Sanitary Landfill
47. Fadrowski Drum Disposal
48. Laskin/Poplar Oil Co.
49. Northern Engraving Co.
50. Southside Sanitary Landfill
51. Tri-State Plating
52. University of Minnesota (Rosemount Research Center) OU1
53. Waste, Inc., Landfill
54. Windom Dump

Region 6

55. Bailey Waste Disposal
56. Cleveland Mill
57. Double Eagle Refinery Co.
58. Dutchtown Treatment Plant
59. Fourth Street Abandoned Refinery
60. Gurley Pit

61. Industrial Waste Control
62. Mallard Bay Landing Bulk Plant
63. Odessa Chromium #2 (Andrews Highway)
64. Old Inger Oil Refinery
65. South 8th Street Landfill

Region 7

66. Farmers' Mutual Cooperative
67. Kem-Pest Laboratories
68. Labounty
69. Waverly Ground Water Contamination
70. White Farm Equipment Co. Dump

Region 8

71. Rose Park Sludge Pit
72. Whitewood Creek

Region 9

73. Del Norte Pesticide Storage
74. Firestone Tire & Rubber Co. (Salinas Plant)
75. Schofield Barracks
76. Southern California Edison Co. (Visalia Poleyard)
77. Western Pacific Railroad Co.

Region 10

78. Bonneville Power Administration Ross Complex
79. Northwest Transformer
80. Union Pacific Railroad Company

TABLE C-1 80 Delisted NPL Facilities with Contaminated Groundwater by "Site Type"

Landfills	Industrial Facilities	Potable Well Fields	Military Bases	Other
9. Love Canal	1. Cheshire Groundwater Contamination	7. Katonah Municipal Well	44. Avenue "E" Groundwater Contamination	8. Krysowaty Farm
10. Mannheim Avenue Dump	2. Pinette's Salvage Yard	14. Suffern Village Well Field	75. Schofield Barracks	12. Pijak Farm
11. Pfohl Brothers Landfill	3. SACO Tannery Waste Pits	19. Vestal Water Supply Well 4-2		13. Spence Farm
17. Upper Deerfield Township Sanitary Landfill	4. Tansitor Electronics, Inc.			18. V&M/Albaladejo
21. Berks Landfill	5. Anchor Chemicals			45. Berlin & Farro
27. Sussex County Landfill No. 5	6. Jones Sanitation			52. University of Minnesota (Rosemount Research Center)
30. Cedartown Municipal Landfill	15. Tabernacle Drum Dump			66. Farmers' Mutual Cooperative
31. Davie Landfill	16. Tronic Plating Co., Inc.			78. Bonneville Power Administration Ross Complex
35. Lee's Lane Landfill	20. Aladdin Plating			
36. Lewisburg Dump	22. Brodhead Creek			
39. Sixty-Second Street Dump	23. Matthews Electroplating			
42. Agate Lake Scrapyard	24. Mid-Atlantic Wood Preservers, Inc.			
46. Dakhue Sanitary Landfill	25. New Castle Spill			
47. Fadrowski Drum Disposal	26. Southern Maryland Wood Treating			

continued

TABLE C-1 Continued

Landfills	Industrial Facilities	Potable Well Fields	Military Bases	Other
50. Southside Sanitary Landfill	28. BMI-Textron			
53. Waste, Inc., Landfill	29. Cedartown Industries, Inc.			
54. Windom Dump	32. Dubose Oil Products Co.			
65. South 8th Street Landfill	33. Gold Coast Oil Corp.			
	34. Independent Nail Co.			
	37. Palmetto Recycling, Inc.			
	38. Rochester Property			
	40. Standard Auto Bumper Corp.			
	41. Woodbury Chemical Co. (Princeton Plant)			
	43. Alsco Anaconda			
	48. Laskin/Poplar Oil Co.			
	49. Northern Engraving Co.			
	51. Tri-State Plating			
	55. Bailey Waste Disposal			
	56. Cleveland Mill			
	57. Double Eagle Refinery Co.			
	58. Dutchtown Treatment Plant			

59. Fourth Street Abandoned Refinery
60. Gurley Pit
61. Industrial Waste Control
62. Mallard Bay Landing Bulk Plant
63. Odessa Chromium #2 (Andrews Highway)
64. Old Inger Oil Refinery
67. Kem-Pest Laboratories
68. Labounty
69. Waverly Ground Water Contamination
70. White Farm Equipment Co. Dump
71. Rose Park Sludge Pit
72. Whitewood Creek
73. Del Norte Pesticide Storage
74. Firestone Tire & Rubber Co. (Salinas Plant)
76. Southern California Edison Co. (Visalia Poleyard)
77. Western Pacific Railroad Co.
79. Northwest Transformer
80. Union Pacific Railroad Co.

TABLE C-2 Contaminants of Concern at the 80 Delisted NPL Facilities

VOCs Only	VOCs and SVOCs	Metals Only	Metals and VOCs	Metals, VOCs, and SVOCs	Other
1. Cheshire Groundwater Contamination	2. Pinette's Salvage Yard	18. V&M/Albaladejo	5. Anchor Chemicals	41. Woodbury Chemical Co. (Princeton Plant)	3. SACO Tannery Waste Pits
4. Tansitor Electronics, Inc.	8. Krysowaty Farm	20. Aladdin Plating	6. Jones Sanitation	48. Laskin/ Poplar Oil Co.	11. Pfohl Brothers Landfill
7. Katonah Municipal Well	9. Love Canal	23. Matthews Electroplating	16. Tronic Plating Co., Inc.	57. Double Eagle Refinery Co.	13. Spence Farm
10. Mannheim Avenue Dump	67. Kem-Pest Laboratories	24. Mid-Atlantic Wood Preservers, Inc.	17. Upper Deerfield Township Sanitary Landfill	59. Fourth Street Abandoned Refinery	22. Brodhead Creek
12. Pikaj Farm		29. Cedartown Industries, Inc.	21. Berks Landfill	61. Industrial Waste Control	28. BMI-Textron
14. Suffern Village Well Field		30. Cedartown Municipal Landfill	31. Davie Landfill	62. Mallard Bay Landing Bulk Plant	34. Independent Nail Co.
15. Tabernacle Drum Dump		37. Palmetto Recycling, Inc.	35. Lee's Lane Landfill	80. Union Pacific Railroad Company	43. Alsco Anaconda

19. Vestal Water Supply Well 4-2	39. Sixty-Second Street Dump	36. Lewisburg Dump	49. Northern Engraving Co.
25. New Castle Spill	40. Standard Auto Bumper Corp.	38. Rochester Property	71. Rose Park Sludge Pit
26. Southern Maryland Wood Treating	47. Fadrowski Drum Disposal	46. Dakhue Sanitary Landfill	79. Northwest Transformer
27. Sussex County Landfill No. 5	50. Southside Sanitary Landfill	53. Waste, Inc., Landfill	
32. Dubose Oil Products Co.	51. Tri-State Plating	55. Bailey Waste Disposal	
33. Gold Coast Oil Corp.	56. Cleveland Mill	58. Dutchtown Treatment Plant	
42. Agate Lake Scrapyard	63. Odessa Chromium #2 (Andrews Highway)	60. Gurley Pit	
44. Avenue "E" Groundwater Contamination	65. South 8th Street Landfill	64. Old Inger Oil Refinery	
45. Berlin & Farro	72. Whitewood Creek	68. Labounty	
52. University of Minnesota (Rosemount Research Center)		70. White Farm Equipment Co. Dump	
54. Windom Dump			

continued

TABLE C-2 Continued

VOCs Only	VOCs and SVOCs	Metals Only	Metals and VOCs	Metals, VOCs, and SVOCs	Other
66. Farmers' Mutual Cooperative					
69. Waverly Ground Water Contamination					
73. Del Norte Pesticide Storage					
74. Firestone Tire & Rubber Co. (Salinas Plant)					
75. Schofield Barracks					
76. Southern California Edison Co. (Visalia Poleyard)					
77. Western Pacific Railroad Co.					
78. Bonneville Power Administration Ross Complex					

TABLE C-3 Remedial Objectives at the 80 Delisted NPL Facilities.

Achieve MCLs or Another Cleanup Level	Objectives Unclear other than Risk Reduction	No Explicit Objective
2. Pinette's Salvage Yard	3. SACO Tannery Waste Pits	1. Cheshire Groundwater Contamination
4. Tansitor Electronics, Inc.	6. Jones Sanitation	5. Anchor Chemicals
7. Katonah Municipal Well	26. Southern Maryland Wood Treating	8. Krysowaty Farm
10. Mannheim Avenue Dump	53. Waste, Inc., Landfill	9. Love Canal
11. Pfohl Brothers Landfill	57. Double Eagle Refinery Co.	12. Pijak Farm
15. Tabernacle Drum Dump	58. Dutchtown Treatment Plant	13. Spence Farm
19. Vestal Water Supply Well 4-2	71. Rose Park Sludge Pit	14. Suffern Village Well Field
23. Matthews Electroplating		16. Tronic Plating Co., Inc.
24. Mid-Atlantic Wood Preservers, Inc.		17. Upper Deerfield Township Sanitary Landfill
28. BMI-Textron		18. V&M/Albaladejo
29. Cedartown Industries, Inc.		20. Aladdin Plating
30. Cedartown Municipal Landfill		21. Berks Landfill
31. Davie Landfill		22. Brodhead Creek
32. Dubose Oil Products Co.		25. New Castle Spill
33. Gold Coast Oil Corp.		27. Sussex County Landfill No. 5
35. Lee's Lane Landfill		34. Independent Nail Co.
36. Lewisburg Dump		41. Woodbury Chemical Co. (Princeton Plant)
37. Palmetto Recycling, Inc.		42. Agate Lake Scrapyard
38. Rochester Property		48. Laskin/Poplar Oil Co.
39. Sixty-Second Street Dump		50. Southside Sanitary Landfill
40. Standard Auto Bumper Corp.		55. Bailey Waste Disposal
43. Alsco Anaconda		56. Cleveland Mill
44. Avenue "E" Groundwater Contamination		59. Fourth Street Abandoned Refinery

continued

TABLE C-3 Continued

Achieve MCLs or Another Cleanup Level	Objectives Unclear other than Risk Reduction	No Explicit Objective
45. Berlin & Farro		60. Gurley Pit
46. Dakhue Sanitary Landfill		61. Industrial Waste Control
47. Fadrowski Drum Disposal		67. Kem-Pest Laboratories
49. Northern Engraving Co.		78. Bonneville Power Administration Ross Complex
51. Tri-State Plating		79. Northwest Transformer
52. University of Minnesota (Rosemount Research Center) OU1		
54. Windom Dump		
62. Mallard Bay Landing Bulk Plant		
63. Odessa Chromium #2 (Andrews Highway)		
64. Old Inger Oil Refinery		
65. South 8th Street Landfill		
66. Farmers' Mutual Cooperative		
68. Labounty		
69. Waverly Ground Water Contamination		
70. White Farm Equipment Co. Dump		
72. Whitewood Creek		
73. Del Norte Pesticide Storage		
74. Firestone Tire & Rubber Co. (Salinas Plant)		
75. Schofield Barracks (USARMY)		
76. Southern California Edison Co. (Visalia Poleyard)		

TABLE C-3 Continued

Achieve MCLs or Another Cleanup Level	Objectives Unclear other than Risk Reduction	No Explicit Objective
77. Western Pacific Railroad Co.		
80. Union Pacific Railroad Company		

TABLE C-4 Extent to which MCLs were Achieved at the 80 Delisted NPL Facilities

MCLs Achieved			MCLs Not Achieved
Active Remedy/No LTM	Active Remedy/ LTM	No Active Remedy/ No LTM	LTM
15. Tabernacle Drum Dump	7. Katonah Municipal Well	1. Cheshire Groundwater Contamination	2. Pinette's Salvage Yard
19. Vestal Water Supply Well 4-2	10. Mannheim Avenue Dump	12. Pijak Farm	3. SACO Tannery Waste Pits
33. Gold Coast Oil Corp.	39. Sixty-Second Street Dump	13. Spence Farm	6. Jones Sanitation
38. Rochester Property	54. Windom Dump	14. Suffern Village Well Field	11. Pfohl Brothers Landfill
44. Avenue "E" Groundwater Contamination		16. Tronic Plating Co., Inc.	21. Berks Landfill
45. Berlin & Farro		23. Matthews Electroplating	30. Cedartown Municipal Landfill
48. Laskin/Poplar Oil Co.		24. Mid-Atlantic Wood Preservers, Inc.	31. Davie Landfill
51. Tri-State Plating		25. New Castle Spill	35. Lee's Lane Landfill
52. University of Minnesota (Rosemount Research Center)		28. BMI-Textron	36. Lewisburg Dump
63. Odessa Chromium #2 (Andrews Highway)		29. Cedartown Industries, Inc.	47. Fadrowski Drum Disposal

		Other		
Risk Assessment/ No LTM	Risk Assessment/ LTM	RAO Didn't Include Achievement of MCLs	MCL Achievement Unknown	No Apparent GW Contamination
5. Anchor Chemicals	17. Upper Deerfield Township Sanitary Landfill	4. Tansitor Electronics, Inc.[a]	8. Krysowaty Farm	18. V&M/ Albaladejo
41. Woodbury Chemical Co. (Princeton Plant)	20. Aladdin Plating	9. Love Canal	46. Dakhue Sanitary Landfill	55. Bailey Waste Disposal
	27. Sussex County Landfill No. 5	22. Brodhead Creek[a]	57. Double Eagle Refinery Co.	60. Gurley Pit
	34. Independent Nail Co.	26. Southern Maryland Wood Treating	59. Fourth Street Abandoned Refinery	78. Bonneville Power Administration Ross Complex (USDOE)
		42. Agate Lake Scrapyard	61. Industrial Waste Control	79. Northwest Transformer
		49. Northern Engraving Co.	64. Old Inger Oil Refinery	
		53. Waste, Inc., Landfill		
		56. Cleveland Mill		
		58. Dutchtown Treatment Plant		
		71. Rose Park Sludge Pit		

continued

TABLE C-4 Continued

MCLs Achieved			MCLs Not Achieved
Active Remedy/No LTM	Active Remedy/ LTM	No Active Remedy/ No LTM	LTM
69. Waverly Ground Water Contamination		32. Dubose Oil Products Co.	66. Farmers' Mutual Cooperative
70. White Farm Equipment Co. Dump		37. Palmetto Recycling, Inc.	68. Labounty
74. Firestone Tire & Rubber Co. (Salinas Plant)		40. Standard Auto Bumper Corp.	72. Whitewood Creek
76. Southern California Edison Co. (Visalia Poleyard)		43. Alsco Anaconda	77. Western Pacific Railroad Co.
80. Union Pacific Railroad Company		50. Southside Sanitary Landfill	
		62. Mallard Bay Landing Bulk Plant	
		65. South 8th Street Landfill	
		67. Kem-Pest Laboratories	

LTM = long-term monitoring.
[a]Indicates a facility with a TI waiver.

Risk Assessment/ No LTM	Risk Assessment/ LTM	Other: RAO Didn't Include Achievement of MCLs	MCL Achievement Unknown	No Apparent GW Contamination
		73. Del Norte Pesticide Storage[a]		
		75. Schofield Barracks[a]		

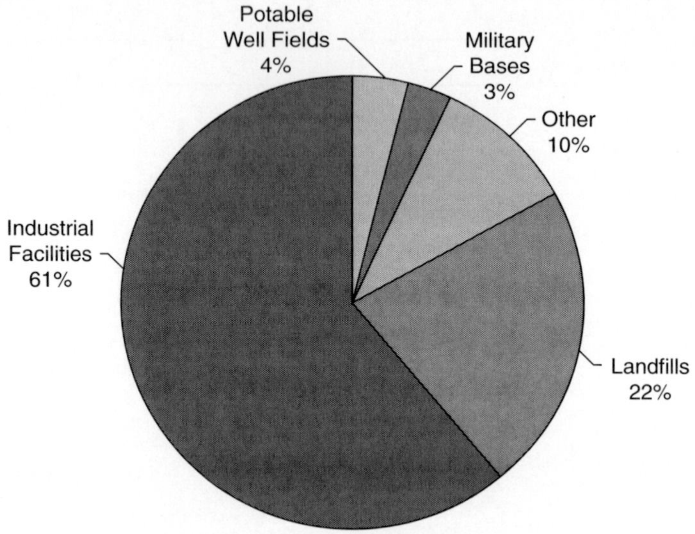

FIGURE C-1 80 Delisted NPL facilities by "site type."

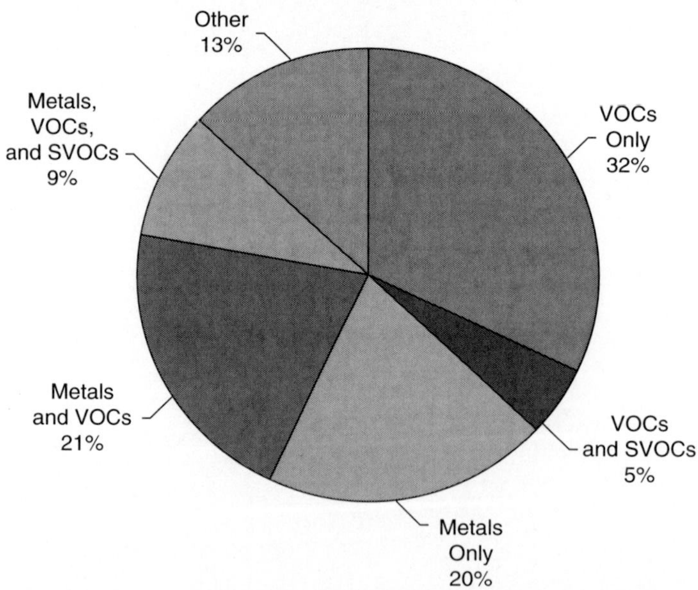

FIGURE C-2 Contaminant groups found at the 80 delisted NPL facilities.

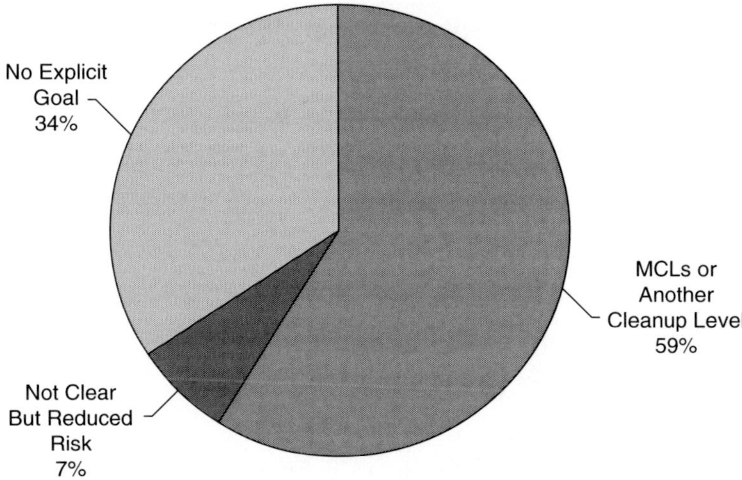

FIGURE C-3 Types of remedial action objectives for the 80 delisted NPL facilities.

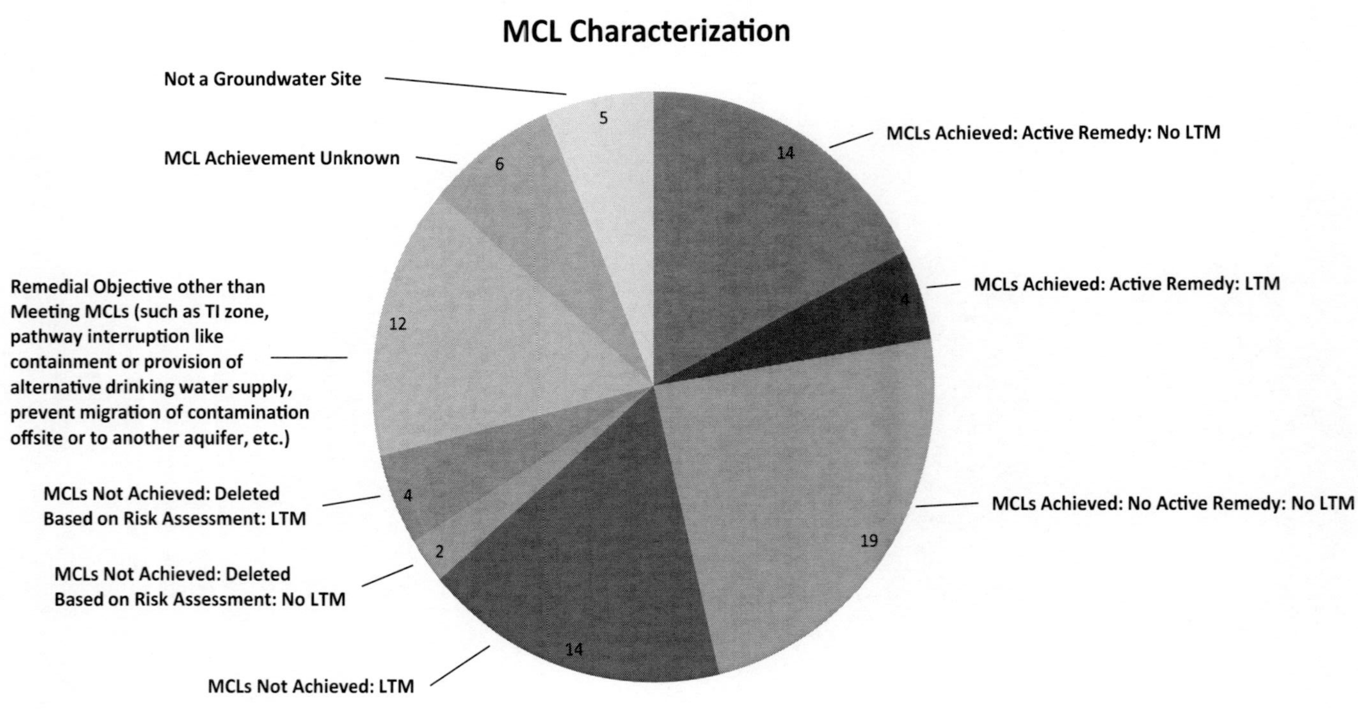

FIGURE C-4 Extent of achievement of MCLs for the 80 delisted NPL facilities.

1. Cheshire Groundwater Contamination

State: CT

Site Lead: State

Date Deleted from NPL: July 2, 1997

Date of Last Five-Year Review: N/A

COCs: Many VOCs including trichloroethane (TCA), dichloroethene (DCE), tetrachloroethene (PCE), and xylenes

RAOs: Protection of human health and the environment; risk assessment results found risk was acceptable, between 1:10,000 and 1:1 million, and the hazard quotient was sufficiently low.

Technologies Applied: Soil excavation

Reported Results: Trichloroethane 3 µg/L, dichloroethane 14 µg/L, dichloroethene 0.9 µg/L, bis(2-ethylhexyl)phthalate 15 µg/L, bromodichloromethane 0.6 µg/L, dichlorobenzene 1 µg/L, dibromochloromethane 0.6 µg/L, tetrahydrofuran 5.2 µg/L, and toluene 1.4 µg/L. Concentrations greater than 10,000 µg/L for PCE, 3,500 µg/L for trichloroethene (TCE), and 5,000 µg/L for chloroform have been found in groundwater from location OW-1 at various times from June 1989 to February 1991.

MCLs Achieved? Yes

ICs in Place? None listed

Vapor Intrusion: Unclear. Soil vapor sampling was performed by a third party but results were inconclusive. Risk assessment for inhalation pathway of some contaminants found negligible risk.

Long-Term Monitoring? No

2. Pinette's Salvage Yard

State: ME

Site Lead: EPA/State

Date Deleted from NPL: September 30, 2002

Date of Last Five-Year Review: September 28, 2010

COCs: Base neutral acids, polychlorinated biphenyls (PCBs), VOCs

RAOs: Reduce the following contaminants in groundwater to below the designated MCL: benzene 5 μg/l, chlorobenzene 47 μg/l, 1-4-dichlorobenzene 27 μg/L, and PCBs 0.5 μg/L.

Technologies Applied: Soil excavation and pump and treat

Reported Results: In October 2009, all wells tested had contaminant concentrations in groundwater below the MCL except DMW-5 (PCB 2.1 μg/L).

MCLs Achieved? No, PCB concentrations in DMW-5 are still above the MCL.

ICs in Place? Yes, they are restricting site and aquifer use through a restrictive covenant.

Vapor Intrusion: Not a concern because: "(1) the remaining levels of VOCs in the groundwater at the Site have been measured to be very low; (2) there are no occupied buildings currently within the Restricted Area; and (3) the Restrictive Covenant prohibits the construction or placement of any buildings within the Restricted Area without prior written permission of the Maine DEP." (http://www.epa.gov/superfund/sites/fiveyear/f05-01016.pdf)

Long-Term Monitoring? Yes

3. SACO Tannery Waste Pits

State: ME

Site Lead: EPA

Date Deleted from NPL: September 29, 1999

Date of Last Five-Year Review: December 30, 2008

COCs: Arsenic and monochlorobenzene

RAOs: Arsenic concentrations less than 10 µg/L and monochlorobenzene concentrations less than 100 µg/L. MCLs were set as the action levels, or standards, for all groundwater contaminants, except for arsenic at four locations. EPA established Alternate Concentration Limits (ACLs) for four site monitoring wells [MW-l0l (70 µg/L), MW-103 (123 µg/L), MW-111B (64 µg/L), and MW-114B (77 µg/L)] based on the maximum concentrations observed in the four wells during the RI.

Technologies Applied: GW monitoring, source control, excavation

Reported Results: Arsenic concentrations continue to exceed the ACL of 123 µg/L in one of four locations (MW-103) and the 2001 MCL in four of the other five wells in the long-term monitoring program. There does not appear to be a consistent trend in concentrations, with arsenic concentrations at some wells stable over the past twelve years and others fluctuating.

Following the 2005 hydraulic assessment, ME DEP discontinued monitoring for monochlorobenzene except at MW-114A. This was done because for the other eight wells in long-term monitoring, six never exceeded the MCL 100 µg/L and the other two only once each. The MCL has been exceeded slightly at MW-l14A seven times (100-130 µg/L), all since April 2000.

MCLs Achieved? No, arsenic and monochlorobenzene concentrations are still above the MCL, but the restriction on use of site groundwater prevents any exposures.

ICs in Place? Yes, there is restricted site access through state legislation converting it into a wildlife preserve.

Vapor Intrusion: Not a concern because the COCs are primarily metals and monochlorobenzene is essentially limited to one area of the site, and institutional controls that prevent development are in place.

Long-Term Monitoring? Yes

4. Tansitor Electronics, Inc.

State: VT

Site Lead: State

Date Deleted from NPL: September 29, 1999

Date of Last Five-Year Review: September 23, 2009

COCs: VOCs

RAOs: 1,1,1-TCA less than 200 μg/L, 1,1-DCE less than 7 μg/L, TCE less than 5 μg/L, PCE less than 5 μg/L, which are the federal MCLs. A technical impracticability (TI) waiver was granted for a 10-acre area where MCLs do not have to be met.

Technologies Applied: Groundwater monitoring and monitored natural attenuation

Reported Results: Groundwater monitoring within the TI Zone has shown gradual reductions in concentrations of contaminants. Groundwater monitoring beneath and outside the TI Zone has demonstrated that there continues to be no migration beyond the TI Zone or the Site.

MCLs Achieved? No, because a TI zone was established.

ICs in Place? Yes, the use of groundwater is prevented through deed restrictions

Vapor Intrusion: With respect to potential vapor intrusion within the manufacturing building, information provided by the facility indicates that the HVAC systems create an ongoing air exchange of 8–24 times per work-day to address the use of solvents within the manufacturing process and soil vapor data levels were below OSHA time weighted average levels. As the contaminated groundwater is a potential vapor intrusion source, EPA will continue to evaluate this pathway in future reviews, particularly if land use of the Site changes.

Long-Term Monitoring? Yes

5. Anchor Chemicals

State: NY

Site Lead: State

Date Deleted from NPL: September 30, 1999

Date of Last Five-Year Review: September 29, 1995

COCs: Lead, chromium, and VOCs

RAOs: Protection of human health and the environment; risk assessment showed that the current risk via multiple pathways was within the acceptable range

Technologies Applied: Excavation of drywell sediments

Reported Results: Highest detected concentrations: 1,1,1-tricholoroethane 10 μg/L, chromium 1440 μg/L, lead 240 μg/L

MCLs Achieved? No, but risk assessment found the risk to fall within an acceptable range. "Groundwater samples were collected in April 1996 to assess the effectiveness of the removal action. No contaminants were detected above drinking water maximum contaminant levels (MCLs). A second round of groundwater samples were taken in July 1997. Nickel and chromium were detected in the second round; however, the levels do not justify a change in EPA's no-further-action decision for the site." (http://www.epa.gov/region02/superfund/npl/0201324c.pdf).

ICs in Place? None listed

Vapor Intrusion: "All drywell clean-outs should be performed using the requirements for potential vapor emissions which are outlined in a New York State Department of Health (NYSDOH) procedures manual. EPA has reviewed the referenced requirements for responding to potentially harmful vapor emissions and will require that they be followed during the drywell removals." (http://www.epa.gov/superfund/sites/rods/fulltext/r0295254.pdf).

Long-Term Monitoring? No

6. Jones Sanitation

State: NY

Site Lead: EPA

Date Deleted from NPL: September 23, 2005

Date of Last Five-Year Review: June 27, 2011

COCs: VOCs and metals

RAOs: *Onsite:* provide additional data on the chemical composition, especially VOCs and metals, of groundwater on-site; evaluate whether the landfill closure has substantially changed groundwater flow patterns and chemistry; and observe the natural attenuation of contaminants in the groundwater.
Offsite: ensure no site-related contaminants are impacting local drinking water sources.

Technologies Applied: Excavation/capping, groundwater monitoring, and MNA

Reported Results: *Onsite:* 2006 – benzene 2.8 µg/L (RAO 0.7), chlorobenzene 19 µg/L, TCE 1.4 µg/L, PCE 5.4 µg/L (MCL 5), vinyl chloride 0.35 µg/L, cis -1 ,2-DCE 5.5 µg/L, 1,2-DCE 0.52 µg/L, 1,2-Dichlorobenzene 1.8 µg/L, 1,3-Dichlorobenzene 0.68 µg/L, 1,3-Dichlorobenzene 0.96 µg/L, manganese 1830 µg/L (RAO 300), iron 43100 µg/L (MCL 300), and lead 29 µg/L (MCL 15).

2005-2011: Benzene has decreased and exhibited concentrations less than the cleanup goal/RAO of 0.7 µg/L, the current NYSDEC standard of 1.0 µg/L, and EPA's MCL of 5 µg/L. Chlorobenzene concentrations have been increasing during the last three rounds of sampling but have been found to be decreasing when compared to the concentrations detected between 2002 and 2005. The concentrations continue to exceed the NYSDEC standard of 5 µg/L but are below EPA's MCL of 100 µg/L. Benzene and chlorobenzene concentrations have only been detected in monitoring well JSMW-3B located downgradient of the original excavated trench areas and directly upgradient of the capped area. This well also exhibits concentrations of iron and manganese that are not consistent with background. In addition, several other VOCs have been detected in the monitoring wells within the past five years above their respective NYSDEC WQS and MCLs;

however, the concentrations have decreased and are currently (2011) below their MCLs and NYSDEC WQS.

Offsite: No site-related contaminants were detected in any of the groundwater wells tested. A review of the sampling results from the residential potable groundwater wells demonstrate that off-site groundwater has not been impacted by the site.

MCLs Achieved? No, concentrations of some VOCs and metals are still above standards.

ICs in Place? Yes, there are deed restrictions and/or well permitting restrictions to prevent human contact with contaminated groundwater at the site.

Vapor Intrusion: Vapor intrusion was not previously evaluated as a potential future exposure pathway based on the (health protective) assumption that buildings are not located above the contaminants of concern in the groundwater.

Long-Term Monitoring? Yes

7. Katonah Municipal Well

State: NY

Site Lead: EPA

Date Deleted from NPL: March 20, 2000

Date of Last Five-Year Review: September 25, 2007

COCs: PCE

RAOs: Achieve PCE concentration less than 5 µg/L

Technologies Applied: Pump and treat

Reported Results: 2007 monitoring well results show PCE concentrations at 0.2 and 0.4 µg/L

MCLs Achieved? Yes

ICs in Place? No. EPA believes actions identified in the Record of Decision (ROD) along with NYC programs protecting its reservoirs and water supplies are sufficient.

Vapor Intrusion: 2006 GW monitoring showed likelihood of vapor intrusion to be low.

Long-Term Monitoring? Yes

8. Krysowaty Farm

State: NJ

Site Lead: State

Date Deleted from NPL: February 22, 1989

Date of Last Five-Year Review: N/A

COCs: VOCs, pesticides, base neutral compounds, PCBs

RAOs: Protection of human health and the environment

Technologies Applied: Excavation/offsite disposal

Reported Results: None provided

MCLs Achieved? Unknown

ICs in Place? Yes, an alternative water supply was provided.

Vapor Intrusion: Not mentioned

Long-Term Monitoring? No. Semiannual monitoring was planned for five years, but no results are available.

9. Love Canal

State: NY

Site Lead: EPA

Date Deleted from NPL: September 30, 2004

Date of Last Five-Year Review: September 29, 2008

COCs: VOCs, SVOCs, and pesticides

RAOs: Protection of human health and the environment

Technologies Applied: Cap and treat along with pump and treat

Reported Results: The 2007 O&M Report shows that there has been no significant change in chemical concentration conditions and that the barrier drain system is successfully capturing leachate from the Site and preventing off-site migration of contamination. Hence, monitoring results continue to confirm that the remediation and containment system, i.e., the leachate collection and treatment system, is functioning properly.

MCLs Achieved? Not applicable because the goal was containment

ICs in Place? Yes, there are zoning and deed restrictions.

Vapor Intrusion: Based on the distance to the nearest residences, further evaluation of vapor intrusion is not deemed necessary. Also, indoor air sampling was performed as part of the 1988 Love Canal EDA Habitability Study which did not find any indoor air issues within the homes in the emergency declaration area.

Long-Term Monitoring? Yes

10. Mannheim Avenue Dump

State: NJ

Site Lead: EPA/State

Date Deleted from NPL: August 28, 2007

Date of Last Five-Year Review: September 23, 2004

COCs: TCE

RAOs: TCE in groundwater aquifer below 1 μg/L which is the NJ Safe Drinking Water Act level (Federal MCL is 5 μg/L).

Technologies Applied: Pump and treat

Reported Results: Monitoring has indicated that the goal of restoring the groundwater aquifer to meet all appropriate drinking water standards specified in the ROD has been achieved. The highest TCE concentration found during 2003 sampling was 0.21 μg/L.

MCLs Achieved? Yes

ICs in Place? None mentioned

Vapor Intrusion: Not mentioned

Long-Term Monitoring? Yes

11. Pfohl Brothers Landfill

State: NY

Site Lead: EPA

Date Deleted from NPL: September 22, 2008

Date of Last Five-Year Review: March 18, 2011

COCs: VOCs, SVOCs, PCBs, dioxins, furans, cyanide

RAOs: Reduce contamination to drinking water standards

Technologies Applied: Pump and treat

Reported Results: Since 2006 data show that no VOCs, PCBs, dioxins, furans, or cyanide were detected above the Class GA water quality standards. SVOCs were detected sporadically above water quality standards in one well, but four groundwater wells continue to be purged dry during sampling, possibly affecting the VOC sampling results.

MCLs Achieved? No, SVOC concentrations are still above water quality standards

ICs in Place? Yes, in the form of declarations of covenants and restrictions on access

Vapor Intrusion: VOCs in the groundwater are located within the containment system and are at a great distance from the residences, so the potential for soil vapor intrusion issues related to this site is low.

Long-Term Monitoring? Yes

12. Pijak Farm

State: NJ

Site Lead: EPA

Date Deleted from NPL: March 3, 1997

Date of Last Five-Year Review: N/A

COCs: Benzene and phthalates

RAOs: Protection of human health and the environment

Technologies Applied: Waste, soil and groundwater removal, site regrading and surface restoration, groundwater removal, berm along site perimeter, and groundwater monitoring

Reported Results: 1984 ROD states that samples from MW-2S contained 14 base-extractable compounds ranging in concentrations from 220 µg/L to 3600 µg/L, and averaging about 1000 µg/l. Included are four priority pollutants: ethylbenzene (MCL 700 µg/L), 1,2-dichlorobenzene (MCL 1.9 µg/L), di-n-butyl phthalate (MCL 2.5 µg/L) and bis (2-ethylhexyl)-phthalate (6 µg/L) in concentrations of 940 µg/L, 220 µg/L, 310 µg/L and 320 µg/L, respectively.

MCLs Achieved? Yes. The potentially responsible party (PRP) Morton International, under State monitoring, completed the removal and disposal of on-site contaminated drums, lab packs, and visibly contaminated soil to an approved facility, thereby eliminating risks posed by exposure to these materials. Soil sampling conducted subsequent to these activities indicated the presence of residual PCB contamination in soil at the site. Therefore, between 1989 and 1994 Morton International, under State supervision, provided for the excavation and off-site disposal of the contaminated soil. In all, approximately 4,000 cubic yards (6,400 tons) of waste material and contaminated soil were excavated and disposed of offsite. The results of groundwater monitoring conducted over a five-year period indicate that the groundwater is clean.

ICs in Place? None listed

Vapor Intrusion: Not mentioned

Long-Term Monitoring? No

13. Spence Farm

State: NJ

Site Lead: EPA

Date Deleted from NPL: March 3, 1997

Date of Last Five-Year Review: N/A

COCs: Acetone, phthalate, phenol, zinc, chromium, and mercury

RAOs: Protection of human health and the environment

Technologies Applied: Excavation and groundwater monitoring

Reported Results: 16 monitoring wells were sampled as part of the investigative program. All organic compounds detected were generally in the concentration range of from 10 to 50 ppb. Two volatile organic compounds were detected, neither of which are priority pollutants. Of the base-extractable organics detected, only two are priority pollutants: Di-n-octyl phthalate being detected at mw-1s, 1d (dup) and mw-8 and bis (2-ethylhexyl) phthalate being detected at mw-3s, mw-4d, and mw-6s. As a result of the random dumping at this site, the organic compounds identified in the groundwater samples are almost entirely different from those compounds found in the test pit, split-spoon, and stream sediment samples.

For inorganic priority pollutant compounds, very low concentrations of mercury (approximately 0.0004 mg/L) and zinc (approximately 0.10 mg/L) were found. Phenols were detected at mw-5, arsenic at mw-8, and chromium at mw-9. Occurrences of contamination at MW-5, the site upgradient well, and MW-8, which was located in an area thought to be uncontaminated, were not anticipated and appear to be the results of high background levels.

According to the Federal Register notice, "The groundwater monitoring conducted annually from 1989 through 1994 did not detect Site-related contaminants above criteria established for the protection of groundwater. NJDEP and EPA have determined that the remedy implemented at the Site is protective of human health and the environment and that no further cleanup by responsible parties is appropriate. Hazardous substances on Site were cleaned up to levels that would allow for unlimited use and unrestricted exposure, therefore the five-year review requirement of section 121(c) of CERCLA, as amended, is not applicable."

MCLs Achieved? Yes

ICs in Place? None listed

Vapor Intrusion: Not mentioned

Long-Term Monitoring? None after 1994

14. Suffern Village Well Field

State: NY

Site Lead: unclear

Date Deleted from NPL: May 28, 1993

Date of Last Five-Year Review: September 30, 1998

COCs: Trichloroethane

RAOs: Protection of human health and the environment

Technologies Applied: Groundwater monitoring, excavation, natural attenuation

Reported Results: 1996 sampling indicated that the level of trichloroethane in the distribution water is well below the NYS Department of Health drinking water standard of 5 µg/L.

MCLs Achieved? Yes

ICs in Place? None listed

Vapor Intrusion: Not mentioned

Long-Term Monitoring? No

15. Tabernacle Drum Dump

State: NJ

Site Lead: unclear

Date Deleted from NPL: May 8, 2008

Date of Last Five-Year Review: September 10, 1998

COCs: 1,1,1-trichloroethane and 1,1-dichloroethene

RAOs: Groundwater concentrations of 26 µg/L for TCA and 2 µg/L for DCE

Technologies Applied: Pump and treat, reinjection into the ground

Reported Results: From May 1996 to July 1997 TCA ranged from 0.5 to 2.0 µg/L and DCE ranged from 0.2 to 1.0 µg/L.

MCLs Achieved? Yes

ICs in Place? None listed

Vapor Intrusion: Not mentioned

Long-Term Monitoring? No

<u>**16. Tronic Plating Co., Inc.**</u>
State: NY

Site Lead: State

Date Deleted from NPL: October 5, 2001

Date of Last Five-Year Review: N/A

COCs: 1,1-DCE, TCA, TCE, PCE, 1,2-DCE, antimony; beryllium; cadmium; total chromium; lead; nickel; silver; thallium; and, hexavalent chromium.

RAOs: Protection of human health and the environment. Risk assessment showed risks from exposure to groundwater/soil are within EPA's acceptable range.

Technologies Applied: Excavation to address localized plume of cadmium in the groundwater

Reported Results: No results given for after the cleanup.

MCLs Achieved? Yes, the groundwater does not exceed MCLs. Although groundwater sampling indicated some MCL exceedances for some contaminants, the distributions of these contaminants suggested either off-site sources or localized contamination and could not be associated with potential sources at the Site.

ICs in Place? None listed

Vapor Intrusion: The remote subsurface location of the contaminants precludes fugitive particulate or vapor emissions from the Site. Low levels of VOCs in the subsurface soil samples also indicate that the volatilization to ambient air will be negligible. During the remedial investigation, ambient air in the work space was monitored using the HNU analyzer. During the drilling of the industrial leaching pool, storm drain, and monitoring well borings, VOCs were detected only once at 0.5 ppm level.

Long-Term Monitoring? No

17. Upper Deerfield Township Sanitary Landfill

State: NJ

Site Lead: EPA

Date Deleted from NPL: June 10, 2000

Date of Last Five-Year Review: September 18, 2009

COCs: VOCs such as benzene, tricholorethene, and vinyl chloride; mercury

RAOs: Protection of human health and the environment. Baseline risk assessment found that under current and likely future land use scenarios, total carcinogenic risks are within EPA's acceptable risk range for the site.

Technologies Applied: Groundwater monitoring

Reported Results: 2009 well sampling results (contaminants above MCL highlighted in italics):

Contaminant	Max Detected Concentration (μg/L)	MCL (μg/L)
2-chlorophenol	*50.2*	*40*
1,4dichlorobenzene	65.5	75
n-nitroso-di-npropylamine	66.4	NS
1,2,4-Trichlorobenzene	69.4	70
2,4-Dinitrotoluene	67.8	NS
Pentachlorophenol	*2.58*	*1*
Aluminum	*2090*	*200*
Arsenic	*2070*	*10*
Barium	*2320*	*2000*
Cadmium	*52.6*	*5*
Cobalt	527	NS
Chromium	*211*	*100*
Iron	*1060*	*300*
Manganese	*555*	*50*
Lead	*529*	*15*
Antimony	*521*	*6*
Selenium	*1860*	*0.50*
Thallium	*1870*	2
Vanadium	531	NS

NS = no standard.

MCLs Achieved? No, but risk assessment determined that existing risk was within the acceptable range.

ICs in Place? Yes, an alternate water supply was provided.

Vapor Intrusion: Not mentioned

Long-Term Monitoring? Yes, the township will monitor groundwater until 2024.

18. V&M/Albaladejo

State: Puerto Rico

Site Lead: State

Date Deleted from NPL: October 22, 2001

Date of Last Five-Year Review: N/A

COCs: Metals

RAOs: Protection of human health and the environment. An investigation into the nature and extent of groundwater contamination was conducted in 1998 and 1999. Six monitoring wells were installed to determine if surface soil contamination had migrated into the groundwater. Four rounds of groundwater sampling indicated that no site-related contamination is present in the groundwater at the site.

Technologies Applied: Soil removal

Reported Results: There were no exceedances of inorganic analytes during the Round 4 (most recent) sampling.

MCLs Achieved? Not applicable because no groundwater contamination ever found

ICs in Place? None listed

Vapor Intrusion: Not mentioned

Long-Term Monitoring? No. EPA stated that the source of groundwater contamination was removed with the successful completion of the soils removal action. No future monitoring of the groundwater is anticipated.

19. Vestal Water Supply Well 4-2

State: NY

Site Lead: State

Date Deleted from NPL: September 30, 1999

Date of Last Five-Year Review: N/A

COCs: 1,1,1-trichloroethane and tricholorethene

RAOs: State groundwater standards

Technologies Applied: Wellhead treatment and excavation

Reported Results: Levels of contaminants in the influent (untreated groundwater) to Well 4-2 have declined to levels which are below drinking water standards.

MCLs Achieved? Yes

ICs in Place? No

Vapor Intrusion: Not mentioned

Long-Term Monitoring? No

20. Aladdin Plating

State: PA

Site Lead: EPA

Date Deleted from NPL: November 16, 2001

Date of Last Five-Year Review: September 14, 2009

COCs: Chromium and hexavalent chromium

RAOs: Protection of human health and the environment. A risk assessment performed during the RI/FS concluded that groundwater contamination posed no threat to human health or the environment.

Technologies Applied: Groundwater monitoring to watch for migration of contamination into drinking water aquifers

Reported Results: The maximum monitoring-well concentrations for chromium appear to be holding steady through successive sampling events at around 40 ppm total chromium (MCL 0.1 ppm). No evidence of migration of chemicals to residential wells has been detected.

MCLs Achieved? No, but risk assessment determined risk to be in the acceptable range

ICs in Place? Yes, a Unilateral Administrative Order which requires the property owner to file a Notice of Use Restriction with the county recorder of deeds.

Vapor Intrusion: Not applicable

Long-Term Monitoring? Yes, groundwater monitoring until 2023

21. Berks Landfill

State: PA

Site Lead: EPA

Date Deleted from NPL: November 14, 2008

Date of Last Five-Year Review: July 30, 2010

COCs: VOCs including vinyl chloride, trichloroethene, and cis-1,2-dichloroethene; metals including aluminum, iron, and manganese

RAOs: Protection of human health and the environment

Technologies Applied: Groundwater monitoring, leachate collection system and landfill caps

Reported Results: Maximum concentrations in onsite groundwater wells from 2005 to 2009: DCE 470 μg/L, TCE 480 μg/L, vinyl chloride 65 μg/L, PCE 36 μg/L, aluminum 6,690 μg/L, iron 6,100 μg/L, and manganese 6,100 μg/L.

MCLs Achieved? No

ICs in Place? Yes, there is a prohibition of consumption of onsite groundwater

Vapor Intrusion: Not considered a pathway of concern

Long-Term Monitoring? Yes

22. Brodhead Creek

State: PA

Site Lead: EPA

Date Deleted from NPL: July 23, 2001

Date of Last Five-Year Review: May 22, 2009

COCs: The polycyclicaromatic hydrocarbons (PAHs) benzo(a)anthracene, benzo(b)fluoranthene, benzo(k)fluoranthene, chrysene, dibenz(a,h)anthracene, and indenopyrene.

RAOs: Protection of human health and the environment. EPA performed a risk assessment and determined there was no significant risk because groundwater from the site was not being used. A TI Zone was established.

Technologies Applied: Excavation

Reported Results: Groundwater at the Site remains significantly contaminated

MCLs Achieved? No, as a TI zone was established

ICs in Place? Yes, institutional controls restrict the use of groundwater at the Site and prohibit excavation unless prior written approval is provided by EPA, PADEP, and the property owner.

Vapor Intrusion: Vapor intrusion would be a potential concern to be quantitatively evaluated if a building were constructed on-site. However, it is extremely unlikely that a building, commercial or residential, will ever be constructed on-site due to the makeup of the land (floodplains).

Long-Term Monitoring? Yes

23. Matthews Electroplating

State: VA

Site Lead: EPA

Date Deleted from NPL: January 19, 1989

Date of Last Five-Year Review: June 21, 2004

COCs: Chromium

RAOs: Reduce chromium concentrations to below the MCL of 100 μg/L

Technologies Applied: None

Reported Results: Total and hexavalent chromium concentrations had dropped below 50 μg/L in all residential wells by February 1992.

MCLs Achieved? Yes

ICs in Place? Yes, alternate drinking water supply was provided (community water supply line was extended to a new water line).

Vapor Intrusion: Not applicable

Long-Term Monitoring? No. EPA has determined that hazardous substances, pollutants, or contaminants no longer remain at the site above levels that allow for unlimited use and unrestricted exposure. Therefore, no further five-year reviews are needed for this site.

24. Mid-Atlantic Wood Preservers, Inc.

State: MD

Site Lead: EPA

Date Deleted from NPL: July 18, 2000

Date of Last Five-Year Review: September 26, 2008

COCs: Chromium

RAOs: Reduce chromium concentrations to below the MCL of 100 μg/L

Technologies Applied: Groundwater monitoring, natural attenuation

Reported Results: The groundwater monitoring program demonstrated to EPA's and MDE's satisfaction that groundwater met water quality standards, and the program was discontinued prior to the Site's deletion from the NPL in 2000.

MCLs Achieved? Yes

ICs in Place? Yes, deed restrictions

Vapor Intrusion: Not applicable

Long-Term Monitoring? No

25. New Castle Spill

State: DE

Site Lead: State

Date Deleted from NPL: June 12, 1996

Date of Last Five-Year Review: June 12, 1996

COCs: Tris

RAOs: Protection of human health and the environment. EPA developed a health-based drinking water cleanup level of 4.4 mg/L for Tris.

Technologies Applied: Groundwater monitoring and natural attenuation

Reported Results: The fact sheet says that Tris, a flame retardant, was detected in soils near the drum storage area and in shallow groundwater beneath the site. The ROD states that there would be a five-year follow-up, but the document listed as the five-year review on the CERCLIS website is only a link to the Federal Register notice of the site being deleted from the NPL.

MCLs Achieved? Yes, by 1996 data showed that Tris levels in the shallow aquifer had reached the established cleanup goals.

ICs in Place? Yes, restriction of well installation in the Columbia aquifer (in the vicinity of the site)

Vapor Intrusion: Not mentioned

Long-Term Monitoring? No

26. Southern Maryland Wood Treating

State: MD

Site Lead: EPA

Date Deleted from NPL: April 5, 2005

Date of Last Five-Year Review: September 30, 1999

COCs: Toluene, naphthalene, benzene, 2-methylnaphthalene, ethylbenzene, acenaphthylene, styrene acenaphthene, total xylenes, fluorene, phenol, phenanthrene, 2-methylphenol, fluoranthene, 4-methylphenol pyrene, 2,4-dimethylphenol, benzo(g,h,i)perylene, pentachlorophenol

RAOs: Protect groundwater as a current or potential future drinking water supply by containing or treating subsurface soil that contains in excess of 1.0 ppm B(a)P equivalence

Technologies Applied: Sheet pile wall, excavation, pump and treat

Reported Results: No results reported

MCLs Achieved? No, but containment was the goal. In December 2000 (the year after the most recent five-year review) the equipment that was used to treat the soils was decontaminated, dismantled, and removed from the site. Water Treatment Plant No. 2 was also dismantled and removed within this timeframe. Groundwater wells placed around the site have been monitored from October 2000 until September 2002 and have continued to confirm that the cleanup was a "success." All wells, including the 600 ft water production well, have been removed and sealed.

ICs in Place? Yes, restricted access. The landfill is secured by a fence with a guard posted at the front gate on a 24-hour basis.

Vapor Intrusion: Not mentioned

Long-Term Monitoring: Yes

27. Sussex County Landfill No. 5

State: DE

Site Lead: EPA

Date Deleted from NPL: September 28, 2001

Date of Last Five-Year Review: February 11, 2005

COCs: Benzene, vinyl chloride, 1,2-dichloropropane, 1,4-dichlorobenzene, tricholorethene

RAOs: As per the most recent five-year review, the goal was to reduce contaminants to the following MCLs: benzene 5 μg/L, vinyl chloride 2 μg/L, 1,2-dichloropropane 5 μg/L, 1,4-dichlorobenzene 75 μg/L, and tricholorethene 5 μg/L. MCLs were not included in the original RAO.

Technologies Applied: None. "In October 1993, Sussex County completed the RI which included EPA-prepared Baseline Human Health Risk Assessment and Ecological Risk Assessment. The Risk Assessment indicated that very low levels of contaminants of concern existed in the groundwater which translated into correspondingly low risk levels at the Site. Based on the results of the RI and the Risk Assessments, EPA determined that a feasibility study was not necessary to evaluate remedial alternatives." (http://www.epa.gov/superfund/sites/rods/fulltext/r0295254.pdf).

Reported Results: As of November 2003, all COCs were detected below their MCL except benzene which was found at concentrations of 8.2 and 6.2 μg/L. However, benzene was not detected above its MCL in any of the residential wells currently used for drinking water.

MCLs Achieved? No, benzene concentration is still above the MCL.

ICs in Place? Yes, provision of an alternate drinking water supply (bottled water and a water purification system), restriction of well installation, and deed restrictions. Residents have since been hooked up to the county water supply and domestic wells were closed.

Vapor Intrusion: Not mentioned

Long-Term Monitoring: Yes

28. BMI-Textron

State: FL

Site Lead: EPA

Date Deleted from NPL: September 19, 2002

Date of Last Five-Year Review: August 25, 2000

COCs: Arsenic, sodium, cyanide and fluoride

RAOs: Drinking water standards (MCLs are supposedly in "table 2" of the ROD, but there appear to be no tables in the document posted).

Technologies Applied: Groundwater monitoring, natural attenuation and soil excavation.

Reported Results: In January 2000 cyanide levels (both total and amenable) were below drinking water standards. However, fluoride was not sampled, as recommended by both agencies. No other results are reported.

MCLs Achieved? Yes. Sampling at the Site continued until MCLs were met for all COCs at all eight wells. The last sampling event occurred in July 2000.

ICs in Place? Yes. Permits to dig wells must be obtained from FL DEP

Vapor Intrusion: Not mentioned

Long-Term Monitoring? No

29. Cedartown Industries, Inc.

State: GA

Site Lead: EPA

Date Deleted from NPL: September 19, 2006

Date of Last Five-Year Review: August 24, 2006

COCs: Cadmium, lead, antimony, arsenic, and beryllium

RAOs: Reduce contaminants to the following MCLs: cadmium 5 µg/L, lead 15 µg/L (MCGL is 0), antimony 6 µg/L, arsenic 10 µg/L (MCGL is 0), and beryllium 4 µg/L

Technologies Applied: Excavation/GW monitoring

Reported Results: With the exception of an apparently spurious result in February 2005, cadmium concentrations in the collected samples have not exceeded the MCL of 5.0 µg/L since the first quarter of 1998. In addition, the results of the groundwater monitoring program have verified that the treatment of soils has not released significant concentrations of the other COCs (lead, antimony, arsenic, and beryllium) to the groundwater.

MCLs Achieved? Yes

ICs in Place? Yes, deed notation

Vapor Intrusion: Not applicable

Long-Term Monitoring? No. The five-year review states that since the objectives of the groundwater remedial action have been achieved, long-term groundwater monitoring should be discontinued.

30. Cedartown Municipal Landfill

State: GA

Site Lead: EPA

Date Deleted from NPL: March 10, 1999

Date of Last Five-Year Review: September 21, 2006

COCs: Beryllium, cadmium, chromium, lead, manganese

RAOs: Reduce contaminants to the following concentrations: beryllium 4 μg/L, cadmium 5 μg/L, chromium 100 μg/L, lead 15 μg/L (MCGL 0), and manganese 50 μg/L (there is no MCL for manganese, this is a National Secondary Drinking Water Regulation).

Technologies Applied: Groundwater monitoring and natural attenuation

Reported Results: July 2006 results show all concentrations below the MCLs except chromium (MCL 100 μg/L) in one well (130 μg/L) and manganese (MCL 840 μg/L) in two wells (1430 and 967 μg/L).

MCLs Achieved? No, chromium and manganese are still above MCLs in some locations

ICs in Place? Yes, record notices and deed, zoning, and land use restrictions

Vapor Intrusion: Not applicable

Long-Term Monitoring? Yes

31. Davie Landfill

State: FL

Site Lead: EPA

Date Deleted from NPL: August 21, 2006

Date of Last Five-Year Review: March 16, 2011

COCs: Vinyl chloride and antimony

RAOs: Vinyl chloride goal of 1 μg/L

Technologies Applied: Groundwater monitoring and natural attenuation

Reported Results: Groundwater monitoring data from the past five years (2005-2010) have shown slight exceedances of the vinyl chloride cleanup goal (1 μg/L).

MCLs Achieved? No, vinyl chloride concentration is still above the cleanup goal.

ICs in Place? Yes, provision of an alternate water supply (residents were connected to the public water line)

Vapor Intrusion: Not mentioned

Long-Term Monitoring? Yes

32. Dubose Oil Products Co.

State: FL

Site Lead: EPA

Date Deleted from NPL: August 4, 2004

Date of Last Five-Year Review: September 25, 2008

COCs: VOCs

RAOs: Reduce COC concentrations below the following MCLs: benzene 1 μg/L, 1,1-dichloroethene 7 μg/L, pentachlorophenol 30 μg/L, PAHs 10 μg/L, trichloroethene 2 μg/L, and xylenes 50 μg/L.

Technologies Applied: Excavation and bioremediation of soils; draining and backfilling of on-site ponds; removal of on-site structures; site grading and revegetation; installation of surface water runoff controls; and groundwater monitoring.

Reported Results: No remaining hazardous contaminants above levels that allow for unlimited use and unrestricted exposure.

MCLs Achieved? Yes

ICs in Place? Yes, the Site is in a Florida Delineated Area where groundwater well drilling is restricted.

Vapor Intrusion: Not mentioned

Long-Term Monitoring? No

33. Gold Coast Oil Corp.

State: FL

Site Lead: EPA

Date Deleted from NPL: October 9, 1996

Date of Last Five-Year Review: September 20, 2001

COCs: VOCs

RAOs: Reduce COC concentrations below the following MCLs: 1,1-dichloroethane 50 µg/L, t-1,2-dichloroethane 700 µg/L, methylene chloride 50 µg/L, tetrachloroethene 7 µg/L, toluene 3400 µg/L, and tricholorethene 30 µg/L.

Technologies Applied: Excavation and a groundwater recovery, treatment, and disposal system

Reported Results: October 1996 sampling showed all COCs below MCLs

MCLs Achieved? Yes

ICs in Place? None listed

Vapor Intrusion: Not mentioned

Long-Term Monitoring? No

34. Independent Nail Co.

State: SC

Site Lead: State

Date Deleted from NPL: October 9, 1996

Date of Last Five-Year Review: April 5, 2000

COCs: Chromium, cyanide, and zinc

RAOs: Protection of human health and the environment. Risk assessment found that contaminant levels in the groundwater presented no imminent or substantial threat to human health or the environment.

Technologies Applied: Excavation and groundwater monitoring

Reported Results: "Chromium, cyanide, and zinc have been detected in groundwater at the Independent Nail Company Site at maximum concentrations of 0.058 mg/L, 0.110 mg/L and 0.098 mg/L, respectively. A comparison of these contaminant concentrations with drinking water MCLs, SMCLs, and health advisories indicates that total chromium is the only groundwater contaminant at the Independent Nail Company Site which exceeds any of these standards or criteria." (http://www.epa.gov/superfund/sites/rods/fulltext/r0488040.pdf).

MCLs Achieved? No, total chromium is still above the MCL. Risk assessment found risk to fall within an acceptable range.

ICs in Place? None listed

Vapor Intrusion: Not applicable

Long-Term Monitoring? Yes

35. Lee's Lane Landfill

State: KY

Site Lead: EPA

Date Deleted from NPL: April 25, 1996

Date of Last Five-Year Review: September 25, 2008

COCs: Lead, arsenic, benzene and chromium

RAOs: Reduce contaminants to the following standards: lead 15 µg/L (action level, like an MCL), arsenic 10 µg/L (the MCL), benzene 5 µg/L (the MCL), and chromium 100 µg/L (the MCL).

Technologies Applied: Groundwater and ambient air monitoring

Reported Results: December 2007: Arsenic 0.011 and 0.033 mg/L, chromium <0.001 mg/L, and lead <0.005 mg/L. Analytical results for benzene have not been reported for groundwater at the site during this review period.

MCLs Achieved? No, arsenic is still above MCL. Lead results were unclear, and there are no benzene results in the most recent five-year review.

ICs in Place? Yes, an alternate water supply was provided, and there is access control via fencing and gating.

Vapor Intrusion: Not mentioned

Long-Term Monitoring? Yes

36. Lewisburg Dump

State: TN

Site Lead: EPA

Date Deleted from NPL: February 21, 1996

Date of Last Five-Year Review: September 19, 2007

COCs: Aluminum, barium, iron, manganese, methylene chloride, and bis(2-ethylhexyl)phthalate

RAOs: Meet the following MCLs: Aluminum 50-200 μg/L (secondary drinking water standard), barium 200 μg/L, iron 300 μg/L (secondary standard), manganese 50 μg/L (secondary drinking water standard), methylene chloride 5 μg/L, bis(2-ethylhexyl)phthalate 6 μg/L

Technologies Applied: Groundwater monitoring

Reported Results: No organic compounds were detected during Nov. 2005 sampling. Inorganics were detected, but only lead (120 μg/L) exceeded the MCL for drinking water, which is 15 μg/L.

MCLs Achieved? No, lead is still above the MCL.

ICs in Place? Yes, there are deed restrictions, security fencing, and prevention of using land for residential construction.

Vapor Intrusion: As long as the institutional controls (which prevent the use of property for housing and prohibit activities that could impair the landfill cover) remain in effect, the site should remain stable and free of soil vapor intrusion or any other future contaminant migration in the future.

Long-Term Monitoring? Yes

<u>37. Palmetto Recycling, Inc.</u>

State: SC

Site Lead: EPA

Date Deleted from NPL: October 13, 2000

Date of Last Five-Year Review: June 7, 2004

COCs: Lead

RAOs: Reduce lead concentrations to below the MCL of 15 µg/L

Technologies Applied: Groundwater monitoring/excavation of soils

Reported Results: 2004 groundwater monitoring indicates that the groundwater concentrations for lead are below 15 µg/L.

MCLs Achieved? Yes

ICs in Place? None listed

Vapor Intrusion: Not applicable

Long-Term Monitoring? No

38. Rochester Property

State: SC

Site Lead: EPA

Date Deleted from NPL: October 9, 2007

Date of Last Five-Year Review: February 25, 2005

COCs: TCE, bis(2-ethylhexyl)phthalate, and manganese

RAOs: Reduce concentrations of the following COCs: trichloroethene 5 µg/L (the MCL), manganese 180 µg/L (no MCL, but the secondary drinking water regulation is 50 µg/L), and bis(2-ethylhexyl) phthalate 6 µg/L (the MCL). A chromium remedial goal was been removed from the RAO due to new IRIS information regarding its hazard quotient posted between the first and second five-year reviews.

Technologies Applied: *In situ* air sparging system

Reported Results: All wells and performance monitoring points were sampled in November 2006 (the final confirmation sampling event) and contaminant concentrations were found to be below the cleanup goals for the site.

MCLs Achieved? Yes

ICs in Place? Yes. A deed restriction is in place in order to deter the installation of a private well.

Vapor Intrusion: Not mentioned

Long-Term Monitoring? To be determined. According to the most recent five-year review, EPA and the South Carolina Department of Health and Environmental Control are currently reviewing the most recent groundwater monitoring report in order to determine the appropriate response with regard to future system operations and groundwater monitoring.

39. Sixty-Second Street Dump

State: FL

Site Lead: EPA

Date Deleted from NPL: October 1, 1999

Date of Last Five-Year Review: September 21, 2009

COCs: Cadmium, chromium, lead

RAOs: Reduce concentrations of chromium to 100 μg/L, lead to 15 μg/L, and cadmium to 5 μg/L. Note that standards for chromium and cadmium changed in 2009 from the original ROD goals.

Technologies Applied: Pump and treat

Reported Results: Jan 2009 sampling results: chromium 1.8 μg/L, cadmium $<$ 0.15 μg/L, and lead 0.16 μg/L.

MCLs Achieved? Yes

ICs in Place? Yes. The site is within a Florida Ground Water Delineation Area so there is a restriction on new drinking water wells.

Vapor Intrusion: Not mentioned

Long-Term Monitoring? Yes

40. Standard Auto Bumper Corp.

State: FL

Site Lead: EPA

Date Deleted from NPL: October 29, 2007

Date of Last Five-Year Review: March 15, 2011

COCs: Chromium, nickel, and thallium

RAOs: Restore groundwater to health-based levels (chromium 100 µg/L, nickel 100 µg/L, and thallium 2 µg/L). (There is no federal MCL for nickel, so the cleanup standard for nickel is likely from FL.)

Technologies Applied: Groundwater monitoring and natural attenuation

Reported Results: The latest groundwater monitoring data show that nickel, chromium, and thallium were not detected in any of the three groundwater wells above the MCL. Groundwater monitoring ended in 2001 in accordance with the ROD.

MCLs Achieved? Yes

ICs in Place? Yes, there is an alert system that notifies the Miami-Dade County DERM of any construction permit requests for the locations adjacent to the Site where soil contamination remains below the surface. Owners must also contact EPA and other authorities prior to any construction, excavation, or removal of any part of a building or road.

Vapor Intrusion: Not mentioned

Long-Term Monitoring? No

41. Woodbury Chemical Co. (Princeton Plant)

State: FL

Site Lead: State

Date Deleted from NPL: November 27, 1995

Date of Last Five-Year Review: N/A

COCs: Pesticides (aldrin, chlordane, DDT, dieldrin, and heptachlor epoxide) and metals (arsenic and chromium) have been detected at low levels in off-site groundwater to the south and east. Nitrates detected in groundwater have been linked to the heavy use of fertilizers in the area surrounding the site.

RAOs: Protection of human health and the environment. Exposure and toxicity assessment showed no unacceptable risk.

Technologies Applied: Soil excavation and groundwater monitoring

Reported Results: None listed. According to the Notice for Site Deletion, "Confirmational monitoring of groundwater demonstrates that no significant risk to public health or the environment is posed by the Site."

MCLs Achieved? No, but risk assessment showed that the current risks are acceptable

ICs in Place? None listed

Vapor Intrusion: Not mentioned

Long-Term Monitoring? No. Groundwater monitoring was conducted for one year from the signing of the ROD (1992).

42. Agate Lake Scrapyard

State: MN

Site Lead: State

Date Deleted from NPL: August 1, 1997

Date of Last Five-Year Review: August 24, 2001

COCs: PCE

RAOs: Protection of human health and the environment; no explicit goals for reducing contaminant concentrations are available.

Technologies Applied: Groundwater monitoring and natural attenuation

Reported Results: In October of 1996, analytical results showed that only two wells had detectable concentrations of PCE. The concentration at OW6 was 6 µg/L, which is below the Health Risk Limit (HRL) of 7 µg/L, and they have shown a steady decrease since January 1992. The PCE concentrations at OW7 in October of 1997 were 3.6 µg/L. This further supports the conclusion that there is not a significant source of PCE impacting the groundwater and that natural attenuation has reduced the PCE concentrations.

MCLs Achieved? Not applicable, as such goals were never set. Reports discuss meeting "health based goals" but these are not specified.

ICs in Place? Yes, deed restrictions

Vapor Intrusion: Not mentioned

Long-Term Monitoring? No

43. Alsco Anaconda

State: OH

Site Lead: EPA

Date Deleted from NPL: November 2001

Date of Last Five-Year Review: June 23, 1997

COCs: Metals, chlorinated benzene compounds, semi-volatile organics, and cyanide.

RAOs: "Site-related contaminants with an existing MCL shall be reduced to a concentration at or below the MCL. Carcinogenic site-related contaminants shall be reduced to levels that pose a cumulative carcinogenic risk of no greater than 1×10^{-6}. Concentrations of noncarcinogenic site-related contaminants shall be reduced to levels that pose a cumulative HI no greater than 1 for any specific toxicological category." (http://www.epa.gov/superfund/sites/npl/d011105.htm).

Technologies Applied: Groundwater monitoring

Reported Results: None listed

MCLs Achieved? Yes. "With the exception of cyanide and arsenic, cleanup criteria for contaminants established for this site have been met since 1999. Monitoring in May, August, and October 2000 met cleanup objectives for cyanide and arsenic." (http://www.epa.gov/superfund/sites/npl/d011105.htm).

ICs in Place? Yes, deed restrictions

Vapor Intrusion: Not mentioned

Long-Term Monitoring? No, since contaminant levels are below those allowing for unlimited use and unrestricted exposure.

44. Avenue "E" Groundwater Contamination

State: MI

Site Lead: EPA

Date Deleted from NPL: March 20, 2007

Date of Last Five-Year Review: September 28, 2005

COCs: Benzene, toluene, acetone, tricholorethene, carbon tetrachloride, and other organic compounds

RAOs: Prevent the groundwater transport of contaminants from the site and treat the water extracted to remove the fuel-related contaminants. Operate the system until contaminant levels are at or below 5 µg/L for six consecutive months.

Technologies Applied: Pump and treat, groundwater monitoring

Reported Results: "All groundwater contaminants have reached the cleanup standards specified in the 1987 agreement between the Coast Guard and MDNR." (http://www.epa.gov/superfund/sites/fiveyear/f05-05024.pdf).

MCLs Achieved? Yes

ICs in Place? Because the groundwater was to be remediated to unrestricted use, there were no provisions for institutional controls at this site.

Vapor Intrusion "Currently soil vapor intrusion is not considered a possible problem at this site. Down gradient monitoring wells placed along Avenue E in the residential area where the plume was traced had no detectable levels of contaminants of concern associated with the USCG plume." (http://www.epa.gov/superfund/sites/fiveyear/f05-05024.pdf).

Long-Term Monitoring? No, monitoring ended in October 2005.

45. Berlin & Farro

State: MI

Site Lead: State

Date Deleted from NPL: June 24, 1998

Date of Last Five-Year Review: N/A

COCs: Methylene chloride, 1,2-dichlorethane, vinyl chloride, 1,1-dichlorethene, benzene, and trichloroethene

RAOs: Reduce COCs to the following cleanup levels: methylene chloride 5 μg/L, 1,2-dichlorethane 0.4 μg/L (1 μg/L if analytic procedure does not allow for lower MDL), vinyl chloride 0.02 μg/L (1 μg/L if analytic procedure does not allow for lower MDL), 1,1-dichlorethene 7 μg/L, benzene 1 μg/L, and trichloroethene 3 μg/L. Some of these values are the federal MCLs and some are lower.

Technologies Applied: Pump and treat, soil excavation, and groundwater monitoring

Reported Results: As documented by the June 27, 1996, Remedial Action Report and the September 18, 1996, Superfund Site Close Out Report, confirmatory sampling verified that all soil, sediment, and groundwater cleanup standards were met at the Site and that all cleanup actions specified in the amended ROD have been implemented.

MCLs Achieved? Yes

ICs in Place? Yes. There are deed restrictions to prevent installation of drinking water wells during remediation.

Vapor Intrusion: Not mentioned

Long-Term Monitoring? No

46. Dakhue Sanitary Landfill

State: MN

Site Lead: State

Date Deleted from NPL: July 1995

Date of Last Five-Year Review: February 2, 2009

COCs: VOCs, chloroform, and heavy metals, including cadmium and lead.

RAOs: Reduce contaminants to the following MCLs: chloroform 70 µg/L (MCLG), cadmium 5 µg/L, and lead 15 µg/L

Technologies Applied: Groundwater monitoring, source control on the landfill OU

Reported Results: No results listed for COCs

MCLs Achieved? Unknown. The most recent five-year review cannot be downloaded from CERCLIS. As of the second five-year review, they had not met MCLs, but concentrations were trending down.

ICs in Place? Yes, restriction of well development

Vapor Intrusion: Not mentioned

Long-Term Monitoring? Yes

47. Fadrowski Drum Disposal

State: WI

Site Lead: EPA

Date Deleted from NPL: September 6, 2005

Date of Last Five-Year Review: July 30, 2008

COCs: Metals (different metals are listed in different locations in the documents, none of which are ever specifically identified as the contaminants of concern)

RAOs: Reduce contaminant concentrations below 1988 NR 140 Wisconsin groundwater standards [Preventative Action Limits (PALs), Enforcement Standards (ESs), or ACLs]. These cleanup goals are more stringent than federal MCLs.

Technologies Applied: Groundwater monitoring, source control, and natural attenuation

Reported Results: As of 2008 many monitored compounds had declined in concentration via the process of natural attenuation such that they met the cleanup criteria (1988 Chapter NR 140 PALs, ESs or established ACLs).

MCLs Achieved? No. Fluoride, iron, and manganese remain about their PALs; however, due to their natural occurrence in the groundwater, achieving PALs for these three constituents via natural attenuation or related methods is neither technically nor economically feasible.

ICs in Place? Yes, restriction of future land and groundwater use and prohibition of future development of the Site within the Waste Management Boundary.

Vapor Intrusion: Not applicable

Long-Term Monitoring? Yes, annual sampling for an unspecified amount of time.

48. Laskin/Poplar Oil Co.

State: OH

Site Lead: EPA

Date Deleted from NPL: September 5, 2000

Date of Last Five-Year Review: May 15, 2009

COCs: PCBs, PAHs, SVOCs, VOCs, and metals

RAOs: Protection of human health and the environment

Technologies Applied: Excavation, groundwater diversion trenches, and monitoring

Reported Results: "A diversion trench was constructed up-gradient of the capped area, in order to intercept all groundwater flow in the shallow aquifer moving northward toward the Site, and a drain in the trench conducts the intercepted flow directly to Cemetery Creek. Treatment of the diverted water was not required because upgradient groundwater is not contaminated. Although the trench and cap effectively de-watered the Site, groundwater and surface water monitoring were initially conducted as part of the Inspection, Maintenance, and Monitoring Plan (IMMP) because hazardous substances were present on-site. Currently, groundwater levels are monitored in order to determine if the Site remains dewatered." (http://www.epa.gov/region5superfund/fiveyear/reviews_pdf/ohio/laskin_poplar_oil.pdf).

The results of water level measurements conducted since the 2004 five-year review show that water levels have remained below the level of the unweathered shale, and thus no groundwater or surface water samples were required to be collected (http://www.epa.gov/region5superfund/fiveyear/reviews_pdf/ohio/laskin_poplar_oil.pdf).

MCLs Achieved? Probably not in the groundwater directly below the site, but yes in the groundwater discharging to the Cemetery Creek, which is a likely point of compliance.

ICs in Place? Yes, there are restrictions on groundwater use for drinking water purposes. The ROD indicated that after the site is de-watered, there will be essentially no groundwater available for any purpose.

Vapor Intrusion: Not mentioned

Long-Term Monitoring? No

49. Northern Engraving Co.

State: WI

Site Lead: EPA

Date Deleted from NPL: October 1997

Date of Last Five-Year Review: April 6, 2010

COCs: Copper, fluoride, nickel, zinc, 1,1- dichloroethylene, tricholorethene, and vinyl chloride

RAOs: Reduce contaminants below the following ACLs: fluoride 4,800 µg/L, copper 1,000 µg/L, nickel 644 µg/L, zinc 5,000 µg/L, tricholorethene 40 µg/L, vinyl chloride 10 µg/L, and 1, 1-dichloroethylene 10 µg/L. These numbers are much higher than MCLs.

Technologies Applied: Source control via a RCRA cover installed atop the lagoon and groundwater monitoring

Reported Results: The concentrations of the indicator contaminants were below the ACLs during the second five-year monitoring period.

MCLs Achieved? No, but MCLs were not the goal; ACLs were.

ICs in Place? Yes, there is a deed affidavit, and the most recent review recommended deed restrictions.

Vapor Intrusion: Not mentioned

Long-Term Monitoring? No. Because groundwater remediation goals were met, groundwater monitoring was discontinued in 2000.

50. Southside Sanitary Landfill

State: IN

Site Lead: State

Date Deleted from NPL: July 3, 1997

Date of Last Five-Year Review: September 20, 2005

COCs: Heavy metals including arsenic, chromium, cadmium, and nickel

RAOs: Protection of human health and the environment

Technologies Applied: Slurry wall and leachate collection system to isolate the groundwater beneath the landfill from the surrounding groundwater

Reported Results: A review of the water level data indicated that the slurry wall performance is within acceptable limits. The analytical results from 11 monitoring wells located outside of the slurry wall showed that chemical concentrations are also within acceptable limits, except in MW-15R.

MCLs Achieved? Yes, outside the slurry wall, but probably not within the wall.

ICs in Place? No. However, within six months of the most recent five-year review report (2005), the IN Department of Environmental Management will develop and submit an IC Plan that will include a schedule for implementing an easement/restrictive covenant, as well as an evaluation of the need for any additional institutional controls.

Vapor Intrusion: Not applicable

Long-Term Monitoring? No

51. Tri-State Plating

State: IN

Site Lead: State

Date Deleted from NPL: July 14, 1997

Date of Last Five-Year Review: N/A

COCs: Chromium

RAOs: Reduce contamination of chromium to meet State and Federal standards (Could not find IN standard but the federal MCL is 100 µg/L)

Technologies Applied: Pump and treat, and groundwater monitoring

Reported Results: Groundwater remediation goals were met in 1995

MCLs Achieved? Yes

ICs in Place? Yes. Fencing was built around portions of the site.

Vapor Intrusion: Not applicable

Long-Term Monitoring? No

52. University of Minnesota (Rosemount Research Center) OU1

State: MN

Site Lead: State

Date Deleted from NPL: February 6, 2001

Date of Last Five-Year Review: June 15, 2007

COCs: Chloroform and TCE

RAOs: Reduce chloroform contamination below 60 µg/L (the MCLG is 70 µg/L) and TCE below 5 µg/L

Technologies Applied: Pump and treat

Reported Results: Groundwater samples from all five monitoring wells showed concentrations of chloroform ranging from 2.3 to 23 µg/L, which was below than the MDH drinking water criteria of 60 µg/L. Trichloroethene was detected (in well MW-23D) at a concentration of 2.6 µg/L, less than the federal MCL of 5 µg/L. Based on the groundwater sampling results, the MPCA determined that the groundwater was potable and therefore, no further groundwater sampling was necessary.

MCLs Achieved? Yes

ICs in Place? Yes, a declaration of restrictions and covenants and affidavit concerning real property contaminated with hazardous substances documents were filed.

Vapor Intrusion: Not mentioned

Long-Term Monitoring? No

53. Waste, Inc., Landfill

State: IN

Site Lead: EPA

Date Deleted from NPL: August 2008

Date of Last Five-Year Review: June 27, 2011

COCs: Benzene, PAHs, PCBs, arsenic, manganese, chromium, cadmium, mercury, antimony

RAOs: Prevent migration of groundwater contamination

Technologies Applied: Groundwater monitoring; landfill cap

Reported Results: Average contaminant concentrations in landfill leachate are lower than the previous reporting period with limited detections and no exceedances of any associated MCLs. This continues to demonstrate that the installation of the multi-layer cap has effectively reduced the infiltration of rainwater through the landfill contents and reduced the leachate concentrations at the site.

MCLs Achieved? No, since the goal was only to prevent migration of the contamination.

ICs in Place? Yes, deed restrictions

Vapor Intrusion: Not mentioned

Long-Term Monitoring? Yes

54. Windom Dump

State: MN

Site Lead: State

Date Deleted from NPL: 2000

Date of Last Five-Year Review: April 23, 2010

COCs: Cis-1,2-dichloroethene and vinyl chloride

RAOs: Reduce contamination below action levels (MCL of 70 µg/L for cis-1,2-dichloroethene and 2 µg/L for vinyl chloride)

Technologies Applied: Pump and treat, groundwater monitoring

Reported Results: Response actions have successfully reduced contaminant concentrations below action levels.

MCLs Achieved? Yes

ICs in Place? Yes, deed restrictions

Vapor Intrusion: Not mentioned

Long-Term Monitoring? Yes

55. Bailey Waste Disposal

State: TX

Site Lead: EPA

Date Deleted from NPL: October 15, 2007

Date of Last Five-Year Review: September 7, 2010

COCs: Organic compounds and heavy metals

RAOs: Protection of human health and the environment, but the ROD did not contain a groundwater remedy.

Technologies Applied: Pump and treat for the short term to remove groundwater that might rise due to waste consolidation.

Reported Results: No results provided. The most recent review states: "The original ROD for the Site did not contain a ground water remedy, as the RI concluded that the Site had no impact on drinking water, and 'in the unlikely event that site constituents were to migrate via a ground water pathway, it would take more than 800 years for them to reach potable ground water. . . [and] shallow ground water beneath and adjacent to the site is saline and not suitable for human consumption'. Moreover, the area is subject to tidal flow (i.e., significant daily flow in and out of the area), so ground water discharge to surface water has not been found to be an issue. The major components of the amended ground water remedy included no long-term ground water treatment, but in the short-term (i.e. during construction of the cap), a 'consolidation water collection layer [was installed] to intercept and remove ground water that rises due to the consolidation of the waste'. The collected water was then pumped to a wastewater holding tank, treated to the discharge limits in effect during the remedy implementation, and discharged." (http://www.epa.gov/earth1r6/6sf/texas/bailey/Bailey_5-Year_Review_Final_w_Signature_Pages.pdf)

MCLs Achieved? Not applicable, as the groundwater was never found to be contaminated.

ICs in Place? Yes, a deed notice

Vapor Intrusion: Not mentioned

Long-Term Monitoring: No

56. Cleveland Mill

State: NM

Site Lead: EPA

Date Deleted from NPL: July 23, 2001

Date of Last Five-Year Review: August 20, 2007

COCs: Arsenic, beryllium, cadmium, lead, and zinc

RAOs: Protection of human health and the environment; no obvious groundwater-specific goal other than to prevent contamination.

Technologies Applied: Soil excavation in the original ROD. Because EPA anticipated that associated groundwater contamination would naturally attenuate once the source was removed, the 1993 ROD did not include a groundwater remedy.

Reported Results: All the monitoring wells and residential wells used to gather RI data were below MCLs and New Mexico Water Quality Control Commission (NMWQCC) standards. (But this has nothing to do with the remedy or the post-deletion monitoring.

MCLs Achieved? Not applicable, as these were not remedial goals. It could be argued that this was not a contaminated groundwater site.

ICs in Place? Yes, restrictive covenants

Vapor Intrusion: Not applicable

Long-Term Monitoring? Yes

57. Double Eagle Refinery Co.

State: OK

Site Lead: EPA/State

Date Deleted from NPL: August 21, 2008

Date of Last Five-Year Review: May 15, 2007 (combined review done for Fourth St. and Double Eagle sites)

COCs: PAH, chlorinated hydrocarbons and PCB, alkyl benzenes, ketones, lead (primary COC), arsenic, and antimony

RAOs: Ensure that contaminants do not migrate deeper, or to a receptor point off-site, and determine if an off-site source of contamination exists. The plan for the site seems to indicate that no one will ever drink the groundwater, precluding the need to meet MCLs.

Technologies Applied: Groundwater monitoring and natural attenuation

Reported Results: COC concentrations tend to be decreasing, except for chlorinated solvents in a few wells. The off-site wells BMW- 6A and BMWD-1 showed an increase in chlorinated solvent concentrations, which prompted the DEQ to drill additional wells in the area. The results of the off-site study indicate that there are off-site sources of chlorinated solvent contamination. The U.S. Geological Survey (USGS) performed an evaluation of the effectiveness of natural attenuation in 2002 indicating that natural attenuation is taking place at the site.

MCLs Achieved? Not sure; no concentrations were provided

ICs in Place? Yes, a deed notice

Vapor Intrusion: Not mentioned

Long-Term Monitoring? No. The 2006 ESD determined that further groundwater monitoring is not necessary. The wells were plugged by the DEQ.

58. Dutchtown Treatment Plant

State: LA

Site Lead: EPA

Date Deleted from NPL: November 16, 1999

Date of Last Five-Year Review: September 12, 2007

COCs: Benzene, ethylbenzene, toluene, xylene (BTEX), and lead

RAOs: Prevent contamination of underlying 150-foot-deep drinking water aquifer and restore contaminated shallow groundwater, based on its classification, for future use.

Technologies Applied: Groundwater monitoring and natural attenuation

Reported Results: Deep Water Zone: According to the first five-year review report (EPA 2002a), "From 1997-2002, BTEX was not detected in any of the Deep Zone wells for any of the sampling events reported in the monitoring results." From 2003 through 2006, BTEX constituents were not detected in the one remaining Deep Zone well (MW-7) for any of the sampling events reported in the groundwater analytical results. There is no evidence of vertical contaminant migration, thus the remedy continues to be protective of the shallowest drinking water aquifer (encountered at 100 feet bgs and extending to 300 feet bgs).

Shallow Water Zone: According to the first five-year review report (EPA 2002a), "For the monitoring results review (1997-2002), no concentrations of BTEX exceeded the Class III groundwater corrective action levels (taking into account the natural attenuation factor of 173) set forth in the O&M work plan (G&M 1997) for any of the sampling events reported." From 2003 through 2006, BTEX constituents continue to remain below the Class III (not an underground source of drinking water) groundwater corrective action levels with the Updated O&M Plan's (AGM 2002a) identified natural attenuation factor (of 173) included.

MCLs Achieved? No, but shallow groundwater would not be used as drinking water so cleanup to MCLs was not required. Deep water wells have never had site related contaminants detected (based on data from 1997-2006).

ICs in Place? Yes, access restrictions, deed notices, and restriction on groundwater use from site wells.

Vapor Intrusion: The risk assessment did not consider vapor intrusion to indoor air. Although there are residences located within 100 feet of the site boundary (i.e., to the east), there are no residences located within 100 feet of the impacted shallow groundwater zone, and groundwater is not migrating toward the residences (i.e., it is migrating towards the northwest). Therefore, the vapor intrusion pathway is not considered a complete pathway.

Long-Term Monitoring? No

59. Fourth Street Abandoned Refinery

State: OK

Site Lead: EPA/State

Date Deleted from NPL: August 21, 2008

Date of Last Five-Year Review: May, 15, 2007 (combined review done for Fourth St. and Double Eagle sites)

COCs: PAH, chlorinated hydrocarbons and PCB, alkyl benzenes, ketones, lead (primary COC), arsenic, and antimony

RAOs: Protection of human health and the environment

Technologies Applied: Groundwater monitoring

Reported Results: COC concentrations tend to be decreasing, except for chlorinated solvents in a few wells. The off-site wells BMW-6A and BMWD-1 showed an increasing trend in chlorinated solvent concentrations, which prompted the DEQ to drill additional wells in the area. The results of the off-site study indicate that there are off-site sources of chlorinated solvent contamination. The U.S. Geological Survey (USGS) performed an evaluation of the effectiveness of natural attenuation in 2002 indicating that natural attenuation was taking place at the site.

MCLs Achieved? Unknown because no concentration data were provided

ICs in Place? Yes, a deed notice

Vapor Intrusion: Not mentioned

Long-Term Monitoring? No

60. Gurley Pit

State: AR

Site Lead: EPA

Date Deleted from NPL: November 6, 2003

Date of Last Five-Year Review: September 27, 2007

COCs: Lead, barium, zinc, and TOC

RAOs: The Remedial Investigation indicated that the groundwater contained no site-related contaminants and that the potential for future groundwater contamination would be prevented by the implementation of the source control remedy.

Technologies Applied: Source control and groundwater monitoring

Reported Results: The results from the 2006 sampling event showed pit concentrations of barium, lead, zinc, and TOC, the COCs for the site, consistent with previous sampling rounds. The results provided no indication that the site was negatively impacting groundwater.

MCLs Achieved? Not Applicable

ICs in Place? Yes, deed notice

Vapor Intrusion: Not applicable

Long-Term Monitoring? No

61. Industrial Waste Control *(There is limited documentation for this site)*

State: AR

Site Lead: EPA

Date Deleted from NPL: April 2008

Date of Last Five-Year Review: September 26, 2007

COCs: Methylene chloride, toluene, PAHs, heavy metals including nickel, chromium, and lead

RAOs: Protection of human health and the environment

Technologies Applied: Off-site disposal; solidification of soil; slurry wall to prevent contamination of groundwater; natural attenuation; groundwater monitoring

Reported Results: No numeric results are given. During September 2003, two monitoring wells recorded higher than baseline levels of contaminants. A Site Assessment Study concluded in November 2006 that the increase above baseline values was related to natural attenuation locally around the monitor wells and that contaminants of concern would not migrate off-site. No offsite migration of COCs has been detected in mine void at the downgradient monitor wells (MW-10, 11, and 103D) or the property line monitor well (MW-15).

MCLs Achieved? Unknown, as no numeric results are given.

ICs in Place? Yes, deed restrictions

Vapor Intrusion: Not mentioned

Long-Term Monitoring? No

62. Mallard Bay Landing Bulk Plant

State: LA

Site Lead: EPA

Date Deleted from NPL: September 19, 2005

Date of Last Five-Year Review: N/A

COCs: Arsenic, benzene, bromodichloromethane, carbon tetrachloride, chloroform, and manganese

RAOs: Reduce contamination to meet the following levels: arsenic (10 μg/L, the MCL); bromodichloromethane/chloroform (there is no specific MCL, but the total trihalomethane amount should not exceed 80 μg/L); carbon tetrachloride (5 μg/L, the MCL); manganese (50 μg/L, the MCL); benzene (5.0 μg/L, the LDEQ Risk Evaluation Corrective Action Program Screening Standard).

Technologies Applied: Emergency removal action

Reported Results: No numeric results given, but the fact sheet states that no hazardous substances remain at the Site above levels that prevent unlimited use and unrestricted exposure.

MCLs Achieved? Yes, but no data were provided.

ICs in Place? No. The waste will ultimately be removed from the site so it is anticipated that no institutional controls will be needed.

Vapor Intrusion: Not mentioned

Long-Term Monitoring? No

63. Odessa Chromium #2 (Andrews Highway)

State: TX

Site Lead: EPA/State

Date Deleted from NPL: July 19, 2004

Date of Last Five-Year Review: September 25, 2001

COCs: Chromium

RAOs: Reduce chromium to MCL (100 μg/L)

Technologies Applied: Extraction and electrochemical treatment of contaminated groundwater

Reported Results

MW-214A		**MW-223R**		**MW-213**	
Jun 98	ND (0.02 mg/L)	Jun 98	ND (0.02 mg/L)	Jun 00	ND (0.02 mg/L)
Oct 98	ND (0.02 mg/L)	Oct 98	ND (0.02 mg/L)	Oct 00	ND (0.02 mg/L)
Dec 98	ND (0.02 mg/L)	Dec 98	ND (0.02 mg/L)	Jan 01	0.01 mg/L

MW-216		**MW-209**		**MW-221**		**MW-231**	
Oct 98	0.03 mg/L	Apr 00	0.09 mg/L	Jun 00	0.1 mg/L	Jun 00	0.01 mg/L
Nov 98	0.05 mg/L	May 00	0.09 mg/L	Jul 00	0.1 mg/L	Oct 00	ND (0.02 mg/L)
Dec 98	0.06 mg/L	June 00	0.08 mg/L	Aug 00	0.1 mg/L	Jan 01	0.02 mg/L

MCLs Achieved? Yes

ICs in Place? None listed

Vapor Intrusion: Not applicable

Long-Term Monitoring? No

64. Old Inger Oil Refinery

State: LA

Site Lead: EPA

Date Deleted from NPL: August 12, 2008

Date of Last Five-Year Review: July 23, 2007

COCs: Heavy metals, phenols, benzene, naphthalene, benzo (a) pyrene, and benzo (a) anthracene

RAOs: Reduce contamination to levels meeting LDEQ's RECAP GWSS standards: benzene 5.0 μg/L (same as MCL), naphthalene 10 μg/L (no MCL), benzo(a)pyrene 0.2 μg/L (same as MCL), benzo(a)anthracene 7.8 μg/L (no MCL).

Technologies Applied: Five emergency removal actions; Pump and treat, groundwater monitoring

Reported Results: The data review determined that the shallow groundwater, when compared to the Risk Evaluation and Corrective Action Program Groundwater Screening Standards values, is not contaminated above levels that pose an unacceptable risk. *The five-year review document was not accessible from CERCLIS.*

MCLs Achieved? Unclear, as numeric results were provided.

ICs in Place? Yes, conveyance notice

Vapor Intrusion: Not applicable

Long-Term Monitoring? No

65. South 8th Street Landfill

State: AR

Site Lead: EPA

Date Deleted from NPL: September 28, 2004

Date of Last Five-Year Review: June 16, 2009

COCs: Lead, arsenic, and manganese

RAOs: Restore groundwater to MCLs (lead 15 µg/L, arsenic 10 µg/L, and manganese 50 µg/L)

Technologies Applied: Monitored natural attenuation

Reported Results: No specific results listed but the fact sheet says that the groundwater is no longer contaminated. According to the last five-year review says "For the ground water OU, nine monitoring wells were sampled during eight sampling events in 2002 and the analytical results confirmed that the source area treatment and natural attenuation processes in the aquifer have reduced the metal concentrations below the remedial goals specified in the 1998 ROD Amendment. The nine groundwater monitoring wells were plugged and abandoned in June 2003."

MCLs Achieved? Yes, but no numeric results provided in the final five-year review.

ICs in Place? Yes. There is a prohibition on further excavation.

Vapor Intrusion: Not applicable

Long-Term Monitoring? No

66. Farmers' Mutual Cooperative

State: IA

Site Lead: EPA

Date Deleted from NPL: November 13, 2001

Date of Last Five-Year Review: September 7, 2000

COCs: Alachlor, atrazine, cyanazine, metolachlor, metribuzin, trifluralin, benzene, toluene, ethyl benzene, and total xylenes

RAOs: Reduce contamination to meet State action levels: alachlor 0.4 μg/L, atrazine 3 μg/L, cyanazine 1 μg/L, metolachlor 70 μg/L, metribuzin 100 μg/L, trifluralin 5 μg/L, benzene 5 μg/L, toluene 2,420 μg/L, ethyl benzene 700 μg/L, and total xylenes 12,000 μg/L.

Technologies Applied: Groundwater monitoring and natural attenuation

Reported Results: Pesticide contaminant concentrations have been shown to be below standards for three consecutive sampling events (November 1997 through November 1999) but benzene concentrations in MW-13 and MW-22 continue to indicate concentrations above action levels.

MCLs Achieved? No, benzene was still above action levels in two wells in 2000.

ICs in Place? Yes. The site remains on the State Registry of Hazardous Waste or Hazardous Substance Disposal Sites and cannot be sold or undergo a significant change in use without the approval of the IDNR.

Vapor Intrusion: Not mentioned

Long-Term Monitoring? In a letter dated April 3, 2000, IDNR agreed with the recommendation to discontinue pesticide monitoring, to abandon 12 monitoring wells, and to continue annual sampling of monitoring wells MW-13 and MW-22 for BTEX, with the understanding that subsequent modification to the monitoring could be required if the magnitude and/or extent of BTEX contamination were found to increase significantly.

67. Kem-Pest Laboratories

State: MO

Site Lead: EPA

Date Deleted from NPL: September 20, 2001

Date of Last Five-Year Review: September 21, 1999

COCs: Pesticides including heptachlor, chlordane, and endrin; VOCs; and SVOCs

RAOs: Protection of human health and the environment. The baseline risk assessment indicated groundwater contamination did not pose a significant threat to human health.

Technologies Applied: Groundwater monitoring will be conducted for the first five years to verify that no unacceptable exposures posed, by conditions at the site, occur.

Reported Results. 1999 review of groundwater monitoring to date indicates that the remedy is protective.

MCLs Achieved? Yes, according to the Fact Sheet MCLs were met.

ICs in Place? None listed

Vapor Intrusion: Not mentioned

Long-Term Monitoring? No

68. Labounty

State: IA

Site Lead: EPA

Date Deleted from NPL: October 6, 1993

Date of Last Five-Year Review: July 22, 2010

COCs: Arsenic and 1,1,2-trichloroethane

RAOs: Reduce arsenic concentration in the Cedar River to acceptable levels (MCL 10 μg/L) by reducing surface infiltration through wastes. The MCL for 1,1,2-trichloroethane is 5 μg/L.

Technologies Applied: Groundwater monitoring and an upgradient groundwater diversion wall

Reported Results: Tables in the most recent five-year review (2010) show contaminant concentrations above MCLs in some wells during 2004-2009 monitoring (for example, M0479-A: arsenic 653 μg/L and 1,1,2-TCA 10,600 μg/L in May 09).

MCLs Achieved? No, contaminant concentrations are still above MCLs in some wells.

ICs in Place? Yes. There are perpetual easements, a restrictive covenant, and prohibitions against the use of groundwater or intrusive activities.

Vapor Intrusion: There is a building immediately north of monitoring well M00379A and upgradient of the constructed chemical fill cover which raises the possibility of vapor intrusion. The concentration of 1,1,2-TCA in this well is less than 5 μg/L (the screening level for 1,1,2-TCA in the 2002 EPA Vapor Intrusion Guidance).

Long-Term Monitoring? Yes

69. Waverly Ground Water Contamination

State: NE

Site Lead: EPA

Date Deleted from NPL: November 20, 2006

Date of Last Five-Year Review: August 13, 2009

COCs: Carbon tetrachloride and chloroform

RAOs: Restore the contaminated aquifer for future use as a source of drinking water by reducing the carbon tetrachloride and chloroform concentrations below health-based criteria described in the ROD (5 μg/L carbon tetrachloride and 100 μg/L chloroform).

Technologies Applied: Pump and treat

Reported Results: Most recent sampling (second quarter FY2009) indicated that carbon tetrachloride and chloroform were not detected in monitoring wells.

MCLs Achieved? Yes

ICs in Place? No

Vapor Intrusion: No. Soil gas compliance results indicated that the contaminants were at such a depth (30 to 35 feet below ground surface) that it was highly unlikely that any vapors would reach any potential receptors.

Long-Term Monitoring? No

70. White Farm Equipment Co. Dump

State: IA

Site Lead: EPA

Date Deleted from NPL: October 30, 2000

Date of Last Five-Year Review: June 22, 2009

COCs: Benzene, cadmium, chromium, and lead

RAOs: Prevent further migration of contaminated groundwater and reduce levels of contaminants below established health-based standards for drinking water [benzene 1 μg/L (MCL is 5), cadmium 5 μg/L, chromium 100 μg/L, and lead 50 μg/L—action level is 15].

Technologies Applied: Pump and treat

Reported Results: As of 2009, the levels of benzene, cadmium, chromium, and lead in the groundwater remain below the performance standards set for the site. It should be noted that the concentrations of all four analytes were below detection limits during both the 1999 and 2004 monitoring efforts. During the 2008 monitoring effort, there were detectable concentrations of cadmium, chromium, and lead, but still below performance standards.

MCLs Achieved? Yes

ICs in Place? Yes, there is a restrictive covenant

Vapor Intrusion: Not mentioned

Long-Term Monitoring? Yes

71. Rose Park Sludge Pit

State: UT

Site Lead: State

Date Deleted from NPL: June 3, 2003

Date of Last Five-Year Review: September 25, 2007

COCs: PAHs

RAOs: "Prevent exposure to the acid waste sludge, eliminate potentially unhealthy odors and vapors, and prevent off-site migration of the sludge through surface water and groundwater." No specific cleanup goals were established for the groundwater at the Site (http://epa.gov/superfund/sites/npl/d030630.htm).

Technologies Applied: Slurry wall/cap

Reported Results: Results from groundwater monitoring suggest that contamination from the waste material remains contained.

MCLs Achieved? Not applicable, as no specific cleanup goals were set for groundwater this Site. The goal was containment.

ICs in Place? Yes. There is a five-party agreement between Salt Lake City, Salt Lake City/County Health Department, Utah State Health Department, EPA, and Amoco Oil Company to provide Site access for all five entities and to prohibit any excavation activities on-site.

Vapor Intrusion: Although the RAOs state that a primary objective is to eliminate potentially unhealthy odors and vapors the five-year review does not elaborate on vapor intrusion.

Long-Term Monitoring? No

72. Whitewood Creek

State: SD

Site Lead: EPA

Date Deleted from NPL: August 13, 1996

Date of Last Five-Year Review: September 27, 2007

COCs: Arsenic, copper, zinc, selenium, and mercury

RAOs: Reduce contaminant concentrations below MCLs (arsenic 10 μg/L, copper 1,000 μg/L, zinc 5,000 μg/L, selenium 50 μg/L, and mercury 2 μg/L). Five-year review suggests a waiver from complying with the arsenic MCL.

Technologies Applied: Groundwater monitoring

Reported Results: No groundwater results provided in the most recent five-year review.

MCLs Achieved? No. The fact sheet says that "Since waste is left in place which precludes unlimited use and unrestricted exposure, EPA will continue to conduct five-year reviews."

ICs in Place? Yes, ordinance restrictions on future digging in remediated areas

Vapor Intrusion: Not applicable

Long-Term Monitoring? Yes

73. Del Norte Pesticide Storage

State: CA

Site Lead: EPA

Date Deleted from NPL: September 18, 2002

Date of Last Five-Year Review: June 14, 2010

COCs: 1,2-Dichloropropane and 2,4-dichlorophenoxyacetic acid

RAOs: Restoration of contaminated on-site groundwater to 100 µg/L for 2,4-dichlorophenoxyacetic acid (federal MCL is 70) and to the health-based level of 10 µg/L for 1,2-dichloropropane (federal MCL is 5)

Technologies Applied: Pump and treat and groundwater monitoring

Reported Results: By October 1997, both the groundwater and soil cleanup levels for 2,4-D had been achieved. Although the 10 µg/L heath-based level for 1,2-DCP has not been achieved, 1,2-DCP concentrations in groundwater continue to decline slowly and the plume is contained within the original contaminated area.

MCLs Achieved? No, the MCL for 1,2-DCP has not yet been met. "EPA's 2000 ROD Amendment concluded that the 5,000 foot plume was not migrating and that it was technically impracticable to restore the 1,2-DCP plume to the 5 µg/L MCL. The 2000 ROD Amendment therefore waived this ARAR on the basis of Technical Impracticability" (http://www.epa.gov/fedrgstr/EPA-WASTE/2002/August/Day-08/f20099.htm).

ICs in Place? Yes, ordinance restrictions on future digging in remediated areas

Vapor Intrusion: Not mentioned

Long-Term Monitoring? Yes

74. Firestone Tire & Rubber Co. (Salinas Plant)

State: CA

Site Lead: EPA

Date Deleted from NPL: April 21, 2005

Date of Last Five-Year Review: September 28, 2001

COCs: VOCs

RAOs: Reduce GW contamination to the following levels:

Contaminant	MCL (μg/L)	federal MCL
1,1 dichloroethylene	6	7
1,1 dichloroethane	5	none
1,1,1 trichloroethane	70	200
1,2 dichloroethane	0.5	5
Tricholorethene	0.7	5
Benzene	0.7	5
Toluene	20	1000
Ethylbenzene	10	700
Xylene	70	10,000

Technologies Applied: Pump and treat, groundwater monitoring, and natural attenuation

Reported Results: The final sampling occurred on July 28, 1998 with only two wells showing contamination above cleanup standards: well S09 containing 55 μg/L of 1,1-DCE and OW4 containing 11 μg/L of 1,1-DCE. These wells are in the shallow zone and are located approximately 250-300 feet from the facility. No contamination above cleanup standards was found in the intermediate or deep zones. The Regional Water Quality Control Board noted a declining trend in 1,1-DCE concentrations in the shallow aquifer (from 130 μg/L in 1995 to 55 μg/L in 1998) and stated that, due to the low concentrations and significant separation by tens of feet from underlying water zones, the shallow zone contamination would not be expected to impact the downgradient groundwater and deeper aquifers in the future. They also stated that they believed that the wells would slowly attenuate to below the cleanup standard. After the last sampling event, wells were destroyed and sealed. No further monitoring has occurred at the site.

MCLs Achieved? Yes, MCLs were achieved in the deep aquifer (but not the shallow).

ICs in Place? No, there are no institutional controls required as part of the remedy for the Firestone site.

Vapor Intrusion: Not mentioned

Long-Term Monitoring? No

75. Schofield Barracks

State: HI

Site Lead: U.S. Army

Date Deleted from NPL: August 30, 2000

Date of Last Five-Year Review: September 24, 2007

COCs: Trichloroethene

RAOs: Reduce level of TCE contamination to drinking water standards only at point of use. A technical impracticability (TI) waiver was prepared, which supports the idea of point-of-use treatment. Because of the TI waiver, the cleanup goals apply only at the wellhead and not throughout the aquifer.

Technologies Applied: Point-of-use treatment

Reported Results: No results provided

MCLs Achieved? No, because water is only being treated at the point of use

ICs in Place? Yes. There are prohibitions on the use or disturbance of groundwater, prohibitions on excavation activities, disturbance of the landfill cover, and any other activities that might interfere with the implemented remedy.

Vapor Intrusion: Not mentioned

Long-Term Monitoring? No

76. Southern California Edison Co. (Visalia Poleyard)

State: CA

Site Lead: EPA

Date Deleted from NPL: August 27, 2009

Date of Last Five-Year Review: July 21, 2010

COCs: Pentachlorophenol, benzo(a)pyrene, and TCDDeqv

RAOs: Reduce contamination to the following levels: pentachlorophenol 1 µg/L, benzo(a)pyrene 0.2 µg/L, and TCDDeqv 30 pg/L.

Technologies Applied: Physical and chemical groundwater treatment system (steam)

Reported Results: The table below shows site-wide average groundwater concentrations over a three-year compliance demonstration period.

	Pentachlorophenol	Benzo(a)pyrene	TCDD eqv.
Cleanup Levels	1.0 µg/L	0.2 µg/L	0.03 ng/L
UCL95 Int. Aquifer	0.075 µg/L	0.055 µg/L	0.019 ng/L
UCL95 Deep Aquifer	0.054 µg/L	0.03 µg/L	0.0053 ng/L

The statistical analysis of the groundwater data demonstrated that the cleanup levels had been met in both the intermediate and deep monitoring zones at the Site, except for two outliers, which were found during a period with low water table elevations.

MCLs Achieved? Yes

ICs in Place? Yes, deed restrictions

Vapor Intrusion: Not mentioned

Long-Term Monitoring? No

77. Western Pacific Railroad Co.

State: CA

Site Lead: EPA

Date Deleted from NPL: August 29, 2001

Date of Last Five-Year Review: September 18, 2008

COCs: 1,1-DCE, 1,1,1-TCA, 1,1-DCA, and TCE

RAOs: Reduce contamination to the following levels: 1,1-DCE to 7 μg/L; 1,1,1-TCA to 200 μg/L; 1,1-DCA to 5 μg/L; and TCE to 5 μg/L.

Technologies Applied: Groundwater monitoring and pump and treat

Reported Results: Analytical results indicated that 1,1-DCE, 1,1,1-TCA, and TCE continue to be present in the groundwater at concentrations below the Federal and State MCLs. The concentration of 1,1-DCA in well MW8902 was 5.4 μg/L, which is above the cleanup level of 5.0 μg/L.

MCLs Achieved? No. Concentrations of 1,1-DCA are still slightly above the MCL.

ICs in Place? Yes, a land use covenant

Vapor Intrusion: Not mentioned

Long-Term Monitoring? Yes

78. Bonneville Power Administration Ross Complex

State: WA

Site Lead: U.S. Department of Energy

Date Deleted from NPL: September 23, 1996

Date of Last 5 Year Review: September 2, 2009

COCs: dichloroethylene and chloroform, found in on-site groundwater at slightly elevated levels. *CERCLIS does not consider this to be a site with contaminated groundwater.*

RAOs: Protection of human health and the environment. Results of the site-specific risk assessment indicated that exposure to either on- or off-site groundwater would not pose a risk to human health.

Technologies Applied: None for groundwater

Reported Results: None reported

MCLs Achieved? Not applicable, as there was likely little groundwater contamination to begin with

ICs in Place? Yes, but not for groundwater

Vapor Intrusion: Not mentioned

Long-Term Monitoring? No

79. Northwest Transformer

State: WA

Site Lead: State

Date Deleted from NPL: September 28, 1999

Date of Last Five-Year Review: July 27, 1999

COCs: PCBs in soil

RAOs: None for groundwater

Technologies Applied: None for groundwater

Reported Results: The PRPs continued to monitor groundwater at the site for five years after cleanup was completed and no contaminants of concern were detected in perimeter or off-site wells.

MCLs Achieved? Not applicable, as groundwater was never contaminated

ICs in Place? A Consent Decree with the owners executed prior to cleanup required a deed notice with requirements to notify the Washington State Department of Ecology before excavating below 15 feet and to notify EPA before using groundwater at the site. A deed notice to that effect was recorded August 30, 1999.

Vapor Intrusion: Not mentioned

Long-Term Monitoring? Yes, for five years after cleanup to make sure there was no contaminant migration from soil to groundwater.

80. Union Pacific Railroad Company

State: ID

Site Lead: State

Date Deleted from NPL: September 22, 1997

Date of Last Five-Year Review: N/A

COCs: Heavy metals, polycyclic aromatic hydrocarbons, and organic compounds

RAOs: Reduce contamination to federal drinking water standards

Technologies Applied: Pump and treat and groundwater monitoring

Reported Results: Since the completion of the cleanup, groundwater monitoring results have met or were lower than federal drinking water standards.

MCLs Achieved? Yes

ICs in Place? Yes, there are deed, land, and groundwater use restrictions.

Vapor Intrusion: Not mentioned

Long-Term Monitoring? No